Lecture Notes on Data Engineering and Communications Technologies

Volume 162

Series Editor

Fatos Xhafa, Technical University of Catalonia, Barcelona, Spain

The aim of the book series is to present cutting edge engineering approaches to data technologies and communications. It will publish latest advances on the engineering task of building and deploying distributed, scalable and reliable data infrastructures and communication systems.

The series will have a prominent applied focus on data technologies and communications with aim to promote the bridging from fundamental research on data science and networking to data engineering and communications that lead to industry products, business knowledge and standardisation.

Indexed by SCOPUS, INSPEC, EI Compendex.

All books published in the series are submitted for consideration in Web of Science.

Rajalakshmi Krishnamurthi · Adarsh Kumar ·
Sukhpal Singh Gill · Rajkumar Buyya
Editors

Serverless Computing: Principles and Paradigms

 Springer

Editors
Rajalakshmi Krishnamurthi ⓘ
Department of Computer Science
and Engineering
Jaypee Institute of Information Technology
Noida, India

Sukhpal Singh Gill
School of Electronic Engineering
and Computer Science
Queen Mary University of London
London, UK

Adarsh Kumar ⓘ
School of Computer Science
University of Petroleum and Energy Studies
Dehradun, Uttarakhand, India

Rajkumar Buyya
School of Computing and Information
Systems
The University of Melbourne
Melbourne, VIC, Australia

ISSN 2367-4512   ISSN 2367-4520 (electronic)
Lecture Notes on Data Engineering and Communications Technologies
ISBN 978-3-031-26632-4   ISBN 978-3-031-26633-1 (eBook)
https://doi.org/10.1007/978-3-031-26633-1

This Springer imprint is published by the registered company Springer Nature Switzerland AG
The registered company address is: Gewerbestrasse 11, 6330 Cham, Switzerland

Preface

Serverless computing is a paradigm shift in cloud computing. Recently, many companies rely on serverless computing for their product application development and deployment, market analysis and customer relationship without investing excess on infrastructure development and maintenance. This book brought a single point of resource for researchers and practitioners on wide aspects of serverless technologies. The book presents serverless computing, data-centric serverless computing, distributed serverless computing and the road ahead and future of serverless computing. Further, it focuses on the fundamental of serverless computing such as the evolution of computing technologies, architecture, benefits, applications, issues and solutions in serverless computing, open challenges and future scope. Further, the book will present critical issues such as fine granularity and performance achievement in serverless computing. Next, the role of hyperscalers in serverless computing in terms of application development, business and economic perspective will be targeted.

The key performance concepts such as no operational costs, scheduling and resource management, performance modelling, fairness, interoperability, virtualisation, data centres and portability are addressed. The merits of conventional serverless computing include autoscaling and pay-as-you-go mode. It lacks efficient data processing due to the shipping of data to the code, isolated VM for serverless functions, and non-addressable and limited internal cache state. However, modern computing in serverless requires data-intensive, distributed applications, open-source platforms and customisable hardware. The topics under serverless data lake architecture include functionalities such as data ingestion, data catalog, data discovery/searching, ETL and ELT methodologies in serverless data lake architecture. Next, the containers orchestration on containers such as Docker, Kubernetes, and Linux Containers will be addressed. The commercial data-centric serverless platforms frameworks such Amazon, Google, Azure and Oracle are covered. This book also discusses the need for hardware-level enhancement for data-centric serverless computing. For this purpose, the impact of multicore CPUs, cluster/grid computing, graphic processing units, tensor processing units and FPGA accelerators for serverless computing will be targeted. Further, the several big data format, storage and services mechanisms for serverless computing are presented. Here, the various modern data types and storage mechanisms such as spatial–temporal data, time series data, key-value data, and graph-based data storage, columnar data storage, real-time data streaming are addressed. The data-centric serverless services include interactive queuing, real-time logging and monitoring, querying, and notification services. Intensive data processing in serverless technology such as prediction, intelligent decision-making, real-time, big data analytics, and data science support for AI, ML and DL models in serverless computing is addressed.

This book focuses on distributed serverless computing. Here, the state management, network file systems, communicating agents, autoscalability, P2P communication, generic- and application-specific frameworks, multi-tenancy and existing distributed serverless computing frameworks are addressed. Further, the performance issues in distributed serverless computing such as reliability, serviceability, high availability, aggregation and broadcasting patterns, consistency, concurrency, consensus, and fault-tolerant mechanism are addressed. Next, the data handling in distributed serverless environments such as data sharing, replication, redundancy, partitioning and indexing are addressed. This book addresses serverless technology and primarily provides efficient mechanisms towards data privacy in terms of access control auditing, attack and abuses. This book will also address the multiple serverless computing and event-driven distributed systems. As a cutting-edge trend, serverless computing is integrated with high-end computing technologies such as blockchain, IoT, cloud computing, fog and edge computing, big data, artificial intelligence, SDN and NFVs. This book serves as a platform for providing key insight and foreseen open challenges towards serverless computing.

Chapters in this book are organised as follows:

The first chapter titled "Serverless Computing: New Trends and Research Directions" discussed that the serverless computing is an innovative method for the production and distribution of software that does not rely on a centralised server management infrastructure. Instead, the cloud service provider must ensure that the code will execute as intended in the cloud environment. This frees up more time for the developers to work on their projects. This chapter introduces serverless computing and associated technologies. Additionally, this work provides future directions as well as a summary of the research done for this book.

The second chapter titled "Punching Holes in the Cloud: Direct Communication Between Serverless Functions" introduced a temporary architecture for function-to-function communication in serverless systems through the use of direct network connections. The framework has been successfully implemented on real, production-ready serverless computing services, most notably AWS. To permit outgoing connections from functions while restricting inbound connections, contemporary serverless computing systems frequently employ a networking configuration called network address translation (NAT). Further, this chapter details the planning and development of a library for transient communication in AWS Lambda. The network connection between serverless applications is simplified by the library's inclusion of function and server components.

The third chapter titled "Hybrid Serverless Computing: Opportunities and Challenges" studied the extent to which serverless computing may be expanded to become hybrid serverless computing. Further, the authors have defined hybrid serverless computing, detailed the methods towards attaining it and highlighted the potential and problems that it presents.

The fourth chapter titled "A Taxonomy of Performance Forecasting Systems in the Serverless Cloud Computing Environments" focused on the classification scheme used to characterise the parallel file system (PFS) structure. To understand how existing PFSs are implemented in distributed computing environments and how they might be adapted for usage in serverless (edge) cloud computing, a taxonomy has been developed.

The fifth chapter titled "Open-Source Serverless for Edge Computing: A Tutorial" investigated the options for deploying a serverless application at the edge of the network using open-source software and Internet of things (IoT) data. Due to its focus on resource economy and flexibility, the serverless method may be especially useful for edge computing-based applications, in which the hosting nodes are deployed close to the consumers and comprise devices and workstations with minimal resources.

The sixth chapter titled "Accelerating and Scaling Data Products with Serverless" covered the framework and tools (data visualisation, pipelines, models, and APIs) that help speed up and control data offers. APIs for data and model serving with containerised solutions as a building block for data products that are driven by machine learning techniques, and for serving a unified data ontology; data visualisation in the context of containerised web applications that deliver excellent methods for data explorations, model predictions, visualisation and consumer insights.

The seventh chapter titled "QoS Analysis for Serverless Computing Using Machine Learning" discussed the importance of artificial intelligence (AI) and machine learning (ML) to make predictions regarding the system configurations that are utilised in serverless computing. In addition, a model that does not incur any costs is proposed to investigate and evaluate the many possible configurations of workstations in an environment that lacks servers.

The eighth chapter titled "A Blockchain-Enabled Serverless Approach for IoT Healthcare Applications" explored how blockchain technology might complement serverless computing to address reliability issues with functions and resource allocation for IoT healthcare applications. The proposed method aims to react to customers' demands in a trustworthy and dependable manner by taking their privacy concerns into account, allocating resources efficiently, and meeting their needs promptly. It is obvious that this paves the way for efficient use of resources, which in turn may boost consumer happiness and service quality.

The ninth chapter titled "Cost Control and Efficiency Optimization in Maintainability Implementation of Wireless Sensor Networks Based on Serverless Computing" provided a conceptual approach to the implementation of maintainability for wireless sensor network (WSN) using serverless computing. To further decouple the device operation and functional development, considerably optimise the reuse of resources and remove the hardware interference, it has been proposed that serverless computing may be accomplished at the software functional level of WSN. To reduce design, manufacturing and operational costs, WSN platforms may be built using the idea of serverless computing, which can support the functions of data collecting and data management into functional development that may benefit from exploration via upfront investments. Finally, a case study is provided that uses existing technology and smart city scenarios to propose a WSN platform for serverless computing.

The tenth chapter titled "Scheduling Mechanisms in Serverless Computing" examined the benefits, drawbacks and uses of the most popular schedulers in serverless computing. The current study's goal is to give a thorough analysis of different and efficient scheduling methods that can be used as a foundation for choosing the right scheduling procedure based on the providers' perspective.

The eleventh chapter titled "Serverless Cloud Computing: State of the Art and Challenges" provided a thorough overview of these restrictions and to showcase state-of-the-art research on ways to address the issues that are preventing serverless from becoming the standard in cloud computing. The primary difficulties of deploying such a cloud platform are examined, and potential research avenues are outlined.

The book provides the best learning resource for researchers, graduates, undergraduates, business people and common readers in the field of serverless computing. When we talk about serverless cloud computing, it brings about tremendous changes in the post-virtual-machine environment. Companies other than technology sectors are using serverless platforms and frameworks at all production levels due to their economic pay-per-use approach. Also, businesses of various shapes and sizes have started to adopt serverless computing because of its scalability. Furthermore, the technology's use has enhanced IT infrastructures in the functions-as-a-service (FaaS) sector. This enables a whole new range of workloads that are capable of benefiting from the same capabilities of stateless programmes. It is now managed by a serverless platform, so the burden of data management is removed for developers. This feature helps business application development in a cloud-native way. This book acts as a bridging information resource between basic concepts and advanced-level content from technical experts to computing hobbyists towards enhancing their knowledge and proficiency.

Noida, India   Rajalakshmi Krishnamurthi
Dehradun, India   Adarsh Kumar
London, UK   Sukhpal Singh Gill
Melbourne, Australia   Rajkumar Buyya

Contents

Serverless Computing: New Trends and Research Directions

Rajalakshmi Krishnamurthi⑩**, Adarsh Kumar**⑩**, Sukhpal Singh Gill**⑩**, and Rajkumar Buyya**⑩

Abstract Serverless computing is an innovative method for the production and distribution of software since it does not rely on a centralised server management infrastructure. As a result of this, serverless computing is becoming more widespread. Instead, the cloud service provider must ensure that the code will execute as intended in the cloud environment. Because everything is taken care of automatically, developers are free to concentrate on creating code rather than establishing and maintaining the infrastructure that is necessary for their programmes to execute. This frees up more time for the developers to work on their projects. This chapter introduces serverless computing and associated technologies. Further, this work summarizes the work done in this book, recent developments and presents future directions.

Keywords Application development · Serverless computing · Serverless functions · Services · Security

R. Krishnamurthi
Department of Computer Science and Engineering, Jaypee Institute of Information Technology, Noida, India
e-mail: k.rajalakshmi@jiit.ac.in

A. Kumar
School of Computer Science, University of Petroleum and Energy Studies, Dehradun, Uttrakhand, India
e-mail: adarsh.kumar@ddn.upes.ac.in

S. S. Gill (✉)
School of Electronic Engineering and Computer Science, Queen Mary University of London, London, UK
e-mail: s.s.gill@qmul.ac.uk

R. Buyya
Cloud Computing and Distributed Systems (CLOUDS) Laboratory, School of Computing and Information Systems, The University of Melbourne, Melbourne, Australia
e-mail: rbuyya@unimelb.edu.au

© The Author(s), under exclusive license to Springer Nature Switzerland AG 2023
R. Krishnamurthi et al. (eds.), *Serverless Computing: Principles and Paradigms*,
Lecture Notes on Data Engineering and Communications Technologies 162,
https://doi.org/10.1007/978-3-031-26633-1_1

1

1   Introduction

Serverless computing eliminates the requirement for a dedicated server farm, making it possible to manage enormous dispersed workloads. Large, geographically scattered workloads may be managed using this kind of computing, eliminating the need for a centralised data centre [1]. This computing method eliminates the need to set up a specialised network of computers to coordinate the efforts of many workers in different places. By using this technique, a cluster of computers is not needed to handle globally distributed computations. In the computer world, eliminating servers has allowed us to manage massive workloads that are distributed across several locations. It wasn't possible until now to do this. Lambda, provided by Amazon Web Services (AWS), is one of the most well-known examples of serverless computing offered by a major cloud provider. This is the case, for example, with cloud systems like Microsoft Azure and Google Cloud. The Google Cloud Platform and the Microsoft Azure cloud storage service are two examples of where this kind of technology is put to use. In this technology, large-scale computer systems provide a substantial challenge that must be controlled when it comes to the process of moving data from one function to another. This problem need to be addressed. The creation of a possible solution is required for this matter.

The major advantages of serverless technology include [1–3]:

(i)   This kind of computing is gaining traction in the corporate world because it may relieve programmers of their server maintenance obligations. So, developers may build and expand their apps without worrying about exceeding the server's resources. This paves the way for the development of innovative app features.

(ii)   When it comes to assuring the continuous good company's information technology infrastructure, business executives often run across challenges. Through the use of serverless computing, it is no longer necessary for programmers to manage the machines on which their programmes are executed. This involves keeping an eye on the server, ensuring that the operating system is up to date, and generating user accounts for all of the different user groups who will be using the server.

(iii)   The advent of cloud computing has made it possible to share software without the need for a single centralised server. It frees up money that may be used toward other endeavours, such as the development of a product that has a greater capacity for usefulness and distinction.

(iv)   Serverless cloud computing frees users from the obligation of managing their own servers, enabling them to focus their attention where it will be of the greatest benefit: on the development and improvement of valuable applications.

(v)   As serverless apps make use of an architecture external to your company, you can take advantage of only benefiting from the functions that you need to develop it. It adapts to your budgets, since its functions scale according to the number of requests that are made.

(vi)   One company that provides this kind of technology is Amazon Web Services, and one of the services it provides is called Lambda. With Lambda, you have the tools at your disposal to turn your ideas into applications that people will find intuitive and easy to use. This is likely made possible by the presence of in-app purchases, geolocation features, user-friendly file and photo sharing, and maybe many more features. Now that serverless computing is a thing, cloud-based IT service concepts that were previously unimaginable are within reach.

(vii) To speed up the process of product distribution, you should attempt to spend less time and effort on administrative responsibilities. Serverless computing provides this opportunity.

1.1 Motivation

The serverless paradigm makes it possible to make software in many different ways, which means it can be used in a wide range of high-performance, and always-available apps. The growth of the Internet of Things (IoT) devices, online and mobile apps, Big Data, and machine learning are just a few examples of the many different domains where serverless computing is finding expanding usage. There are many more sectors as well [3]. This is because serverless apps have their own specific needs, each of these spheres will have its own set of nuances that set it apart from the others, improved resource management practises that take into account this reality are necessary as quickly as possible. A serverless architecture may be useful for workloads that often expand and contract but still need a significant amount of computing power. The concept that will drive the next generation of cloud computing services, known as serverless computing, is garnering an increasing amount of attention. Even when their functionality and popularity increase, it is essential for serverless systems to keep the important qualities and characteristics that set them apart in the first place.

1.2 Traditional Versus Serverless Computing

An increasing number of people are looking towards serverless computing as a viable option to the conventional server and cloud-based designs. This trend may be attributable to the fact that serverless computing has become more popular since it does away with the requirement for traditional server infrastructure. Taking this kind of action is counter to accepted procedures in the world of web design. There is no longer any need for developers to set up and manage backend servers because of serverless architectures. This eliminates the need for programmers to do these mundane chores. Less money will be spent on creating and maintaining the product. The term "serverless computing" is gaining popularity because it appeals to programmers who would rather not have to worry about the care of server infrastructure. This is one of the reasons why people are using the term more often. As a direct consequence of this advancement, programmers are no longer restricted by the capabilities of the servers they make use of in their work. As a result, investing in DevOps is a decision that will prove to be profitable, and using this strategy has the potential to save costs associated with the investment.

This chapter is organized as follows. Section 1 introduces the background of serverless computing, compares the features of serverless computing with traditional computing and discussed the need for serverless computing in present and future. Section 2 introduces the important serverless functions, architectures and computing feasibilities. Section 3 introduces the integration of serverless computing with advanced technologies. Section 3 presents resource management in serverless computing environments. Section 4 presents the integration of serverless computing with other advanced technologies. Section 5 presents the open challenges. Section 6 discusses the future directions. Finally, the conclusion is presented in Sect. 7.

2 Serverless Functions, Architectures and Computing

FaaS provides capability to execute the applications without dependent on any infrastructure and effortless managing of services to the customers. The key characteristics of the FaaS includes (i) support for wide programming languages, (ii) no overhead of administration, (iii) scalability and availability, (iv) APIs and Cloud services based triggers for execution of codes. The conglomeration of "Functions as a Service" together with the "Backend as a Service" leads to emerging of the serverless computing. The leading Serverless solutions include Amazon Web Services (AWS), Azure and Google Cloud [4].

Amazon Web Services: AWS is the top leading marketer of serverless products and cloud space provider [5]. AWS cloud services provide developers to build and execute their applications independent of infrastructure, and computing resources. The basic functionalities such as traverse, deploy and publish of serverless applications within fractions of time. AWS allows developers to (i) incorporate multiple serverless services and applications, (ii) customize the computing resources as per the user requirements and (iii) integrate variety of serverless applications [6]. In addition, visual based workflow creation, coordination, inconnection between Lambdas can be incorporated using AWS step functions.

Amazon provides several real time data processing solutions such as AWS Lamba, Amazon S3, Amazon Kinesis and Dynamo DB. Amazon kinesis provides real time streaming of data and data analytics. The Amazon databases supports NoSQL database functionalities for two formats of data storage namely (i) key value model and (ii) document type data. Further, AWS supports messaging services namely (i) Amazon SNS for Publish/Subscriber and (ii) Amazon SQS for Message Queuing.

Figure 1 depicts the serverless computing based basic web applications and its associated backends. Here, S3 can be utilized for web hosting, Lambda functions can be used to data processing and Amazon API gateways for configure the environment and Dynamo DB can be used for retrieving data.

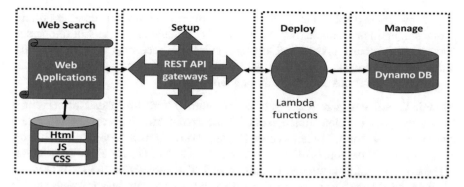

Fig. 1 Basic Amazon Web Services in serverless computing for web application

In terms of intelligent functionalities, Amazon machine learning services enables the real time prediction of hidden pattern and knowledge discovery from the processed data [7]. Similarly Amazon SageMaker provides facility to manage and deploy large scale machine learning models. Other services include Amazon Recognition for digital video and audio processing, Amazon Lex for semantic analytics and Amazon Polly for text to speech conversions [8]. Several tools are available for developers, such as AWS serverless Application Model (SAM), AWS CodeBuild and Code Deploy, CodeStar and Code Pipeline. In AWS, the Integrated Development Environment (IDE) supports several enhanced editing and project management platforms such as Cloud9IDE, Visual Studio. It also supports software development tool kits such as Python, Go, Scala, Node.js, .NET, Java programming [9].

Microsoft Azure: Second most popular Serverless computing platform is the Microsoft Azure and it provides large scale serverless computing space [10]. Azure functions are based on event driven FaaS. Particularly, Azure IoT Edge enables to excite program within the IoT edge devices even in unrealisable network connectivity. Azure storage space supports voluminous data storage, high scalability, availability and efficient durability [11].

Azure Active Directory provides strong security mechanism and access control methods for remote cloud systems. In distributed environments, the interconnectivity between public and private cloud platforms can be performed using Azure Service Bus Messaging functionalities, and Azure Event Grid can be utilized for the event based routing services. In terms of API management, the Azure Logic enables the integration of data with developer applications, and provides transparent code integration between systems without complex coding by developers. In addition, Azure Functions Proxies are capable of providing microservices through monolith APIs using single API interface.

Microsoft Azure provides intelligence support system by means of Azure Bot services for various platform namely Twitter, Microsoft teams, Slack, Skype etc. Several cognitive functionalities such as audio, video, image, speech, text based interpretation and processing are handled by Azure Intelligence services [12, 13].

From developers point of view, Microsoft Azure provides several serverless framework plugins. For example, Visual studio IDE framework permits developers to incorporate various functions and logical applications according the user specifications and requirements. In addition, Azure SDKs support almost all major software development platforms and programming languages.

Google Cloud Platform: GCP sets as the world's third top player of serverless computing [14]. The cloud function of GCP incorporates the event drive computation on serverless platform. The object based storage is involved on the GCP cloud storage units. The cloud datastores includes services such as NoSQL Database as a Service (DBaaS). For real time storage of data from IoT edge devices, the GCP provides Firebase Real time Database platform. In terms of security and privacy, Google Firebase platform supports wide variety of authentication APIs and allows user to customize their mobile applications. GCP provides visualization and management of workflow through GCP FantasM platform. Google Cloud Endpoints and APIgee API management allows developers to design, scale applications, predictive analytics and deploy securely on unrealiable multiregional distributed environments [15]. GCP enables to utilize intelligence through serverless machine learning using Cloud ML Engine, vision processing through Cloud Vision API, Speech processing through Cloud Speech API, Text processing through Cloud Translational APIs. Google Cloud Function provides developers with coding event trigged applications, deploying, management and operational infrastructure.

3   Resource Management in Serverless Computing Environments

A serverless environment's resource management is the process of balancing the needs of an application's workload with those of the system, with the customer's participation kept to a minimum [16]. Given the autonomous character of the anticipated resource management process in such contexts, careful attention to every stage of this procedure is necessary to improve application and system efficiency [17]. We single out three key areas of resource management that must be addressed in a way that is appropriate for the serverless computing paradigm.

1. To characterise and anticipate workload performance, programmers would want as little effort as possible when utilising a serverless deployment paradigm. If an application's deployment requires the specification of a resource configuration and other features, the process may be laborious. Because of this, it is preferable for a serverless platform to be able to forecast performance by inferring features of applications and workloads using simulation and benchmarks methodologies [18]. Users' quality of service needs may be met with the use of an efficient strategy for establishing this comprehension, which in turn leads to improved scheduling and scalability decisions for available resources.

2. Resource scheduling is a major difficulty for both developers and cloud providers or system owners since it involves effectively allocating workloads to host nodes with the appropriate resources [16]. When the need for resources is greater than the supply, scheduling also entails deciding which programmes will run first. Even if the developer requires certain quality of service guarantees, the supplier must prioritise resource efficiency [17].

3. The serverless architecture dynamically spins up environments and distributes their resources to apps in response to incoming demands. This guarantees increased efficiency and adaptability in the use of available resources [19]. Scaling at such a fine granularity necessitates the use of sophisticated and dynamic resource scaling approaches to preserve the expected level of application performance.

4 Serverless Computing and Advanced Technologies

Using serverless computing has now become extremely prevalent for developing cloud-native apps. When it comes to cloud computing, the serverless paradigm is all about removing the burden of managing servers. Serverless computing is expected to grow at a far faster rate than traditional cloud services since developers no more need to be concerned about keeping up with infrastructure [3]. With serverless computing, cloud service providers might have an easier time handling infrastructure management and automatic provisioning. The amount of work and materials needed to maintain the infrastructure are cut down as well [20]. The goal of serverless computing is to leverage the most cutting-edge serverless technology while minimising costs and maximising returns.

AI is the potential of technology, so it's no surprise that platforms are starting to include it. Due to these AI-driven platforms, we've been able to make more accurate, timelier decisions [21]. Their impact may be seen in the altered methods of doing company, communicating with consumers, and analysing financial information. Software engineers' output and effectiveness are drastically impacted by complex machine learning algorithms. However, most of the problems that programmers face may be solved by switching to a serverless architecture. By employing a serverless architecture, both the machine learning models and their associated resources may be controlled with more efficiency and precision. Thanks to this architecture, programmers may be able to devote more time and energy to training AI models and less to maintaining servers.

Building machine learning systems is often necessary when confronting difficult problems. They perform tasks such as data analysis and pre-processing, model training, and AI model tuning [22]. Therefore, APIs should function without a hitch. The usage of serverless computing and artificial intelligence can guarantee the constant storage and transmission of data and messages. Since serverless architecture provides a number of benefits and advantages, it may be a good fit for machine learning models. Almost no management is needed to support the operation of any kind of application or back-end service [23]. The supplier of the underlying infrastructure efficiently distributes its own CPU execution power in response to incoming requests of varying traffic loads [24]. The advantages of serverless architecture [25–28] are as follows:

1. Serverless architecture enables usage-based pricing, so you'll only ever have to fork out cash for the services you actually need. Because of this, the pricing structure is more adaptable, and the overall price is decreased.
2. Because of serverless computing, software developers may work independently and quickly. Because of this, models are treated separately from other functions. Activating this feature at any moment is completely safe and will not affect the rest of the system in any way.
3. With the advent of the "pay-per-use" model, clients are charged solely for the resources they actually employ. In serverless computing, you pay for the services you employ instead of the number of servers you utilize.
4. Serverless computing eliminates the requirement for consumers to monitor and maintain servers by making available back-end services on demand. Users of a serverless service don't have to worry about setting up or maintaining the underlying infrastructure. When using serverless architecture, service providers may easily scale up or down their bandwidth needs without making any changes to their current infrastructure.
5. Serverless programmes have gained traction as a result of their inherent reliability and fault tolerance. Because of this, you won't have to build any services to provide these capabilities to your application.

5 Open Challenges of Resource Management in Serverless Computing

We've discovered that serverless architectures provide unique difficulties in terms of resource management [16–19, 29]. The following are important challenges of resource management in serverless.

1. Cold Start Latency: Auto-scaling systems introduce delay since resources must be created dynamically, delaying the start of a function's execution by a significant amount of time [30]. Especially for routines with relatively brief execution periods, this early delay might have a considerable impact on application performance. Nevertheless, in order to solve this problem, several providers keep reserves of available resources.
2. Resource Utilization: Serverless platforms are efficient in terms of resource usage since they only pay for what is actually used by an application through its execution, as opposed to a more generic cloud computing pricing approach [17]. Nevertheless, the providers could be keeping the underlying infrastructure operational for longer. Since this is the case, it's crucial to pay extra attention to developing techniques for optimal resource utilisation by the host nodes. Customers often

overbook resources for function executions in an effort to prevent their applications from performing poorly [18]. Regularly underutilizing these resources may cause the user to receive bad price for their money and lose faith in the reliability of these services.

3. Heterogenous Workloads: Controlling a wide variety of workloads with little input from the user is made possible by serverless architectures. Thus, in order to provide a desirable result, such systems must learn about the specifics of the application and workload on their own [16, 29]. This is complicated by the wide variety of serverless apps in use today. It's possible that customer discontent will come from delays in resource installation time, increased resource interference impacts, and other similar issues due to a lack of knowledge of the application's requirements and features.

4. QoS: There are no assurances of QoS because the serverless architecture hides most internal workings from the customers [30]. While most consumers would appreciate this, platforms without appropriate performance assurances may be useless for high-precision, latency-sensitive workloads [31]. Providing a consistent level of service to all of the users in a distributed system is a hard and time-consuming undertaking for the provider.

6 Future Directions

The following are promising future directions in the area of serverless computing:

1. Delay: Response time in a serverless architecture is the sum of request queuing, resource setup, and function execution [31]. Although most independent functions have execution duration that are less than a second, or of just few seconds, the capability to keep low latency for function executions is a crucial challenge in serverless deployments [32]. Since the time it takes to build up resources from scratch during a serverless environment's "cold start" is typically far longer than the time it takes to actually run an application, this is the primary reason for excessive delay.

2. Reliability: When anything goes wrong on a serverless platform, the platform will often repeat the operation. In the event that a platform's function execution fails, it will, for instance, resubmit the request immediately [16]. It has been determined that, especially when state management is employed via external storage services, a retry-based approach to fault tolerance might still not yield right output. They stress that precision may be compromised if a partially performed failure attempt of the same execution is viewed by a parallel execution of the function [19].

3. Sustainability: Since serverless computing facilitates the on-demand provisioning and release of resources utilised in the execution of functions, it has been heralded in the sustainability literature as a key technology for advancing green computing [33]. As a plus, the approach of invoicing per execution time encourages programmers to reduce resource use and boost code execution speed. Nevertheless, decomposing an app into functions and the practise of configuring

resources on demand are believed to result in extra delay and an execution cost, both of which impact energy usage [34].

4. Utilization of Resources: Because of the granularity of the serverless billing model, users are only paid for the resources their applications really utilise [17, 18]. However, the provider is still responsible for the whole infrastructure, therefore it is in the provider's long term interest to move as many serverless apps as feasible onto a single host. Unfortunately, performance suffers when there are too many requests competing for a limited resource [31]. This is indicative of the usual tension between the goals of suppliers and customers, who each want to minimise costs while simultaneously maximising benefits [19, 30]. As a result, it's crucial to arrive at a mutually agreeable resource consolidation plan.

5. Security: The use of serverless computing has improved the safety of a variety of different infrastructures and computer systems. For example, Bila et al. [35] provide in-depth information about one method that may be used to secure Linux containers. There are security solutions that can detect intrusions using serverless services [35]. There are more solutions required than existing to secure sensitive information that has been saved in the cloud. This is because advancements have been made in serverless architectures. For example, advancements in authentication and authorization schemes, attacks against common execution environments, resource exhaustion attacks, and privacy concerns are some of the challenges that need to be addressed in future [1].

6. Lack of Tools and Paradigms: There aren't enough tools available right now to make serverless apps. This is a big challenge. Further, the use of insufficient modelling paradigms, which in turn leads to the creation of incoherent methodology, directly contributes to a drop in the overall quality of the code that has been written. Pérez et al. [2] made a new way to write code and middleware for use with serverless apps. This method could help apps that don't need a server. Benchmark suites [1] are important tools that help people figure out how likely it is that a new idea or concept will work. Thus, there is a need to focus on new tools and paradigms for serverless computing and application development.

7. Price Standardization: The market for serverless computing services is now controlled by several significant technology companies, each of which offers its unique price tiers and feature sets. It is expected in future that there will be an expansion not just in the number of companies offering serverless services but also in the range of price choices that are available for such services. Both of these trends are expected to occur shortly. The estimate provided by each company is one-of-a-kind because it takes into consideration a variety of criteria, such as the sort of platform it utilises, the history of client association, transparency in service and the time of day when the request is made. This is because it may be difficult to devise a pricing model for service providers, there is a need to continue looking into the matter.

8. Data Sharing and Infrastructure Requirements: Serverless software is composed of several distinct functions, all of which work together to provide the required functionality. It will be necessary for the functions to have some means of interacting with one another and transferring either their data or the condition they

are now in for this objective to be realised. Thus, the first challenge that has to be conquered is function addressing, and the second challenge that needs to be conquered is intercommunication for functions. Both of these challenges need to be solved. Additionally, data sharing and infrastructure requirements are related to each other because of auto-scaling, short-lived functions, and exponential growth in the usage of serverless services with an increase in the number of function copies.

9. Other Challenges: Serverless describes a scenario in which an application may scale automatically without adding additional servers. This goal may be achieved by completing and delivering copies of the service offerings to the appropriate customers. Given that there are no established rules for determining the places where the real copies of functions are saved, there is no way to offer an answer to this request as it is impossible to do so. Additionally, data caching, service provider management, distributed processing with different modes of execution and customized service scheduling are some of the other concerns that need to be focused upon in detail.

7   Conclusion

This work looks at how serverless computing opens up new perspectives, terminologies, architectures, service options and opportunities. Later in the work, the possibilities, different points of view and recent developments are discussed briefly. Initially, this work outline how the wide use of serverless computing technology has opened up a world of possibilities. Next, this work discussed the open issues and challenges that keep serverless services from being as good as they could be. After the current problems are fixed, serverless computing is likely to become the most popular way to do computing, overtaking cloud computing shortly.

References

1. Shafiei H, Khonsari A, Mousavi P (2022) Serverless computing: a survey of opportunities, challenges, and applications. ACM Comput Surv (CSUR) 54(11):1–32
2. Pérez A, Moltó G, Caballer M, Calatrava A (2019) A programming model and middleware for high throughput serverless computing applications. In: Proceedings of the 34th ACM/SIGAPP symposium on applied computing, Apr 2019, pp 106–113
3. Mampage A, Karunasekera S, Buyya R (2022) A holistic view on resource management in serverless computing environments: taxonomy and future directions. ACM Comput Surv (CSUR) 54(11s):1–36
4. Pierleoni P, Concetti R, Belli A, Palma L (2020) Amazon, Google and Microsoft Solutions for IoT: architectures and a performance comparison. IEEE Access 8:5455–5470. https://doi.org/10.1109/ACCESS.2019.2961511
5. Mathew S, Varia J (2014) Overview of Amazon Web Services. Amazon Whitepap 105:1–22
6. Wittig M, Wittig A (2018) Amazon Web Services in action. Simon and Schuster

7. Newcombe C, Rath T, Zhang F, Munteanu B, Brooker M, Deardeuff M (2015) How Amazon Web Services uses formal methods. Commun ACM 58(4):66–73
8. Chong N, Cook B, Eidelman J, Kallas K, Khazem K, Monteiro FR, Schwartz-Narbonne D, Tasiran S, Tautschnig M, Tuttle MR (2021) Code-level model checking in the software development workflow at Amazon Web Services. Softw Pract Exp 51(4):772–797
9. Jackson KR, Ramakrishnan L, Muriki K, Canon S, Cholia S, Shalf J, Wasserman HJ, Wright NJ (2010) Performance analysis of high performance computing applications on the Amazon Web Services cloud. In: 2010 IEEE second international conference on cloud computing technology and science. IEEE, pp 159–168
10. Bisong E (2019) An overview of Google Cloud Platform services. In: Building machine learning and deep learning models on Google Cloud Platform, pp 7–10
11. Wankhede P, Talati M, Chinchamalatpure R (2020) Comparative study of cloud platforms—Microsoft Azure, Google Cloud Platform and Amazon EC2. J Res Eng Appl Sci 5(02):60–64
12. McGlade J, Wallace L, Hally B, White A, Reinke K, Jones S (2020) An early exploration of the use of the Microsoft Azure Kinect for estimation of urban tree diameter at breast height. Remote Sens Lett 11(11):963–972
13. Kamal MA, Raza HW, Alam MM, Su'ud MM (2020) Highlight the features of AWS, GCP and Microsoft Azure that have an impact when choosing a cloud service provider. Int J Recent Technol Eng 8(5):4124–4232
14. Bataineh AS, Bentahar J, Mizouni R, Wahab OA, Rjoub G, Barachi ME (2022) Cloud computing as a platform for monetizing data services: a two-sided game business model. IEEE Trans Netw Serv Manage 19(2):1336–1350. https://doi.org/10.1109/TNSM.2021.3128160
15. Ariza J, Jimeno M, Villanueva-Polanco R, Capacho J (2021) Provisioning computational resources for cloud-based e-learning platforms using deep learning techniques. IEEE Access 9:89798–89811. https://doi.org/10.1109/ACCESS.2021.3090366
16. Li Z, Guo L, Cheng J, Chen Q, He B, Guo M (2022) The serverless computing survey: a technical primer for design architecture. ACM Comput Surv (CSUR) 54(10s):1–34
17. Suresh A, Somashekar G, Varadarajan A, Kakarla VR, Upadhyay H, Gandhi A (2020) Ensure: efficient scheduling and autonomous resource management in serverless environments. In: 2020 IEEE international conference on autonomic computing and self-organizing systems (ACSOS). IEEE, pp 1–10
18. Großmann M, Ioannidis C, Le DT (2019) Applicability of serverless computing in fog computing environments for IoT scenarios. In: Proceedings of the 12th IEEE/ACM international conference on utility and cloud computing companion, Dec 2019, pp 29–34
19. Cicconetti C, Conti M, Passarella A, Sabella D (2020) Toward distributed computing environments with serverless solutions in edge systems. IEEE Commun Mag 58(3):40–46
20. Mampage A, Karunasekera S, Buyya R (2021) Deadline-aware dynamic resource management in serverless computing environments. In: 2021 IEEE/ACM 21st international symposium on cluster, cloud and internet computing (CCGrid). IEEE, pp 483–492
21. Gill SS, Xu M, Ottaviani C, Patros P, Bahsoon R, Shaghaghi A, Golec M et al (2022) AI for next generation computing: emerging trends and future directions. Internet Things 19:100514
22. Agarwal S, Rodriguez MA, Buyya R (2021) A reinforcement learning approach to reduce serverless function cold start frequency. In: 2021 IEEE/ACM 21st international symposium on cluster, cloud and internet computing (CCGrid). IEEE, pp 797–803
23. Jonas E, Schleier-Smith J, Sreekanti V, Tsai C-C, Khandelwal A, Pu Q, Shankar V et al (2019) Cloud programming simplified: a Berkeley view on serverless computing. arXiv preprint arXiv:1902.03383
24. Golec M, Ozturac R, Pooranian Z, Gill SS, Buyya R (2021) iFaaSBus: a security- and privacy-based lightweight framework for serverless computing using IoT and machine learning. IEEE Trans Ind Inform 18(5):3522–3529
25. Castro P, Ishakian V, Muthusamy V, Slominski A (2019) The rise of serverless computing. Commun ACM 62(12):44–54
26. Zafeiropoulos A, Fotopoulou E, Filinis N, Papavassiliou S (2022) Reinforcement learning-assisted autoscaling mechanisms for serverless computing platforms. Simul Model Pract Theory 116:102461

27. Du D, Liu Q, Jiang X, Xia Y, Zang B, Chen H (2022) Serverless computing on heterogeneous computers. In: Proceedings of the 27th ACM international conference on architectural support for programming languages and operating systems, pp 797–813
28. Aslanpour MS et al (2021) Serverless edge computing: vision and challenges. In: 2021 Australasian computer science week multiconference, pp 1–10
29. Xie R, Tang Q, Qiao S, Zhu H, Yu FR, Huang T (2021) When serverless computing meets edge computing: architecture, challenges, and open issues. IEEE Wireless Commun 28(5):126–133
30. Djemame K (2021) Energy efficiency in edge environments: a serverless computing approach. In: International conference on the economics of grids, clouds, systems, and services. Springer, Cham, pp 181–184
31. Gill SS (2021) Quantum and blockchain based serverless edge computing: a vision, model, new trends and future directions. Internet Technol Lett e275
32. Baldini I, Castro P, Chang K, Cheng P, Fink S, Ishakian V, Mitchell N et al (2017) Serverless computing: current trends and open problems. In: Research advances in cloud computing. Springer, Singapore, pp 1–20
33. McGrath G, Brenner PR (2017) Serverless computing: design, implementation, and performance. In: 2017 IEEE 37th international conference on distributed computing systems workshops (ICDCSW). IEEE, pp 405–410
34. Hassan HB, Barakat SA, Sarhan QI (2021) Survey on serverless computing. J Cloud Comput 10(1):1–29
35. Bila N, Dettori P, Kanso A, Watanabe Y, Youssef A (2017) Leveraging the serverless architecture for securing Linux containers. In: 2017 IEEE 37th international conference on distributed computing systems workshops (ICDCSW), pp 401–404. https://doi.org/10.1109/ICDCSW.2017.66

Punching Holes in the Cloud: Direct Communication Between Serverless Functions

Daniel Moyer and Dimitrios S. Nikolopoulos

Abstract Serverless computing allows Cloud users to deploy and run applications without managing physical or virtual hardware. Since serverless computing can scale easily via function replication, a growing trend is to use serverless computing to run large, distributed workloads without needing to provision clusters of physical or virtual machines. Recent work has successfully deployed serverless applications of data analytics, machine learning, linear algebra, and video processing, among others. Many of these workloads are embarrassingly parallel and follow the stateless function execution paradigm for which serverless computing is designed. However, some applications, particularly those implementing data pipelines, necessitate state sharing between different data processing stages. These workloads have a high degree of parallelism and can also scale easily with the number of concurrent functions but use slow Cloud storage solutions to communicate data between functions. Current serverless application deployments use containers or lightweight virtual machines with limited memory, computation power, and execution time. Therefore, a direct communication path between functions would need to be ephemeral and function under constrained resources. Introducing an ephemeral communication path between functions raises a number of additional challenges. Serverless providers use network firewalls to block inbound connections. Furthermore, the performance and scaling characteristics of a direct communication path would be entirely opaque to users. This chapter presents an ephemeral communication framework for serverless environments that uses direct network connections between functions. The framework has been successfully deployed on actual, production-strength serverless computing offerings, specifically AWS. The insight behind the proposed framework is that current serverless computing environments use a common networking configuration called Network Address Translation (NAT) to allow outbound connections from functions while blocking inbound connections. This work presents the design and implementation of an ephemeral communication library for AWS Lambda. The library

D. Moyer (✉) · D. S. Nikolopoulos (✉)
Department of Computer Science, Virginia Tech, 2202 Kraft Drive, Blacksburg, VA 24060, USA
e-mail: dmoyer@vt.edu

D. S. Nikolopoulos
e-mail: dsn@cs.vt.edu

© The Author(s), under exclusive license to Springer Nature Switzerland AG 2023
R. Krishnamurthi et al. (eds.), *Serverless Computing: Principles and Paradigms*,
Lecture Notes on Data Engineering and Communications Technologies 162,
https://doi.org/10.1007/978-3-031-26633-1_2

includes function and server components so that serverless applications can use network communications easily. It specifies an interface for serverless application code that runs on each function. The communication library supports multi-function jobs and manages communication between functions automatically. This work also implements an orchestrator server to invoke functions and send control messages between them. An external server is necessary to perform NAT traversal, and is also used for coordination. By using network connections, the proposed library achieves high performance and excellent scaling in workloads with over 100 functions. This work measures throughput of 680 Mbps between a pair of functions and verifies that this is the maximum throughput achievable on the current AWS Lambda offering. It also evaluates the framework using a multi-stage reduce-by-key application. Compared to an equivalent implementation using object storage, the library is 4.7 times faster and costs only 52% as much.

Keywords Serverless computing · AWS Lambda · NAT traversal · TCP hole punching · communication framework

1 Introduction

Serverless computing is a service that allows developers to run programs directly without having to manage servers themselves. Also known as Functions as a Service (FaaS), serverless computing is offered by major cloud platforms, for instance, Amazon Web Services (AWS) Lambda. Since serverless computing can scale rapidly, a growing trend is to use it to run large distributed workloads without needing to provision a cluster of machines. Recent works have used serverless for data analytics [1], machine learning [2–4], linear algebra [5], and video processing [6, 7]. However, since serverless programs, which are called functions, each run in their own isolated environment, a key problem for large-scale computing applications is transferring data between multiple instances of a function. Individual serverless functions have limited memory, computation power, and execution time, so inter-function communication is a requirement for many applications. Existing works have used object storage and in-memory databases for communication, but these techniques require trade-offs with regard to performance, cost, and scalability [8, 9]. Using direct network connections between functions is attractive since it would be fast, incur no extra cost, and scale with the number of functions; however, serverless providers use network firewalls to block inbound connections.

1.1 Novelty

This work presents a method to bypass the firewall for AWS Lambda, and develops the first communication framework that uses direct network connections between functions. Lambda uses a common networking configuration called Network Address Translation (NAT) to allow outbound connections from functions while blocking inbound connections. However, there are several standard techniques, such as hole punching or relaying that allow endpoints behind NAT to establish a connection with each other by using an external server for setup. This process is called NAT traversal and the proposed communication library implements TCP hole punching in order to transfer data between serverless functions over TCP connections. This work demonstrates that the communication library can scale to workloads with over 100 functions and show it achieves a significant speedup compared to exchanging data with object storage. There have been several previous serverless execution frameworks [10–12], but to our knowledge, none of them use direct network communication. Also, prior studies [8, 9] on inter-function communication for large serverless workloads have found limitations in existing methods: object storage is slow and in-memory databases are expensive and do not scale easily. Fouladi et al. [13] mention that NAT traversal between functions is possible, but do not evaluate it or discuss it in detail. This work demonstrates that serverless functions can communicate using network connections, which has significant advantages over current techniques.

1.2   Design and Contributions

This work develops a communication library with function and server components so that serverless applications can use network communications easily. It specifies an interface for serverless application code that runs on each serverless function. The proposed library supports multi-function jobs and manages communication between functions automatically. This work also implements an orchestrator server to invoke functions and send control messages between them. An external server is necessary to perform NAT traversal, and it is also used for coordination. By using network connections, the library achieves high performance with a measured throughput of 680 Mbps between a pair of functions. This work also evaluates it using a multi-stage reduce-by-key application with over 100 functions. Compared to an equivalent implementation using object storage, the proposed library is 4.7 times faster and costs only 52% as much.

2   Background

This section provides background on serverless computing, the storage systems used in current stateless serverless computing environments, and NAT traversal, which is the fundamental technique that we are using to develop an ephemeral communication library for serverless functions. This section also discussed the motivation for this work in relation to the state of the art.

2.1 Serverless Computing

Serverless computing, also called Functions as a Service (FaaS), is a paradigm in cloud computing where users can execute short-lived application code in a quickly scalable, stateless environment without needing to deploy or manage servers. A single instance of a serverless application is called a function. Serverless functions can be triggered automatically to respond to events or used to process data in parallel. As an example, uploading an image to a particular bucket in object storage might trigger a serverless function to resize it automatically. Major commercial FaaS providers include Amazon Web Services (AWS) Lambda [14], Azure Functions [15], Google Cloud Functions [16], IBM Cloud Functions [17], and Oracle Cloud Functions [18] and there are also several open-source serverless platforms including OpenFaaS [19], Apache OpenWhisk [20] and Kubeless [21].

A key feature of serverless is that it scales quickly compared to traditional server-based computing where users provision virtual machine servers. It is possible to simultaneously launch hundreds of serverless functions and the initialization time is usually well under a second, whereas it may be up to a minute for server-based machines. However, each serverless function has a maximum execution time, which is 15 min in the case of AWS Lambda. Users can also configure the memory and CPU power allocated to each function, which ranges from 128 to 10,240 MB with AWS Lambda. Serverless computing uses a pay-as-you-go cost model where users are billed based on the total execution time of their functions proportional to the configured memory size as well as the number of function invocations. The rapid scalability and usage-based billing means that serverless is ideal for uneven or bursty workloads since users do not need to provision resources in advance.

Besides scalability, one of the main characteristics of serverless functions are that they run in a stateless, isolated environment called a container. Although containers have a small amount of storage space (512 MB for AWS Lambda) during their execution, it is not persisted so functions must use external storage mechanisms. Containers also block incoming network connections, so it is not possible to run web servers directly in functions or to access running function code in a remote shell for debugging. While different functions are guaranteed to run in separate containers, a FaaS provider may at its discretion, reuse a container from a previous execution of the same function. This reduces startup time since the function does not need to reinitialize and is called a *warm start*, as compared to when a function runs in a new uninitialized container, which is called a *cold start*. Overall, serverless computing has several distinct properties, including minimal management, rapid scalability, and isolated execution environments.

2.2 Data Storage with Serverless

Serverless applications must solve problems related to data transfer and limited function lifetime that do not occur with traditional server-based computing. One challenge in serverless is how to persist data and transfer it between multiple functions, since functions are ephemeral by nature. Although functions do have access to limited storage space during runtime, it is not persistent. Also, functions can be invoked with a small amount of metadata, but this is not sufficient for data processing applications. Thus, it is necessary for serverless functions to input and output data via one or more external data stores, such as database servers or object storage. Because functions are isolated from each other, using external data storage also allows functions to communicate for large computing workloads. This differs from traditional server-based computing frameworks where nodes can easily exchange data over the network. In addition, conventional server-based computing nodes do not have limited lifetimes, but serverless functions do. Serverless computations must account for limited function lifetime when scheduling functions and ensure that functions finish writing their output before they timeout.

There are several different external cloud computing services that serverless functions can use to send and receive data, such as object storage, memory- and disk-backed databases and proxied network connections. Persistent object storage such as AWS Simple Storage Service (S3) can store large amounts of data for a few cents per gigabyte per month. A disadvantage is that it does not support high throughput and rate-limits requests to a few thousand per second [8]. While object storage is cheap, the fastest storage available is memory-backed key-value stores, such as AWS ElastiCache, which provides managed Memcached or Redis instances. These memory-backed databases can support a very fast request rate (over 100,000 requests per second) and are efficient for small objects but are significantly more expensive: ElastiCache is hundreds of times more expensive than S3 when storing the same amount of data [8]. Also, memory-backed stores must have servers provisioned to run on. Of course, any other type of database can also be used in combination with serverless to provide different trade-offs with respect to cost and performance. Most databases still need provisioned servers, however, which limits scalability. In theory, it is also possible to use external proxy servers so serverless functions can communicate over the network. However, this would only be practical if the bandwidth each function needs is small compared to that of a server since provisioning multiple servers just to act as proxies would defeat the purpose of using serverless in the first place. Overall, for large computing serverless workloads, there are several options for external storage, including object storage, databases, and proxied connections, but they all require trade-offs between cost and performance.

2.3 Network Address Translation (NAT)

2.3.1 Definition

Network Address Translation [22] is a commonly used technique to connect an internal network with a different IP addressing scheme from its external network. A network address translator rewrites the IP packets that cross from one network to the other in order to translate between internal and external IP addresses and port numbers, modifying the packet headers to do so. It maintains connection state so that when a client on the internal network makes an outbound connection, it can correctly forward the return packets back to that client. Typically, NAT is used in combination with filtering rules to block unsolicited incoming packets from entering the internal network. In this case, two clients behind different internal networks are unable to connect to each other directly without using a special technique to bypass NAT, which is known as NAT traversal.

2.3.2 NAT Traversal

Transmission Control Protocol (TCP) hole punching is a technique for establishing a TCP connection between two endpoints that are both behind separate NAT networks and thus cannot connect directly [23]. The way TCP hole punching works is that both endpoints simultaneously make an outbound connection to the other endpoint. If the destination IP address and port of each connection matches the source IP address and port of the other, then each NAT device will allow the incoming connection from the other endpoint because it treats it as the response to the outbound connection from its own endpoint. For this to work, each endpoint must know the public source IP address and port of the other endpoint's outbound connection so it can make the destination IP address and port of its connection match. Thus, a prerequisite for hole punching is that both endpoints exchange their external source IP addresses and ports using an out-of-band method such as a relay server. However, the NAT device must use a known or predictable external source IP address and port for TCP hole punching to be possible. If the NAT device uses an unpredictable external port, then the opposite endpoint will not know what destination port to use, and the NAT traversal will not work.

2.3.3 AWS Lambda NAT Behavior

AWS Lambda creates each function container with a private IP address behind a NAT firewall that blocks incoming connections. The NAT firewall translates the function's private IP address into a public IP address so it can access the public internet. However, since the firewall does not allow functions to receive connections, this means that Lambda cannot be used to run externally-accessible servers and that two functions cannot directly communicate with each other over the network without

NAT traversal. AWS does not explicitly state the behavior of the NAT firewall they use for Lambda function containers. However, the observed behavior is that it rewrites the private IP address of the function to a public IP for outgoing connections, but preserves the TCP port used by the function. Also, the public IP address assigned to a container is static throughout the container's lifetime, which means that the public IP address and port of an outgoing connection is predictable as required for TCP hole punching. RFC 3489 [24] classifies NAT implementations as *cone NAT* or *symmetric NAT* based on whether the mappings it assigns are consistent across multiple connections. When a client makes outbound connections from the same private IP address and port, cone NAT consistently maps the connections to the same public IP address and port; however, symmetric NAT assigns a different public IP address or port to subsequent connections. In general, it is not possible to perform hole punching with symmetric NAT since the public IP address and port number cannot be known in advance (except when symmetric NAT uses a sequential or other predictable assignment). However, AWS Lambda uses cone NAT, which makes hole punching feasible. This work demonstrates that it is possible to perform NAT traversal to establish a TCP connection between two concurrently running Lambda functions and incorporates this ability into the proposed communication library.

2.4 Motivation for This Work

Current serverless computing systems lack direct communication and synchronization paths between functions. Functions can only share state via slow Cloud storage systems, but not directly either through datacenter networks or through shared memory on the same server. This limitation prevents the deployment and scaling of numerous data-intensive applications with data processing pipelines that require efficient communication between functions. This work addresses this challenge by presenting the first, to the best of our knowledge network communication library for serverless computing. The library addresses the unique challenges of network communication in serverless environments, namely the ephemeral nature of communicating functions and the arbitrary resource limitations imposed on functions during execution while providing a scalable communication channel that significantly outperforms Cloud storage. Importantly, the library presented in this work is portable and deployed on AWS Lambda, arguably the most widely used serverless computing substrate to date.

3 Design

This section presents the design of an ephemeral communication library for serverless computing. The section explores how to setup, control, and communicate through secure communication channels between ephemeral functions using NAT traversal.

3.1 Communication Library Functionality

In order to use serverless functions to run a distributed application, there is a fair amount of setup involved. Each function must be started and then receive instructions for what input sources and output destinations to use. Also, for direct network communication, functions must exchange the network address information for the peer functions with which they will communicate. This paper presents a communication framework for performing this setup as well as establishing network connections between functions. The implemented communications library consists of two main parts: an orchestrator that invokes functions and manages communication between them, and worker functions that perform NAT traversal and run application code for data processing. The orchestrator must be hosted on a machine that allows inbound network connections since this is a requirement for NAT traversal. When starting an application run, the orchestrator first generates a communication graph of all the functions to deploy that specifies which pairs of functions should establish direct network connections and provides a list of all the inputs and outputs for each function. The orchestrator uses the information in this graph to invoke the necessary functions and then starts a listening server. When each worker function starts, it connects to this server via a WebSocket to register itself with the orchestrator and fetch the list of peer functions to connect to. After a function receives the address information for its peers, it performs NAT traversal to establish direct TCP connections with them. The proposed library uses TCP instead of UDP to provide reliable transmission of data between functions. Once this setup is complete, the worker begins running the actual application code. This initialization process for the first TCP connection between two functions is shown in Fig. 1. Worker applications that use the communication library must implement interface methods to produce input data, consume output data, transfer data between functions, and process input data into output data.

3.2 Client-Server Design

Our framework uses a separate server for communication to bypass the restriction that Lambda functions do not allow inbound connections. Since AWS firewalls Lambda functions, it is not possible for one function to connect to another directly by themselves. However, with a server external to Lambda, it is possible to relay traffic between functions or to facilitate NAT traversal so functions can establish a direct connection. Because having a server is necessary to use NAT traversal in our communication library, we decided to also use it as an orchestrator to invoke functions and send control messages between them. The server starts a service using the WebSocket protocol [25] on a publicly accessible host and port and includes this information as part of the function event data on invocation. Then, when each function starts, it makes a persistent connection to the orchestrator server so it can exchange small

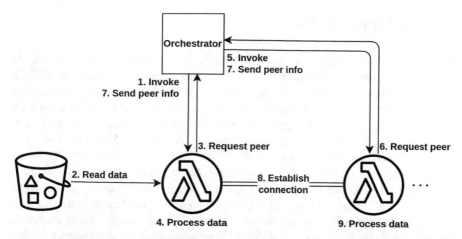

Fig. 1 Communication library initialization procedure for the first two worker functions. The first worker starts reading and processing data while establishing a connection to the second in parallel. If the second worker has any other peers, it will repeat this setup with them once it starts reading data from the first worker

control-plane messages. At this point, the orchestrator sends peer network information for each pair of functions that need to perform NAT traversal.

While implementing the orchestrator on a single server may affect scalability, the per-function resource requirements in our design are small, so we believe that this is an acceptable trade-off. Because performing NAT traversal requires an external server so functions can exchange network address information, the only design variable is whether or not to have that server support additional functionality. The communication library uses that server to implement a function orchestrator, which maintains a persistent connection with each function throughout its lifetime to send and receive control messages. However, an alternate design would be to have the central server only act as a NAT traversal relay and implement the orchestrator as of the serverless functions themselves. Implementing the orchestrator on a traditional server has the benefit of allowing persistent connections to a centralized point. This saves the need for a more complicated peer-to-peer control approach, or running the orchestrator in a separate function and having to handle the case where it must run for longer than the 15-min maximum function lifetime and having to use some replacement scheme. The main potential drawback of moving additional features to the server is that it might decrease the scalability of the design. For each function, the server must maintain some data structures in memory and a single persistent connection with that function. The difference is that in the NAT-traversal-only case, the server can free these resources for a function once it has completed NAT traversal, whereas in our design, the server must maintain them until the function exits. However, a single server should be able to scale to many thousands of concurrently-running functions since the per-function resources are fairly small, and running jobs of that scale would probably also require additional design improvements, regardless.

3.3   NAT Traversal Protocol

Although there are multiple protocols that can be used to perform NAT traversal, we used a custom implementation of TCP hole punching because we wanted to use TCP for data reliability and to use the same system for NAT traversal and exchanging control messages. In general, there are many possible configurations for a network address translator, but our design only needs to work in the specific environment of AWS Lambda. One of our main design decisions was to use TCP instead of UDP. The proposed communication library uses TCP because it automatically detects and corrects transmission errors. Although UDP hole punching may be more common than TCP hole punching, [23] state the latter is just as "fast and reliable" as the former. However, TCP hole punching is slightly more complex to implement from an application perspective, since the application must both listen on and make outbound connections from the same socket. This is not a problem for our communication library since we control the implementation and can configure it to allow socket reuse in this way. Another potential concern with TCP is that its handshake and slow-start behavior may lead to slower data speeds than UDP soon after creating a connection. However, when sending large amounts of data, these startup costs should have minimal effect, but for applications that care about fine performance optimizations, we leave the investigation of UDP as future work. Another design decision we made was not to use existing protocols such as STUN [24] or TURN [26] to help with NAT traversal. The STUN protocol uses a public server to help a client determine its public IP address, whether it is behind NAT, and the NAT type if so. However, since we already knew the NAT behavior for AWS Lambda, it was easier for us to find the public IP address using our orchestrator server than to set up a separate STUN server simply for that purpose. Also, the TURN protocol provides a way to implement relaying when NAT traversal is not possible, but this is not applicable for our case and would be very inefficient anyway. While our hole punching implementation is implemented specifically for AWS Lambda, the TCP hole punching technique works for most NAT configurations such as cone NAT as well as nested NAT when the outer NAT supports hairpin translation, although it does not work with symmetric NAT [23]. Based on our testing, AWS Lambda does not currently support IPv6, which could eliminate the need for NAT if its implementation assigned a public IP address directly to each function. However, there would likely still be a stateful firewall that would block incoming connections with IPv6. If AWS Lambda deployed IPv6 but prevented functions from communicating directly, a similar process to bypass this by making simultaneous outbound connections might still be necessary. Although there are many possible behaviors for NAT, our communication library only needs to support the one used by AWS Lambda, so we decided to use TCP hole punching because it was simple enough to implement as part of our orchestrator server and because we wanted to use TCP for its built-in data reliability.

4 Implementation

This section outlines the implementation of the proposed ephemeral communication library for serverless computing. The section explores in depth the use of NAT traversal, the control of communicating serverless functions, the API of the proposed communication library, and the use of the proposed library for the development of serverless applications.

4.1 NAT Traversal Technique

In order to create a direct connection between Lambda functions, the proposed communication library uses TCP hole punching, which is a known NAT traversal technique [23], to bypass how Lambda blocks incoming connections to functions. On startup, the function component of the library connects to the orchestrator server in order to receive control information. Since the orchestrator server runs outside of Lambda, this outbound connection is permitted by Lambda's firewall. Once the orchestrator accepts connections from both functions that plan to establish a mutual TCP connection, it sends the public IP and a TCP port number of each function to the other. Thus, both functions have the necessary IP address and port number information to make an outbound TCP connection and repeatedly try to connect to the other function using this information. While trying to make the outbound connection, each function also opens a socket to listen on its own fixed port number in case the connection from its peer succeeds first. After both functions attempt outbound connections, eventually, the Lambda NAT firewall will allow traffic from the other function through as it treats this as return traffic from the outbound connection. At this point, from one function's perspective, either its outbound connection attempt succeeds or the listening socket accepts a connection from the other function. Which event succeeds first is non-deterministic, but either way, the function has a valid TCP connection open to its peer function and can begin sending and receiving data.

See an example sequence diagram of TCP hole punching between two AWS Lambda functions in Fig. 2. In this example, Function 1 has public IP 3.0.1.1 and Function 2 has public IP 3.0.2.2. Also, both functions start listening for incoming connections on their corresponding private IP addresses and ports: address 172.16.1.1 port 10001 for Function 1 and address 172.16.2.2 port 20002 for Function 2. At this point, both functions have exchanged public IP addresses and listening port numbers via the orchestrator server. Once they receive this information, both functions try to open an outbound TCP connection to the other. In the diagram, the first connection attempt is made by Function 1 to the public IP address of Function 2. Note that Function 1 sets the source port of the TCP SYN connection request packet to be the same as the port number it is listening on. The NAT firewall for Function 1 rewrites the private source port of this packet to the public IP of Function 1, namely 3.0.1.1, but it keeps the source port number the same. The fact that the AWS Lambda

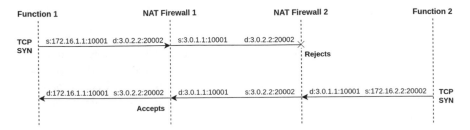

Fig. 2 TCP hole punching example

firewall preserves port numbers makes TCP hole punching possible. However, at this point the NAT firewall for Function 2 sees the incoming packet as unsolicited and drops it. Independently, Function 2 now sends a connection request to Function 1, and its NAT firewall rewrites the source IP address to the public IP address of Function 2. When the connection request reaches the firewall for Function 1, the source and destination addresses and ports are exactly switched compared to the connection previously initiated by Function 1. Thus, the Function 1 firewall treats the second connection request as a response to the first, and allows it. Now, Function 1 receives the connection SYN request from Function 2 on its listening port, and sends a SYN-ACK acknowledgment packet back to continue the TCP connection setup. The firewall for Function 2 will also accept the acknowledgment because it is a valid response to the SYN packet Function 2 sent. Thus, both firewalls will allow traffic through, so the functions can finish the handshake to establish a valid TCP connection between them.

4.2 Server-Function Communication

The proposed communications library uses persistent WebSocket connections implemented with the Python Socket.IO [27] library to exchange control messages between the orchestrator server and worker functions. As discussed in Sect. 3.2, this work uses a client-server design for the communication library where the orchestrator server is responsible for sending and receiving commands and status updates between function. These control messages include the initial information needed for NAT traversal, when a function has finished, and notification of when a function has output available so the orchestrator can invoke functions that depend on it. As it invokes functions, the server creates a listening WebSocket service, and each function establishes a connection to it on startup. The function keeps this connection open throughout its lifetime and only disconnects when it exits. One requirement of this design is that the orchestrator must be able to push messages to functions. Thus, since functions cannot receive inbound connections, they must either poll the server repeatedly for updates or establish a persistent full-duplex communication channel. Polling inherently has a higher latency and introduces more overhead since it uses multiple connections, so the library opts for a single persistent connection. The library uses the WebSocket protocol [25] because it supports full-duplex communication over a single TCP connection and is lightweight for small messages. The advantage of using a preexisting message protocol with an available Python library is that it simplifies implementation compared to a custom protocol written directly on top of TCP. Thus, the library uses the Python Socket.IO library for sending messages as both a client and server WebSocket implementation.

4.3 Communication Library API Overview

The serverless communications library presented in this work presents a high-level Application Programming Interface (API) that supports the development of serverless applications with inter-function communication over TCP. Applications using the library must specify two components: a topology graph of worker functions that dictates how they should communicate with each other and the data processing implementation itself. The communication library is responsible for invoking all of the functions and performing NAT traversal to establish all of the TCP connections for data transfer specified by the application topology. In the topology, applications must define the inputs and outputs for each worker function with two possible types for each: an internal TCP connection or an external application-implemented subroutine. Inputs and outputs with an internal TCP type are managed by the communication library itself, where the output data of one function is sent over TCP to become the input data of another. These internal TCP connections form directed edges between functions in the topology graph. The application is required to make sure the function graph is acyclic so that there are no circular dependencies between functions. While TCP connections can be used to send data between functions, applications must be able to read initial data and write their results to an external source because serverless functions are ephemeral. Thus, the communication library also allows applications to define custom subroutines for input and output and then specify in the graph which functions should run these subroutines and with what parameters. As an example, Fig. 3 defines a topology with three functions. The application's custom input and output subroutines could respectively read from and write to the S3 object specified in the path metadata field. Then, the first two functions, worker-0 and worker-1, download data from S3, process it, and send their immediate results over TCP to worker-2, which combines them and writes the final output to S3.

 In addition to the function topology, applications that use the proposed communications library must implement the actual data processing logic that the serverless functions run. The communications library expects applications to write a Python class that implements a list of required methods. Each of these methods uses Python generators to efficiently produce and/or consume a stream of data objects and run simultaneously in their own thread. As discussed before, the communication library expects the application to implement two methods to handle external inputs and outputs specified by the function topology. In addition, the application must write its

```
[ { "name": "worker-0",
    "inputs":  [ {"type": "s3", "path": "s3://example-bucket/input-0.csv"},
                 {"type": "s3", "path": "s3://example-bucket/input-1.csv"} ],
    "outputs": [ {"type": "tcp", "name": "worker-2"} ] },
  { "name": "worker-1",
    "inputs":  [ {"type": "s3", "path": "s3://example-bucket/input-2.csv"} ],
    "outputs": [ {"type": "tcp", "name": "worker-2"} ] },
  { "name": "worker-2",
    "inputs":  [ {"type": "tcp", "name": "worker-0"},
                 {"type": "tcp", "name": "worker-1"} ],
    "outputs": [ {"type": "s3", "path": "s3://example-bucket/result.csv"} ] } ]
```

Fig. 3 Example function topology definition

main data processing method that transforms objects received as inputs into objects that can be sent as outputs. Finally, the application must implement two methods that can serialize and deserialize the data objects it uses into a binary format for transmission over TCP.

4.4   Function Program Design

The proposed communication library uses a multi-threaded design to run application data processing subroutines and to perform network communication. A major advantage of using TCP for communication is that multiple functions running in parallel can use pipelining to improve throughput. So, as soon as one function is finished processing some block of data, it can immediately send that block to the next function in the pipeline while continuing to process the next data block. Thus, to implement pipelining efficiently, each function should be simultaneously reading input data, processing it, and writing output data. The communication library uses separate threads for each of these tasks to implement parallelism within a function so that a series of functions can achieve data pipelining. In each function, the library uses one thread to run the application's data processing subroutine, as well as one thread for each input or output source or destination, whether it is a TCP connection or a custom application subroutine. An alternative design would be to use an event-based approach using Python coroutines, but this requires non-blocking I/O, which is difficult to implement. Furthermore, the library's NAT traversal implementation must listen on a socket while making outbound connections at the same time, and using a separate thread for each is the most straightforward approach. Putting each application input and output subroutine in a separate thread means that it can block to read or write without affecting data processing. One disadvantage to using Python threads is that the Python interpreter uses a global lock so threads do not gain a performance benefit from running on a multi-core CPU [28]. For small function sizes, this does not matter since AWS Lambda will allocate less than one virtual CPU to the function, but it does prevent the communication library from fully utilizing multiple cores with larger function sizes. However, the communication library's use of multiple threads allows it to still support pipelining with other functions since it can process data at the same time as performing blocking input and output.

5   Evaluation

This section investigates the performance of the proposed communication library in AWS Lambda, one of the most widely used serverless computing environments. The section evaluates latency and throughput of the proposed library in Lambda, how this scales with data transfer sizes and number of communicating functions, and how it compares with communication through the S3 cloud storage. The performance analysis includes both microbenchmarks and templates of data-intensive applications written in MapReduce, a dominant programming paradigm for developing data processing pipelines.

5.1   TCP Performance in Lambda

5.1.1   Throughput

In order to set some expectations for the performance of network communications, this work ran several experiments to measure the performance of TCP connections between Lambda functions. The first metric measured was the raw throughput for a single connection between two functions. This work implemented a simple benchmark application using the proposed communication library, which uses Python TCP sockets with default settings. The benchmark consisted of two components, a producer and a consumer, each of which ran in a separate function. The producer repeatedly transmitted a fixed block of data as fast as possible and the consumer read data from its connection and counted the total size of the data received. To calculate throughput, the consumer also recorded the time between the first and last block of data received. Each of the tests sent 1 GB total and used functions with 3072 MB of memory. The orchestrator server used for the tests was hosted on a Ubuntu 18.04 virtual machine with 4 CPU cores and 8 GB of memory. This work tried different buffer sizes from 1 kB to 1 MB for reading and writing from the network socket and measured each with 5 runs. For each buffer size, Run 1 was a cold start due to the fact that the function code was modified and Lambda was thus forced to provision a new container for execution. The remaining Runs 2–5 were warm starts, but since the experiment only measured time while sending data after both functions had already initialized and established a TCP connection between them; this should not affect the results. The results are graphed in Fig. 4 with the full data listed in Table 1. The maximum throughput of about 680 Mbps is achieved when using buffers of 8 kB or larger. It makes sense that smaller buffer sizes would increase overhead, and the data show that the performance drops off with 4 kB buffers and smaller. While there

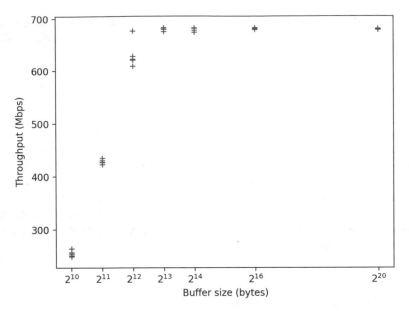

Fig. 4 Throughput (Mbps) between two Lambda functions based on buffer size

Table 1 Throughput (Mbps) between two Lambda functions based on buffer size

Buffer	1 kB	2 kB	4 kB	8 kB	16 kB	64 kB	1 MB
Run 1	264.965	435.219	676.116	682.517	673.219	678.374	679.434
Run 2	250.180	427.054	621.060	674.673	681.465	679.088	679.643
Run 3	257.868	423.257	609.191	680.248	677.673	682.674	680.327
Run 4	254.905	430.758	627.690	680.236	681.192	681.604	679.864
Run 5	252.256	427.881	622.661	679.362	681.917	680.902	677.918
Average	256.035	428.834	631.344	679.407	679.093	680.528	679.437

was some variability in the results; this is understandable because AWS Lambda makes no guarantees on the performance or bandwidth available to functions. An additional comment is that Fouladi et al. report observing speeds of 600 Mbps for direct network communications between functions in their future work discussion, which is similar if not quite as fast as the results in this work.

This work also experimented with two producers and with two consumers to see if a different function topology would increase the throughput. Each experiment used the same methodology as before with two functions, except adding either an additional producer or consumer. It halved or doubled the data sent by each producer so that each consumer would receive exactly 1 GB of data as before. The buffer size was fixed to 8 kB since increasing the size further made no difference in the measured throughput based on the previous experiment. The results for the first test with two producer functions sending to a single consumer function are listed in

Table 2 Throughput (Mbps) with two producer and one consumer functions

	Throughput
Run 1	679.197
Run 2	682.057
Run 3	677.897
Run 4	680.409
Run 5	681.373
Average	680.187

Table 3 Throughput (Mbps) with one producer and two consumer functions

	Consumer 1	Consumer 2	Total
Run 1	309.753	372.020	681.773
Run 2	369.452	307.845	677.297
Run 3	313.596	311.577	625.173
Run 4	306.886	354.791	661.677
Run 5	367.429	318.463	685.892
Average	333.423	332.939	666.362

Table 2. Likewise, the results for one producer function sending to two consumers are presented in Table 3. This shows the throughput for each consumer separately as well as the combined total. In both experiments, the maximum total throughput was approximately 680 Mbps, so the authors conjecture that Lambda imposes a bandwidth cap on each function that limits the observed speed in each case.

After measuring the network throughput for direct connections, the experiment was extended to determine if the throughput would decrease when using a longer chain of functions. One use case for the proposed communication library would be to implement a data pipeline with a chain of functions where each function receives data over TCP, processes it, and transmits its results to be processed by the next function. To see how network speed would affect the performance of the communication library in this case, the number of functions in the benchmark application were increased. This work used a simple copy operation for the data processing stage, so for an execution with a chain of n functions there would be a single producer, $n - 2$ forwarders, and finally a single consumer to measure throughput. This function graph is shown in Fig. 5. Each forwarder in the chain simply copied the data it received on its input connection and wrote it to its output connection as fast as possible without doing any other processing. Note that the case $n = 2$ with no forwarders is identical to the setup for the previous throughput test between two functions. This work used the same methodology as previous tests with the same sizes for function memory (3072 MB), read buffers (8 kB), and data sent (1 GB). It ran five trials for each value $n = 3, 4, 5, 10, 50$ and lists the results in Table 4. This table also includes the results for $n = 2$ from the previous throughput test for comparison. Note that the performance

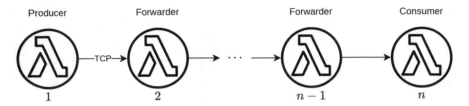

Fig. 5 n-function chain for throughput experiment

Table 4 Throughput (Mbps) for n-function chain

n	2	3	4	5	10	50
Run 1	682.517	613.047	679.856	604.428	679.200	629.934
Run 2	674.673	679.712	636.615	673.883	652.200	610.505
Run 3	680.248	681.710	667.813	620.689	647.970	631.217
Run 4	680.236	677.091	643.818	519.874	672.744	608.416
Run 5	679.362	679.981	665.965	646.269	677.361	618.527
Average	679.407	666.308	658.813	613.029	665.895	619.720

is not very consistent between runs, even for an extremely simple application that does no actual computation. While there is substantial variance between runs and none of the larger cases are as consistent as $n = 2$; notice it is still possible to get data throughput well over 600 Mbps using TCP communication at the end of a long function pipeline. One additional observation is that a complete trial with 50 functions took about 28 s of real time, whereas a trial with two functions only took about 12 s. This is due to the time it takes to initialize functions and perform NAT traversal to establish connections and the fact that the data simply has more functions to flow through. The purpose of this experiment was to measure potential throughput, not latency. The results demonstrate the possibility of using the proposed communication library to implement a multi-stage data processing pipeline with high throughput, assuming a workload where CPU processing is not a bottleneck. This gives an indication of what throughput is possible with AWS Lambda using the communication library, but also shows the inherent variability in its performance.

5.1.2 Latency

While the throughput between Lambda functions is more important for data processing, this work also tried to measure the network latency. There is variance to where Lambda provisions containers for different instances of a function—they could possibly be on the same machine or separate machines either close together or far apart. The latency experiment started two instances of a function where one sent a small message over a TCP connection, and the other waited to receive it and then responded immediately. The first function measured the total round-trip time between when it

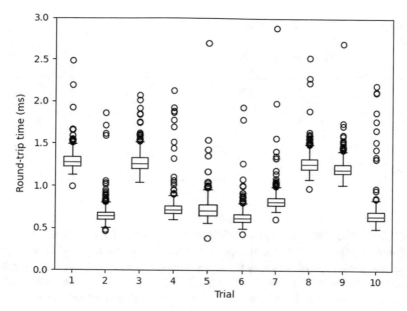

Fig. 6 Round-trip time between pairs of function instances

sent the original message and received a response. A small message size of 16 bytes was chosen so it would be guaranteed to fit in a single packet and because using other sizes made no appreciable difference. Each trial with two functions sent 500 round trip messages at 50 ms intervals. There were ten trials and each of which was a cold start so that Lambda used new containers for every function to get a broader sample. The round-trip times varied from 0.38 to 22.34 ms with an average across trials of 0.98 ms. See box plots for each trial in Fig. 6. A notable finding is high variability between trials. Since using cold starts for each trial forced Lambda to provision new containers each time, this shows that the latency between functions depends on AWS Lambda's internal placement of containers. The results also had a high number of outliers; note that Fig. 6 only shows outliers up to 3 ms, but the maximum overall time was 22.34 ms. Even though the round-trip time for TCP connections between functions has variability, the average latency of around 1 ms will be more than sufficient for most data processing applications, as throughput is more important.

5.2 TCP Versus S3 Comparison

This work developed a distributed serverless application using the proposed communications library to compare the performance of using TCP networking for communication instead of S3 object storage. It implemented a large-scale reduce-by-key

Table 5 Communication method performance for reduce-by-key application

Trial	S3		TCP	
	Billed duration (s)	Real time (s)	Billed duration (s)	Real time (s)
Run 1	821.158	156.972	434.830	33.310
Run 2	758.467	157.013	420.921	33.631
Run 3	808.216	154.258	430.776	32.629
Run 4	837.696	153.305	419.542	32.907
Run 5	810.812	155.919	405.974	32.412
Average	807.270	155.493	422.409	32.978

application using multiple stages of functions for merging based on a map-reduce computation paradigm. The input data set consisted of string keys and floating point values that were merged by summing the values for each distinct key. The application used a binary tree topology as shown in Fig. 7. Each function receives data from two sources: either S3 objects for the first stage or TCP connections for subsequent stages. After processing its input, it sends it to a function in the next stage via a TCP connection, except in the case of the final function, which writes the completely reduced output to S3. The experiments used 101 input files with a total of 44.5 million key-value pairs that were reduced to 1.9 million unique keys. Each run used 104 Lambda functions configured with 3072 MB of memory each. The experiment consisted of 6 executions, but the first was discarded so that all runs used warm starts for consistency. It measured the total billed duration of all functions as well as the real time elapsed from when the communication library orchestrator started the first function to when the last function exited. For comparison, this work ran an experiment using the same data and methodology but using S3 for inter-function communication instead of TCP. Instead of using TCP connections between pairs of functions to send data, the first function wrote all the data to an S3 object, and the receiving function read that object once it was completely written. Table 5 compares the real-time and cost measurement results for the TCP and S3 implementations. Using TCP connections is 4.7 times faster and costs only 52% as much based on Lambda execution time. One reason the proposed communication library achieves a large improvement using network connections is that functions send data incrementally so later stages can start processing data in parallel to previous ones. However, when using S3, an entire object must be created before it could be read, so the implementation with S3 does not achieve this same parallelism. This could be somewhat optimized by writing multiple smaller S3 objects, but this would require significant implementation effort and is not done by Amazon Web Service's sample Lambda map-reduce implementation either [29]. Compared to am S3-only implementation, the proposed communication library achieves significant time and cost savings by using direct TCP connections for inter-function communication.

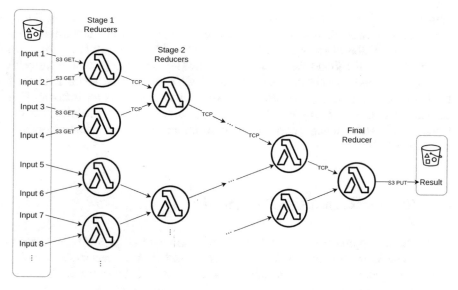

Fig. 7 Function graph for reduce-by-key application

6 Related Work

This section discusses related work in communication, execution frameworks, and application development strategies for serverless computing environments. The section articulates how the work presented in this chapter departs from prior work and the state of the art.

6.1 Serverless Communication

A major challenge and topic of study with using serverless computation for data processing is exchanging data between functions. Many serverless frameworks either use object storage for its low cost and scalability [5, 10], or in-memory stores for their speed [11, 12]. However, [8] develops a hybrid storage system that combines S3 and Redis to balance cost and performance. They evaluate their implementation on analytics applications and their results are significantly faster than S3-only but use less resources than traditional server clusters. Klimovic et al. [9] evaluate different types of storage systems, specifically technologies based on disk (S3), memory (Redis), and NVMe flash (Apache Crail and ReFlex). They look at several workloads and observe that the storage properties serverless applications need for communication are scalability, high IOPS, high throughput, and low latency while noting that high durability is not necessary because computations are short-lived. In addition, they find that NVMe flash storage can meet the performance requirements of most applications at a lower cost than memory-based solutions. Pocket [30] is a distributed elastic data store implementation designed for inter-function communication that uses a combination of storage based on DRAM, NVMe flash, and HDD. It scales automatically and chooses the most cost-effective storage type to meet bandwidth requirements, and matches the performance of AWS ElastiCache Redis at 59% of the cost. Based on these works, using a mixture of storage types for inter-function communication is necessary to balance cost, performance, and scalability, but the results in this work demonstrate that network communication can achieve all of these properties itself.

6.2 Serverless Execution Frameworks

With different approaches to handle scheduling, communication, and failures, there are multiple prior works that implement serverless execution frameworks for running distributed applications. A commonality between the different frameworks is that for communication between functions they all use object storage such as S3, in-memory databases such as Redis, or a combination of the two. One of the foundational execution frameworks in the field is PyWren [10], which provides a simple interface to map data using parallel tasks on serverless functions. Functions read input data from and write their results to S3 object storage, which the authors note is a bottleneck. PyWren is designed to support embarrassingly parallel tasks and does not have built-in support for exchanging data between functions, although applications could implement this themselves. There are several works that extend PyWren, including IBM-PyWren [1], which ports it to IBM Cloud Functions and extends it to run Map-Reduce jobs. In addition, NumPyWren [5] implements parallel linear algebra algorithms on top of PyWren. Perez et al. [31] develop a similar programming model and implementation that make it simple to process S3 files using parallel functions. While serverless functions are very well suited to run embarrassingly parallel applications, there are several works that support more general-purpose computation. For example, the gg tool [13] supports jobs where the computation graph is not available before execution, while managing function failures and stragglers. The authors demonstrate that it can compile real-world software projects up to 5 times faster than a warm AWS EC2 cluster without the need for provisioning. To simplify application implementation, Kappa [12] presents a programming interface using futures similar to standard concurrency programming models. It also provides checkpointing of function results in order to allow failure recovery. Both the gg and Kappa frameworks support either S3 or Redis as storage backends for communicating data between functions. The Wukong [11] framework is distinguished in that it does not use a centralized scheduler. Its worker functions manage scheduling themselves by either becoming or invoking their successor functions, which improves scalability and data locality. Wukong also uses a Redis cluster for inter-function communication. There are a variety of serverless frameworks for parallel computation, but they all rely on object storage or memory-based stores for transferring data between functions.

6.3 Serverless Applications

Serverless computing supports a wide variety of data processing and analytics applications. One strength of serverless computing is for embarrassingly parallel applications that need to scale quickly. Zhang et al. [4] implement an application to find the best hyperparameters for machine learning models, while [7] use serverless for video processing and consider the effect of function size and type. Ali et al. [2] use the inherent scalability of serverless combined with batching logic to efficiently perform machine learning inference with bursty workloads. While there are many other works that use serverless for straightforward data processing, some have more creative use-cases. Singhvi et al. [32] develop a network middle-box framework by using functions to operate on packet streams by dividing network flows into small time-limited chunks. By exploiting container reuse policies, [33] use serverless functions to implement an in-memory cache with a pay-per-use cost model. Overall, the scalability and pricing structure of serverless computing supports a broad range of applications.

6.4 FaaS Platforms

Besides frameworks and applications that run on commercial serverless providers, there are several works that improve the FaaS platform itself. To avoid the problem of copying data to serverless functions, [34] run functions directly on storage nodes to improve locality and reduce latency. Thomas et al. [35] implement an overlay network for serverless containers that reduces initialization time when many functions are started in parallel. Dukic et al. [36] reduce memory usage, and cold starts by sharing the application libraries and runtime between multiple instances of the same function. In addition, [37] provide quality-of-service for invoking and scheduling functions with a front-end framework to commercial FaaS platforms. To support large-scale scientific computing, [38] develop a distributed FaaS platform that uses heterogeneous clusters and supercomputers as function executors. All of these frameworks improve or extend the actual serverless execution platform, which demonstrates some of the current limitations of existing FaaS providers.

7 Conclusion

7.1 Limitations and Future Work

A major area for future work with the proposed communication framework is providing failure detection and recovery. The library currently does not implement a way to recover from connection or function failures and requires restarting the entire job.

Rerunning a computation from the beginning should be sufficient for many serverless applications, especially since they tend to be short-lived due to the 15-min maximum lifetime of Lambda functions. The elasticity of serverless computing makes spawning new functions to restart a job simple. Also, the delay in saving large amounts of intermediate results to durable object storage would be prohibitive and defeat the purpose of using direct connections for communications to begin with. Thus, using network connections prioritizes computation speed at the cost of some reliability. The main times for potential failures in the proposed design are at function initialization, connection establishment, and during data processing. It would be simple to add recovery logic to the proposed library for the first two cases by invoking a replacement function for the one that failed to connect. However, if a function failed during data processing, whether due to a connection reset, runtime error, or exceeding a resource quota, whatever data was in its memory at the time is lost so it would be extremely difficult for the remaining functions to recover. This kind of mid-computation recovery would be application specific and difficult to implement without checkpoint to persistent storage. One approach that uses the proposed communication library as-is would be to write an application in multiple stages that run in series where each stage completely writes its output to persistent storage before the next stage starts. This way, only a failed stage would need to be restarted instead of the whole application. In practice, the main source of failures seen during development were caused by programming mistakes. At the scale tested, there were not any consistent failures related to network connections or Lambda functions themselves. A topic for further research would be to determine how larger or longer-running application are affected by these issues and to implement a checkpointing or other recovery method for them. Although the design with direct network connections loses some reliability because data is not persisted in the case of failures, the elasticity, performance, and simplicity of only using network connections should outweigh this limitation for many applications, especially short-lived or data-intensive ones.

While the design of the proposed communications library does have some limitations, the results provide a lot of potential for future work. Primarily, either the library or direct network connections in general could potentially be used to improve the performance of serverless applications. Also, it would be useful to verify that NAT traversal is possible on serverless providers besides AWS Lambda and to port the communication library to them. A better solution would be for serverless providers to support direct networking between multiple instances of a function, which would eliminate the need for NAT traversal in the first place. One restriction with the proposed framework is that the number of functions and their respective inputs and outputs must be defined in advance for an application before execution. A possible improvement would be to add the ability for functions to dynamically start successor functions and initiate TCP connections with them during runtime, such as in the design used by the gg [13] or Wukong [11] frameworks. Despite these limitations, the proposed communication library introduces a fast, cost-effective, and scalable way for serverless functions to exchange data.

7.2 Summary

This work implements the first serverless communication library that uses direct network connections to transfer temporary computation data between functions. It use TCP hole punching to bypass the AWS Lambda firewall, which restricts incoming connections. The proposed communication library manages NAT traversal and connection setup and provides an interface for developing serverless applications with network connections. Based on this work's experiments, network communication is significantly faster and cheaper than writing and reading from object storage. Using TCP also eliminates the cost and provisioning overhead of in-memory databases. With its benefits for performance, cost, and scalability, the proposed communication framework and NAT traversal technique can be used to improve existing serverless computing applications and open the potential for new ones.

Acknowledgements We declare that this work is based on Daniel Moyer's (primary author) thesis titled "Punching Holes in the Cloud: Direct Communication Between Serverless Functions Using NAT Traversal", submitted in partial fulfillment of the requirements for the award of degree of Master of Science (M.Sc.) in the Virginia Polytechnic Institute and State University (Virginia Tech). This thesis is an authentic record of research work carried out by Daniel Moyer (first author) under the supervision of Dimitrios S. Nikolopoulos (second author), with comprehensive references to related work which are duly listed in the reference section. This Thesis has been verified for originality using Turnitin by the Virginia Polytechnic Institute and State University and the verification of originality has been stored along with the Thesis in the Virginia Tech archival online thesis repository (Vtechworks).

References

1. Sampe J et al (2018) Serverless data analytics in the IBM cloud. In: Proceedings of the 19th international middleware conference industry. Middleware'18. Association for Computing Machinery, Rennes, pp 1–8. ISBN: 9781450360166
2. Ali A et al (2020) Batch: machine learning inference serving on serverless platforms with adaptive batching. In: Proceedings of the international conference for high performance computing, networking, storage and analysis. SC'20. IEEE Press, Atlanta. ISBN: 9781728199986
3. Deese AS (2018) Implementation of unsupervised K-means clustering algorithm within Amazon Web Services Lambda. In: Proceedings of the 18th IEEE/ACM international symposium on cluster, cloud and grid computing. CCGrid'18. IEEE Press, Washington, DC, pp 626–632. ISBN: 9781538658154
4. Zhang M et al (2019) Seneca: fast and low cost hyperparameter search for machine learning models. In: 2019 IEEE 12th international conference on cloud computing (CLOUD), pp 404–408
5. Shankar V et al (2020) Serverless linear algebra. In: Proceedings of the 11th ACM symposium on cloud computing. SoCC'20. Association for Computing Machinery, Virtual Event, pp 281–295. ISBN: 9781450381376
6. Fouladi S et al (2017) Encoding, fast and slow: low-latency video processing using thousands of tiny threads. In: 14th USENIX symposium on networked systems design and implementation (NSDI 17), Mar 2017. USENIX Association, Boston, MA, pp 363–376. ISBN: 978-1-931971-37-9

7. Zhang M et al (2019) Video processing with serverless computing: a measurement study. In: Proceedings of the 29th ACM workshop on network and operating systems support for digital audio and video. NOSSDAV'19. Association for Computing Machinery, Amherst, MA, pp 61–66. ISBN: 9781450362986

8. Pu Q, Venkataraman S, Stoica I (2019) Shuffling, fast and slow: scalable analytics on serverless infrastructure. In: 16th USENIX symposium on networked systems design and implementation (NSDI 19), Feb 2019. USENIX Association, Boston, MA, pp 193–206. ISBN: 978-1-931971-49-2

9. Klimovic A et al (2018) Understanding ephemeral storage for serverless analytics. In: 2018 USENIX annual technical conference (USENIX ATC 18), July 2018. USENIX Association, Boston, MA, pp 789–794. ISBN: 978-1- 939133-01-4

10. Jonas E et al (2017) Occupy the cloud: distributed computing for the 99%. In: Proceedings of the 2017 symposium on cloud computing. SoCC'17. Association for Computing Machinery, Santa Clara, CA, pp 445–451. ISBN: 9781450350280

11. Carver B et al (2020) Wukong: a scalable and locality-enhanced framework for serverless parallel computing. In: Proceedings of the 11th ACM symposium on cloud computing. SoCC'20. Association for Computing Machinery, Virtual Event, pp 1–15. ISBN: 9781450381376

12. Zhang W et al (2020) Kappa: a programming framework for serverless computing. In: Proceedings of the 11th ACM symposium on cloud computing. SoCC'20. Association for Computing Machinery, Virtual Event, pp 328–343. ISBN: 9781450381376

13. Fouladi S et al (2019) From laptop to lambda: outsourcing everyday jobs to thousands of transient functional containers. In: 2019 USENIX annual technical conference (USENIX ATC 19), July 2019. USENIX Association, Renton, WA, pp 475–488. ISBN: 978-1-939133-03-8

14. Amazon Web Services (2021) AWS lambda: serverless compute. Url: https://aws.amazon.com/lambda. Accessed 31 Mar 2021

15. Microsoft Azure (2021) Azure functions serverless compute. Url: https://azure.microsoft.com/en-us/services/functions. Accessed 01 Apr 2021

16. Google Cloud (2021) Cloud functions. Url: https://cloud.google.com/functions. Accessed 01 Apr 2021

17. IBM (2021) IBM cloud functions. Url: https://www.ibm.com/cloud/functions. Accessed 01 Apr 2021

18. Oracle (2021) Cloud functions. Url: https://www.oracle.com/cloudnative/functions. Accessed 01 Apr 2021

19. OpenFaaS Ltd. (2021) OpenFaaS: serverless functions made simple. Url: https://www.openfaas.com. Accessed 01 Apr 2021

20. Apache Software Foundation (2021) Apache OpenWhisk is a serverless, open source cloud platform. Url: https://openwhisk.apache.org. Accessed 01 Apr 2021

21. Kubeless (2021) Kubeless: the kubernetes native serverless framework. Url: https://kubeless.io. Accessed 01 Apr 2021

22. Srisuresh P, Holdrege M (1999) IP network address translator (NAT) terminology and considerations. RFC 2663. RFC Editor, Aug 1999

23. Ford B, Srisuresh P, Kegel D (2005) Peer-to-peer communication across network address translators. In: Proceedings of the annual conference on USENIX annual technical conference. ATEC'05. USENIX Association, Anaheim, CA

24. Rosenberg J et al (2003) STUN—simple traversal of user datagram protocol (UDP) through network address translators (NATs). RFC 3489. RFC Editor, Mar 2003

25. Fette I, Melnikov A (2011) The WebSocket protocol. RFC 6455. RFC Editor, Dec 2011

26. Mahy R, Matthews P, Rosenberg J (2010) Traversal using relays around NAT (TURN): relay extensions to session traversal utilities for NAT (STUN). RFC 5766. RFC Editor, Apr 2010

27. Grinberg M (2021) Python Socket.IO server and client. Url: https://github.com/miguelgrinberg/python-socketio. Accessed 01 Apr 2021

28. Python Software Foundation (2021) Glossary: Python 3.8.8 documentation. Url: https://docs.python.org/3.8/glossary.html#term-globalinterpreter-lock. Accessed 01 Apr 2021

29. Amazon Web Services (2021) Serverless reference architecture: MapReduce. Url: https://github.com/awslabs/lambda-refarch-mapreduce. Accessed 06 Apr 2021
30. Klimovic A et al (2018) Pocket: elastic ephemeral storage for serverless analytics. In: 13th USENIX symposium on operating systems design and implementation (OSDI 18), Oct 2018. USENIX Association, Carlsbad, CA, pp 427–444. ISBN: 978-1-939133-08-3
31. Perez A et al (2019) A programming model and middleware for high throughput serverless computing applications. In: Proceedings of the 34th ACM/SIGAPP symposium on applied computing. SAC'19. Association for Computing Machinery, Limassol, pp 106–113. ISBN: 9781450359337
32. Singhvi A et al (2020) SNF: serverless network functions. In: Proceedings of the 11th ACM symposium on cloud computing. SoCC'20. Association for Computing Machinery, Virtual Event, pp 296–310. ISBN: 9781450381376
33. Wang A et al (2020) InfiniCache: exploiting ephemeral serverless functions to build a cost-effective memory cache. In: 18th USENIX conference on file and storage technologies (FAST 20), Feb 2020. USENIX Association, Santa Clara, CA, pp 267–281. ISBN: 978-1-939133-12-0
34. Zhang T et al (2019) Narrowing the gap between serverless and its state with storage functions. In: Proceedings of the ACM symposium on cloud computing. SoCC'19. Association for Computing Machinery, Santa Cruz, CA, pp 1–12. ISBN: 9781450369732
35. Thomas S et al (2020) Particle: ephemeral endpoints for serverless networking. In: Proceedings of the 11th ACM symposium on cloud computing. SoCC'20. Association for Computing Machinery, Virtual Event, pp 16–29. ISBN: 9781450381376
36. Dukic V et al (2020) Photons: lambdas on a diet. In: Proceedings of the 11th ACM symposium on cloud computing. SoCC'20. Association for Computing Machinery, Virtual Event, pp 45–59. ISBN: 9781450381376
37. Tariq A et al (2020) Sequoia: enabling quality-of-service in serverless computing. In: Proceedings of the 11th ACM symposium on cloud computing. SoCC'20. Association for Computing Machinery, Virtual Event, pp 311–327. ISBN: 9781450381376
38. Chard R et al (2020) FuncX: a federated function serving fabric for science. In: Proceedings of the 29th international symposium on high-performance parallel and distributed computing. HPDC'20. Association for Computing Machinery, Stockholm, pp 65–76. ISBN: 9781450370523

Hybrid Serverless Computing: Opportunities and Challenges

Paul Castro, Vatche Isahagian, Vinod Muthusamy,
and Aleksander Slominski

Abstract In recent years, the adoption of serverless computing has surged due to the ease of deployment, attractive pay-per-use pricing, and transparent horizontal auto-scaling. At the same time, infrastructure advancements such as the emergence of 5G networks, the explosion of devices connected to the Internet known as the Internet of Things (IoT), as well as new application requirements that constrain where computation and data can happen, will expand the reach of Cloud computing beyond traditional data centers into the emergent Hybrid Cloud market that is predicted to expand to over a trillion dollars in next few years. In Hybrid Cloud environments, driven by serverless tenants, there is an increased need to focus on enabling productive work for application builders using a distributed platform consisting of public clouds, private clouds, and edge systems. In this chapter we investigate how far serverless computing can be extended to become Hybrid Serverless Computing,

Paul Castro, Vatche Isahagian, Vinod Muthusamy, Aleksander Slominski—equal contributions to the paper.

All four authors of this chapter have deep expertise in serverless middleware. They were part of the founding team from IBM Research who worked on the Apache OpenWhisk Serverless project and the IBM Cloud Functions platform. Furthermore, they have been constant contributors to raise the awareness and advancement of Serverless computing since its beginning through tutorial presentations (at ICDCS 2017), organization of multiple workshops (https://www.serverlesscompu ting.org/), speaking at seminars, and contributing several papers in top conferences and journals on serverless computing. Some of the noteworthy contributions are a cover article on serverless computing in Communications of the ACM, and a book chapter in Chaudhary, S., Somani, G. and Buyya, R. eds., 2017, *Research advances in cloud computing,* Springer.

P. Castro · V. Isahagian (✉) · V. Muthusamy · A. Slominski
IBM Research, NY, USA
e-mail: vatchei@ibm.com

P. Castro
e-mail: castrop@us.ibm.com

V. Muthusamy
e-mail: vmuthus@us.ibm.com

A. Slominski
e-mail: aslom@us.ibm.com

© The Author(s), under exclusive license to Springer Nature Switzerland AG 2023   43
R. Krishnamurthi et al. (eds.), *Serverless Computing: Principles and Paradigms,*
Lecture Notes on Data Engineering and Communications Technologies 162,
https://doi.org/10.1007/978-3-031-26633-1_3

outline its definition, describe steps towards achieving it, and identify opportunities and challenges.

Keywords Cloud · Computing · Hybrid · Serverless · Standards · Vision

1 Introduction

Cloud computing has evolved from infrastructure services provided by a cloud vendor to a whole array of services needed to develop an application. As more enterprises have migrated their applications to the cloud, there is a trend towards *hybrid cloud* architectures [1], where application components are distributed across multiple cloud providers. This could be to deliberately avoid vendor lock-in, the consolidation of applications that were independently developed across heterogeneous platforms, a result of regulatory constraints that necessitate a hybrid on-prem and cloud architecture, or because developers want to take advantage of differentiated features offered by different platforms. As depicted in Fig. 1, a natural next step in this evolution is to include edge platforms, IoT devices, and personal computing devices, an architecture we refer to as *hybrid computing*.

Today, serverless computing platforms do not work in a hybrid setting. Serverless computing is still largely limited to applications running on a single provider's platform. Our vision is a hybrid serverless computing paradigm where the serverless principles are applied to applications that span multiple clouds and devices.

This approach provides multiple advantages (e.g. application portability), which are discussed in more detail in Sects. 2 and 3.

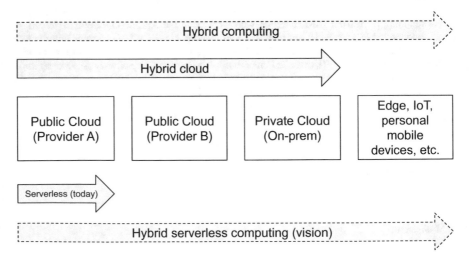

Fig. 1 Evolution of hybrid cloud and related areas

Serverless platforms arrived less than a decade after public clouds and can be viewed as a natural extension of cloud that takes a radical step forward to focus on simplicity, scalability, and pay-per-use with little or no knowledge of cloud servers. In serverless [2], most commonly in the form of Functions-as-a-Service (FaaS) and newer serverless offerings like Knative Serving [3], users ignore most infrastructure concerns and focus on developing and deploying their application logic. This has had a positive effect on developer productivity as it allows developers to focus more on their application logic and less on non-differentiating configuration and operation of a platform. We see a lot of potential benefit in applying serverless principles to a hybrid setting, where distribution and heterogeneity magnify the challenges of cloud native development. A broad view of the "serverless mindset" means designing systems that maximize focus on business value and minimize any overhead [4].

Directly applying serverless principles originally designed for a single vendor platform to a hybrid setting is difficult. However, the same technical and non-technical forces steering the evolution towards hybrid cloud and hybrid computing are going to lead serverless application developers to consider a broader and more heterogeneous view of serverless. The emergent Hybrid Cloud market could potentially expand to over a trillion dollars [5] and in our view, serverless vendors have both business incentives and technical reasons (discussed in Sect. 2) that act as headwinds towards this broad view of hybrid serverless computing. We chart a course for both serverless practitioners and researchers towards hybrid serverless computing architectures.

This chapter is organized into sections with objectives as follows:

- Sect. 2 presents an overview of the evolution of cloud computing
- Sect. 3 proposes a case for hybrid serverless computing
- Sect. 4 designs and draws a path towards hybrid serverless computing
- Sect. 5 lists and discusses the challenges in achieving this vision

The focus of this chapter is on ensuring developer productivity when building hybrid serverless applications. There are also important runtime performance considerations, which this work mentions in Sect. 5, but are not the core focus of this chapter.

2 Trends in Cloud and Serverless Computing

2.1 Cloud and Hybrid Cloud

Cloud evolved from early ideas of Utility [6] and Grid Computing [7]. An early definition of cloud computing has similarities with the definition of serverless computing [8]:

- *The appearance of infinite computing resources available on demand, quickly enough to follow load surges.*
- *The elimination of an up-front commitment by cloud users.*

- *The ability to pay for use of computing resources on a short-term basis as needed (for example, processors by the hour and storage by the day) and release them as needed.*

That definition of Cloud computing [9] (and others such as [10]) evolved to be "a way of using computers in which data and software are stored or managed on a network of servers". Cloud computing as Infrastructure-as-a-Service (IaaS) began in 2006 with the general availability of Amazon Elastic Compute Cloud in Amazon Web Services (AWS).

Today, we are in an era of hyperscalers, where the cloud market is dominated by a few planet-scale, public platform providers. The term Cloud is often synonymous with commercial public clouds like AWS, Google Cloud, and Microsoft Azure, that offer a catalog of hardware and software services that users can provision based on a pay as you go model [11]. Users still have to be aware of, configure, and manage cloud resources, though this can all be done in a self-service manner with no need to manage physical data centers. Each cloud provider platform is a siloed, vertical stack of software services and infrastructure options designed as general-purpose components to build most cloud native applications. Cloud providers offer a broad catalog of services that are unique to each provider. Services across providers do not cleanly interoperate and today there are few, if any, industry standards. While the cloud providers do provide some managed versions of open source software, the tooling and features of these will differ between providers. Applications deployed at one provider are not easily portable to another [12, 13].

In practice, this lack of interoperability may highlight the need to move beyond a few hyperscale cloud providers and make applications more portable across cloud providers. Today, many enterprises already make use of more than one cloud provider [1, 14, 15]. This could be deliberate; users will naturally want to build multi-cloud or hybrid cloud solutions that avoid vendor lock-in and take advantage of the best services available, or just a result of the organic, uncoordinated nature of cloud adoption in some industries [16]. Larger enterprises have significant capital investments in their own private data centers and in many cases it makes sense to adopt a hybrid mixture of on-premises and public clouds. Also, increasing awareness of data privacy, industry specific requirements, and evolving regulatory issues are now a pressing concern—efforts like GAIA-X [17] seek to restrict the use of data and computation in order to preserve data sovereignty. As such, the hyperscale cloud providers and other 3rd party vendors are looking to ease the burden of creating multi-cloud or hybrid cloud solutions.

There are many terms—multi-cloud, hybrid cloud, polycloud, supercloud—being defined that capture different perspectives on this emerging, heterogeneous, distributed platform [18]. In this chapter, we refer to Hybrid Cloud in the broadest sense, encompassing public clouds, on-prem data centers, and edge services. At IBM, hybrid cloud combines and unifies public cloud, private cloud, and on-prem infrastructure to create a single, flexible, cost-optimal IT infrastructure [19]. Microsoft's definition focuses on sharing data and applications across hybrid cloud environments [20]. Some people define hybrid cloud to include "multi-cloud" configurations where

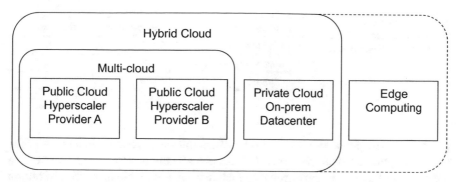

Fig. 2 Relations between public cloud, private cloud, multi-cloud, hybrid cloud, and edge computing

an organization uses more than one public cloud in addition to their on-prem datacenter. Figure 2 shows how different components of hybrid computing relate to each other.

Many in academia and industry are taking the perspective of Hybrid Cloud as a distributed application development platform not tied to a single provider with applications composed of heterogeneous services running anywhere. For example, the vision of Sky Computing is inspired by the evolution of the Internet and the goal would be to connect the offerings of cloud providers and other heterogeneous cloud platforms through the use of industry-wide protocols [21]. The RESTless Cloud [22] invokes the need for an industry-wide POSIX-like standard to cloud APIs to ease interoperability. Similarly, User Defined Cloud [23] looks for high-level declarative techniques for users to define their own clouds based on application requirements. In industry for example Dell Technologies APEX [24] and HPE GreenLake [25] market hybrid cloud solutions. Crossplane [26] is an open source project currently in the incubation stage in the Cloud Native Computing Foundation (CNCF), that allows users to control resources in different clouds using a high-level interface running in Kubernetes.

2.2 Computing Trends Toward Hybrid Computing

It is estimated that there are between 10 and 100 million servers used in all of the public clouds [27] and many more in private clouds and on-premises data centers. There are about 1–3 billion personal computing devices used worldwide [28, 29]. Modern computing devices not only support more advanced computing capabilities but they are also always connected to the Internet. There is a large number of computational resources available today in hybrid clouds in the form of servers, but the number of personal computing devices (PCs) such as laptops, smartphones, and IoT devices, is larger and growing rapidly. According to current research [30], the

Computing paradigm	Number of devices (in billions)
Hybrid cloud	0.1–1B
Personal computing (PC)	1–3B
Smartphones	6–7B
IoT	> 35B

Table 1 Number of computing devices in 2022 based on estimates available

number of smartphone users is over 6.5 billion, translated roughly to be 83% of the world's population. The number of IoT devices already is several times more than the number of people and growing very fast: in 2021 there were more than 35 billion of IoT devices and double of this number is predicted by 2025 [31].

An important part of connecting the cloud with consumer and IoT devices is edge computing providing low-latency high-bandwidth intermediate computing capabilities. Edge computing is also growing fast, according to IDC [32]: "the worldwide edge computing market will reach $250.6 billion in 2024 with a compound annual growth rate (CAGR) of 12.5% over the 2019–2024 forecast period".

Computing devices are getting cheaper, more energy efficient, and almost always connected to the Internet. We summarize the number of different types of computing devices in Table 1 based on cited numbers above.

Hybrid Computing is a superset of Hybrid Cloud to include not only cloud or edge computing but any kind of computing that is connected in some ways to the Internet, even intermittently. Additional types of computing may become more important or emerge in the future such as Fog Computing, Web3 and Blockchain based distributed computing, metaverse with AR/VR devices, and so on. In Fig. 3 it is shown how Hybrid Computing is related to Hybrid Cloud and different types of computing paradigms.

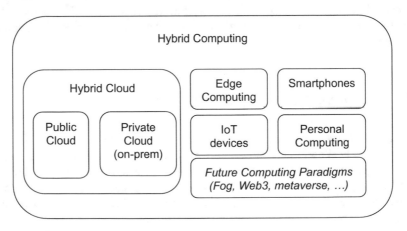

Fig. 3 Hybrid computing as a superset of hybrid cloud

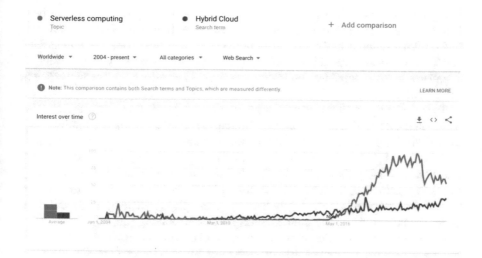

Fig. 4 Popularity of "serverless computing" and "hybrid cloud" in the last 15 years. (Google Trends)

Definition: Hybrid Computing combines any computing paradigm to allow creation of applications that can operate in a seamless way in multiple computing infrastructures connected by the Internet.

Even though the definitions of Hybrid Cloud are evolving, and Hybrid Computing may change in the future, this area is gaining popularity alongside serverless computing. For example, Google Trends shows the growth in popularity of "Serverless computing" and "Hybrid Cloud" as search terms [33] (see Fig. 4.)

It is our hypothesis that future computing paradigms will have serverless characteristics: simplicity, scalability, and developer focus.

3 The Case for Hybrid Serverless Computing

We previously defined serverless computing as *a platform that hides server usage from developers and runs code on-demand, automatically scaled, and billed only for the time the code is running* [2]. This view of serverless computing focuses on simplifying compute and emphasizes the benefits of hiding complexity and offloading much of the work around application scaling, security, compliance, and other infrastructure management concerns to the cloud provider. Recent work [34] predicts that serverless will be the dominant form of cloud computing and that the future of serverless will be abstractions that simplify cloud programming. With this broader interpretation of serverless, and the economics associated with serverless, one can argue that it represents an evolution of the cloud platform itself and is a key element to attract the next generation of developers to the cloud. Simplifying cloud programming has been one of the values promoted by serverless since its inception. Tim Wagner, the former GM for AWS Lambda and considered the father of serverless compute, emphasized that the value of serverless is enhancing productivity because it "gets rid of undifferentiated heavy lifting" for application developers and allows for "capital efficient value creation" [35].

Today, cloud providers are offering services beyond Functions-as-a-Service (FaaS) that are branded as "serverless." Managed serverless container runtimes like IBM Code Engine and Google Cloud Run broaden the executable component to containerized HTTP endpoints, making it easier to run more general purpose code and de-emphasizing the low-latency startup requirements found in FaaS [36, 37]. Accessing state from serverless functions in FaaS runtimes is also getting easier as serverless storage is increasingly becoming an option, where users benefit from minimal configuration, finer grained billing, and auto-scaling [38, 39].

Open source communities provide serverless platforms that can run on any cloud, and are sometimes offered as managed services by cloud providers. For example, IBM Code Engine and IBM Cloud Functions are both based on open source projects. There are active industrial and research communities using serverless open source projects such as Apache OpenWhisk, OpenFaas, and Knative [40, 41, 42]. This has a normalizing effect, where open source software is available across cloud providers and potentially provides a foundation for application portability.

Vendors are also providing cloud agnostic, planet-scale serverless platforms. Newer services like CloudFlare Workers will automatically migrate function code as needed using their existing content distribution network, breaking the regional deployment restrictions found in traditional FaaS platforms [43]. Snowflake is a popular planet-scale data warehouse with serverless features and users don't necessarily need to be aware of where it is running [44]. Cloud agnostic service provider ecosystems may grow more in the future as players can specialize in addressing specific concerns for a single industry more nimbly than today's hyperscalers. The success of these offerings is in part due to the ease of deploying application code globally with less low-level configuration.

In our view, serverless' focus on reducing complexity and allowing the transparent and elastic use of compute resources provides an ideal development platform that federates the heterogeneous, distributed hybrid cloud platform. The largest benefit that serverless provides to improved developer productivity may be raising the level of abstraction for developers so that their interface to the cloud is programmatic and requires minimal configuration to conform to system-defined limits. Applying the serverless design principles beyond just compute and expanded for Hybrid Computing can potentially provide an abstraction layer that separates developer productivity concerns from platform optimizations.

3.1 Hybrid Serverless Architecture Design Principles

The traditional serverless principles, at least as applied to FaaS platforms, provides a simplified but naive view of compute where concerns about the distributed nature of the platform can be abstracted away from the developer. In a hybrid cloud setting, abstracting away the distribution concerns may not always be possible given the greater time scales for latency, the higher complexity of fixing bugs that only appear at scale, or the need to provide specific guidance on the geographical placement of application components. The high-level goals of serverless are still valid, but it is necessary to update the traditional serverless design principles to surface the necessary abstractions that developers may require. We re-cast the traditional serverless design principles as follows:

1. **Don't worry about infrastructure**: Beyond minimizing the need to configure compute, hybrid cloud serverless requires the need to abstract away the additional middleware, often deployed when provisioning primary services, across different cloud providers. For example, networking is often auto-provisioned when requesting a service component but today this only works within a single cloud provider. Identity management across cloud estates is also an open issue.
2. **Don't worry about capacity**: Today, many serverless offerings beyond FaaS will auto-scale based on the workload requirements. For hybrid cloud serverless, developers will need to be aware of the differences in services across clouds, even those services (like ones based on open source) that are purportedly the same across providers. This means automated discovery and the ability to choose from the discovered options based on transparent optimizations becomes important.
3. **Don't worry about idle**: Not paying for idle is beneficial for workloads that don't need resources to be provisioned all the time. This finer grained pricing model is beneficial to hybrid cloud serverless but needs further refinement. First, it should apply to the entire hybrid cloud stack. Beyond charging GB-sec for compute or storage and egress costs for data, it may be necessary for the developer to be more aware of what type of hardware is available, e.g. GPU vs non-GPU, and pricing should reflect this. Also, scale-to-zero has impacts beyond direct costs; improving compute density may impact other factors like sustainable computing [45].

Note that these expanded serverless design principles are a simplification for the developer and are not meant to solve all the problems of running distributed applications. Instead, it provides a perspective on how serverless can help the cloud evolve to simplify cloud native development. This can inform how to construct a hybrid serverless architecture.

Borrowing the hourglass metaphor from computer networking, let's consider the waist of the hourglass to be the interface that the cloud vendor offers to application developers. As illustrated in Fig. 5, in an era when developers were building applications using VM abstractions, the waist was relatively low in the cloud application stack. In the serverless era, however, the vendor takes on many more responsibilities under the serverless interface (such as FaaS).

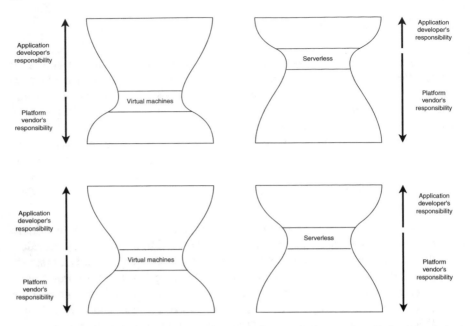

Fig. 5 Building cloud applications over a virtual machine abstraction requires the developer to be responsible for more of the application stack. Conversely, serverless abstractions give the platform vendor opportunities to optimize more of the stack

The fact that the vendor owns a much larger part of the application stack leads to a natural tendency for single-vendor serverless applications. There are technical reasons for this, such as the vendor's visibility into the larger stack affording them the ability to implement platform-level optimizations. There are also business incentives, such as pay-as-you-go the serverless billing models only becoming economical if the vendor has the ability to monitor and manage the underlying resources.

The challenge is how to break out of the ingrained bias towards single-vendor serverless applications. In this chapter, we argue that there are two approaches to building hybrid serverless applications that span the services offered by vendors. The compiler-based approach does not require any special support from vendors, whereas the standards-based one is more robust but requires vendor support.

3.2 Benefits to Application Developers and Platform Providers

A hybrid serverless architecture brings a number of benefits both to the serverless application developers and platform providers.

- Developers benefit from the flexibility to migrate their serverless applications to different platform vendors to take advantage of pricing differences. The choice of vendor is not a one-time decision, since an application may exhibit different load patterns as it evolves, and different vendors may offer pricing plans optimized for different classes of workloads.
- Similarly, an application may require different non-functional guarantees (such as throughput targets) or require different platform services (such as a global load balancer) as it matures from a proof-of-concept to a popular production application. Developers need flexibility to migrate vendors as their requirements evolve.
- A related point is that developers may want to deploy portions of their applications to the edge to improve user-perceived latency or leverage a combination of VM-based and serverless based resources for application execution [46]. A hybrid architecture would facilitate more fine-grained deployment choices for portions of the application.
- Similarly, developers may have constraints on where parts of their application can be deployed. For example, there may be regulations that govern where health records can be stored, whether within a country or an on-prem datacenter managed by the company.
- Another developer benefit is the ability to mix and match services provided by different platform vendors rather than make the choice to deploy their entire application on a single platform. For example, they may choose the FaaS platform of one vendor, the logging service from another vendor, and the messaging service from a third one.

Regarding the last point, vendors may discourage such hybrid architectures with their pricing plans, and there may be legitimate performance reasons to deploy an entire application on a single platform, but we still think that developers should have choice in their deployment options. Furthermore, we believe that giving developers this choice can in turn benefit platform vendors, especially the growing ecosystem of non-hyperscale vendors who are filling gaps in regional and industry specific offerings.

In particular, serverless platforms today need to offer a breadth of services that developers may need. Making hybrid architectures more developer friendly gives an opportunity for vendors to offer specialized services without having to replicate the complete stack of services needed for an enterprise serverless application. For example, FaaS can facilitate the integration of quantum computers as accelerators for existing enterprise applications by decoupling the specification of the quantum circuit from the underlying runtime system and providing transparent request auto-scaling [47]. Similar advantages are seen in other domains like serverless Complex Event Processing [48, 49].

Developer flexibility in this case can lead to more competition among platform vendors, resulting in a virtuous cycle of lower costs, more innovative service offerings, and increased adoption of hybrid serverless architectures.

3.3 Roadblocks to Hybrid Serverless Computing

The limitations of today's serverless computing platforms with respect to hybrid cloud mirror the overall cloud itself. Recent surveys on multi-cloud identify the trend towards hybrid cloud platforms and outline specific challenges. We summarize the latter from two surveys [1, 50] in Table 2.

Current serverless platforms including AWS Lambda, Azure Functions, Google Cloud Functions, IBM Cloud Functions, and offerings based on open source software like OpenFaaS, Apache OpenWhisk, and Knative, are managed services and do not interoperate across cloud provider boundaries. This is reflected by the current use of serverless applications which typically use a single platform because:

1. Serverless relies heavily on a cloud provider's integrated ecosystem of databases, storage, etc. which are not entirely serverless.
2. Serverless applications are mostly built as extensions of already deployed cloud applications, to compose cloud services, or as features of cloud applications.
3. Its relatively easy to end up in anti-patterns of serverless applications design [51, 52].

There is little or no expectation of interoperability across different serverless platforms. This is part of a broader challenge around the operational complexity of today's hybrid environments that force developers to use different tooling for each vendor [1]. Experience in one cloud environment is not transferable across clouds [12, 13]. Current multi-cloud management tools focus more on operational tasks around managing multiple Kubernetes clusters. For example, SUSE Rancher

Table 2 Summary of practitioner challenges for hybrid cloud from recent surveys

Hybrid cloud challenge	Description
Security	Security across multiple clouds remains the highest priority concern as the attack surface is increased, and each cloud vendor has its own identity management and security controls that don't integrate across clouds
Integration complexity	Operating multiple clouds requires specialized skills that don't translate across clouds, and integration of edge components creates additional challenges that are not fully addressed by existing toolsets
Managing costs	Opaque pricing models and lack of comprehensive tools to understand and manage cloud costs makes it difficult to optimize cloud spending
Reduced visibility and control	Observability of deployed resources across clouds reduces the ability to understand what is deployed and how an application is operating
Application modernization	Integrating legacy applications with cloud is an existing problem that is exacerbated by multi-cloud. Even cloud native applications may have inertia and are not portable across different cloud providers

[53] helps manage multiple Kubernetes clusters, while the CNCF incubating project Crossplane [26] focuses on cross-provider resource management through Kubernetes operators.

One exception is Funqy, a Java-based library that allows you to write Java functions using a single API that works across FaaS implementations using the Quarkus runtime [54]. While this potentially improves code portability, it does not address the broader issues of hybrid cloud development.

There is a class of enterprise applications that aren't amenable to run fully on a single public cloud due to economics (e.g. existing investments in an on-premises data center) and regulatory constraints [55]. However, inconsistent development experience across vendors [56], and the vendor and platform lock-in in today's most popular serverless platforms results in reduced developer productivity because of the need to develop non-transferable expertise across cloud estates [51, 57].

4 Towards a Hybrid Serverless Architecture

It is possible today to use multiple clouds and multiple serverless offerings (sometimes called hybrid serverless multi-cloud [58]) but it is not easy. The main limitations are in part due to the difficulty in promoting a set of standards and shared abstractions in an environment as complex as Hybrid Cloud. There is a long history of developing standards to allow interoperability among clouds, but most with limited success [18]. However, as cloud evolves, we believe there will be a growing understanding on how to make hybrid cloud deployments easier, which will include agreement on interoperability standards perhaps based on APIs provided by popular open source projects [21, 59].

Examples from industry validate the trend towards solutions that run across multiple cloud environments [60]. Snap describes their architecture as inherently multi-cloud to encourage better pricing and allow for regionalization [61]. Data warehouse service Snowflake [44] has overtaken AWS RedShift [62] in market share and is deployed across clouds.

Recent research work considers Hybrid Serverless deployments for FaaS architectures; for example Baarzi et al. [63] discuss the merits of a virtual serverless provider to aggregate serverless platforms across clouds. Chard et al. [64] federate endpoints to execute serverless functions in the domain of scientific computing. Perez et al. [49] goes a step further to enable not just the deployment of functions on distributed endpoints, but also scale those endpoints automatically.

More generally, researchers are proposing a new view of cloud as the amalgamation of all cloud estates. Recent proposals emphasize consistent developer experience across multiple clouds either as "supercloud" [65] or Sky computing [21]: "the vision of a Sky full of computation, rather than isolated clouds."

We believe that a combination of evolving computing trends, hybrid cloud approaches and serverless computing may lead to one common computing approach: **Hybrid Serverless Computing**. In this view, serverless principles are inherent across the entire cloud stack from data centers, edge, and IoT devices to help simplify cloud programming and provide a separation of concerns between developer productivity/operator simplicity and the engineering work to optimize cloud platforms.

Definition: Hybrid Serverless Computing extends Hybrid Cloud to include all computing paradigms unified by use of containers and providing serverless approach to computing by using a set of standards that allow better interoperability between clouds and/or a compiler-based approach which can help automate deployment provisioning across clouds.

Serverless promotes a "rental" view of cloud resources [34] that are consumed on-demand and the responsibility of managing the non-differentiating overhead is the responsibility of an automated platform. It is our view that the rental model always makes sense, even for users that own their own platforms:

- Developer productivity is potentially improved because of a simplified model of the "rented" resources and how they are charged
- Serverless, in the form of FaaS, allows developers and platform operators a portable unit of compute that provides a common abstraction where requirements for improved developer productivity and platform operation meet. Whether it be function source code or containers (or any other unit of compute), this allows a higher degree of flexibility when an application is targeting a heterogeneous platform.
- Auto-scaling and in particular, scale-to-zero, provide value even if the platform is not shared among tenants. Higher utilization helps increase the value of a platform and can help lower costs. In addition, as sustainability becomes a higher priority, reducing the carbon footprint of the platform can also be accomplished by being more serverless.

This section describes an approach to build hybrid serverless computing, identifies what capabilities are available today and what needs to be developed.

Observation: Most serverless applications today primarily use the services of a single vendor. This is because of the ease-of-use and performance benefits of using a coherent ecosystem of services in a vendor's platform. Pricing incentives like volume discounts provided by the vendor also encourage lock-in.

Hypothesis: There won't be a single vendor that will provide all the services with the desired functional and non-functional properties that an application developer wants. There is too much ongoing innovation in cloud services to expect that a single vendor will offer the best-of-breed service along all dimensions.

Assertion: There will be a need for both standards-based and compiler-based approaches.

Fig. 6 A simple serverless application running on Initech's platform used as a running example

4.1 Two Approaches to Achieving Hybrid Serverless Computing: Compilers and Standards

4.1.1 Running Example

Consider a simple "Hello, World" serverless application that consists of a function to process images. The function may generate a thumbnail of the image or convert it to another format. In our example, we consider a function to automatically enhance an image, such as adjusting exposure or changing the sharpening settings. Multiple cloud-based photo organization services provide such a feature.

In the Fig. 6, we see that the "enhance_image" serverless function is triggered by new images arriving in an event queue. The output enhanced images are emitted to another event queue, which may trigger other functions to store the image or notify the user of the new image.

This architecture assumes that all the components, including the queues and functions are deployed on a single vendor's serverless platform. Let's call this vendor Initech. Consider the case where another vendor, Acme, offers a more desirable image enhancement service. It may be that Acme's service uses a more advanced image processing AI model, is cheaper to operate, or runs on special hardware to offer lower latency. The following sections will consider increasingly more sophisticated ways in which to incorporate the Acme service into this architecture.

4.1.2 Implementation Step 1: Blocking API Call

Figure 7 shows an implementation where we add a function to the application deployed on Initech's platform to make a blocking call to Acme's image enhancement service.

The advantage of this implementation is that it is a relatively straightforward change that does not require modifications to any other part of the application, nor any special requirements of Initech's platform or Acme's service. The disadvantage is that this is not a pure serverless application. In particular, the blocking function call is an anti-pattern that results in wasted compute and double-billing [66].

4.1.3   Implementation Step 2: Event Bridge

In this next implementation, Acme's service is now more serverless-friendly. We assume that neither vendor allows direct access to the queues in their platform from outside but do allow FaaS functions to be triggered from external calls.

The resulting architecture, shown in Fig. 8, includes a number of additional queues and functions in both vendors' platforms. In particular, there is an event queue in Acme's platform that can trigger a serverless function to perform the image enhancement. And the function output is pushed to an event queue that triggers a function to send the enhanced image back to the rest of the application pipeline in Initech's platform.

The advantage of this implementation is that it is now a pure serverless architecture, taking advantage of event-driven design patterns. The disadvantage, which is apparent when comparing Figs. 7 and 8, is the additional architectural components added to bridge the event queues between Initech and Acme's platforms. These are not part of the core application logic. Another disadvantage is that the additional components are only there in the case of the hybrid platform. If we revert back to using Initech's image enhancement services, these additional components are superfluous. Adding and removing these components depending on which service the developer chooses to use adds to the deployment complexity.

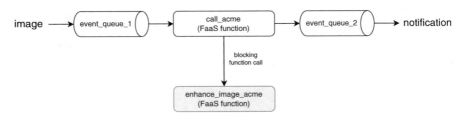

Fig. 7 A hybrid serverless application that spans two vendor platforms used as a running example. The clear diagram elements are those deployed on Initech's platform, while the gray elements are those running on Acme's platform

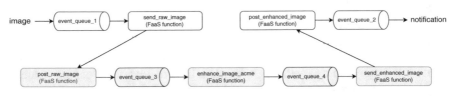

Fig. 8 A hybrid serverless application that adheres to serverless best-practices

Fig. 9 Having vendors support a common eventing standard simplifies the hybrid serverless architecture

4.1.4   Implementation Step 3: Event Standard

In this implementation, we suppose that eventing is a common enough service that there is an eventing standard that all serverless platform providers adhere to. Among other things, this means that Acme's serverless function can subscribe to and be triggered by events in Initech's event queue, and Acme's serverless function can publish events to Initech's event queue.

The advantage of this implementation is that it results in a far simpler architecture, with no extra components that arise from deciding to deploy the application across multiple vendor platforms. The obvious disadvantage of this approach is that vendors need to adopt and adhere to an eventing standard (Fig. 9).

An even more fundamental challenge is that an eventing standard needs to be defined. This may be an existing event interface, such as Knative Eventing [67], that becomes an ad-hoc standard, or an official standard published by a standards body. The interface and semantics of a serverless eventing standard that supports the range of environments we want to support—including resource constrained IoT devices, secure on-prem data centers, and powerful public clouds—remains an open research challenge.

4.1.5   Implementation Step 4: Event Bridge with Compiler

In this implementation, a compiler is used to generate the glue components to wire the application logic between Initech and Acme's platforms.

Notably, as depicted in Fig. 10, the developer's application logic should look like the one in the Event Standard implementation, while the deployed architecture output by the compiler looks like the one in the Event Bridge implementation.

The advantage of this implementation is that it hides from the developer the architectural complexities that are not related to the application logic.

The disadvantage is that it requires someone to build the compiler to perform this task. Furthermore, the compiler needs to have native understanding of the capabilities of each vendor platform, and hence developers are limited to those vendors supported by the compiler.

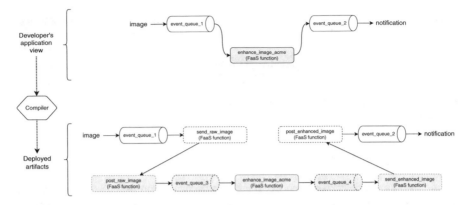

Fig. 10 In lieu of a common eventing standard, a compiler can help generate the artifacts needed to bridge vendor platforms. The generated artifacts, which are not part of the core application logic, are depicted with a dotted outline

4.1.6 Implementation Step 5: Offline Service Selection

In this implementation (c.f. Fig. 11), we now assume that image enhancement is a common enough service that there is a well-defined serverless-friendly standard for this service that vendors adhere to. In other words, this service becomes a first class capability in a serverless platform. Note that in practice there may be other more core backend services, such a key-value storage service, that are likely to become a first class standard before something like an image enhancement service.

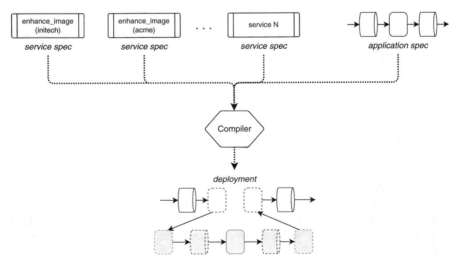

Fig. 11 The presence of serverless-friendly standards for service interface specifications and application requirements allows the compiler to select the optimal service across platforms

The new capability that this implementation offers is the ability to consult a catalog of functionally equivalent services and automatically select the optimal one according to the non-functional requirements of the application. For example, suppose Initech's version of the image enhancement service is slower but cheaper than Acme's. If the developer knows that offline batch processing is acceptable for the application's use case, they can specify a latency requirement in the order of hours, and let the compiler choose the cheaper Initech service. On the other hand, an application with low latency requirements would choose to use Acme's service. In either case, the vendor-specific artifacts are automatically generated and deployed.

The ability to perform service selection requires a standard to describe the services, including functional specifications such as the input and output parameters, and non-functional ones such as the cost per invocation and response time estimates. While there have been service interface standards defined in the past, such as WSDL [68] or recently proposed AsyncAPI [69], they are not in common use today. We believe there is still an open question on what service interface specification should look like, especially serverless-specific concerns such as cold-start latencies.

There is also a requirement for a standard way for application developers to define the requirements of their application. This too is an open research challenge and needs to include aspects such as the pricing or latency constraints.

Once there are standards for the service interfaces and applications, a compiler can make deployment and service selection decisions. For example, suppose Initech's version of the image enhancement service is slower but cheaper than Acme's. If the developer knows that offline batch processing is acceptable for the application's use case, they can specify a latency requirement in the order of hours, and let the compiler choose the cheaper Initech service.

The design of the compiler's service selection algorithm is also an open research challenge. It is not clear how the techniques in the literature [70] need to be extended to consider serverless-specific architectural requirements, such as the overhead of additional queues and functions when crossing vendor boundaries, and the span of hybrid environments, such as possible intermittent connectivity when interacting with IoT devices.

4.1.7 Implementation Step 6: Runtime Optimizer

This implementation is an extension of the one above, but has a component that monitors an application at runtime and makes changes to the deployment to maintain the application requirements.

In addition to the service and application specifications required in the Offline Service Selection implementation from the previous step, we now need a standard way to instrument each vendor's platform to monitor the application behavior at runtime (c.f. Fig. 12).

In addition, this solution requires a standard way to deploy and configure services in a vendor's platform. This can be optional, since there is an alternative where vendor-specific support is added to the compiler. This latter approach does not change

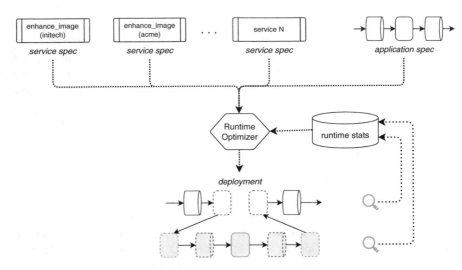

Fig. 12 An optimizer monitors application statistics at runtime and makes redeployment decisions based on changing service profiles and requirements

the application developer's user experience but limits the set of vendors that the runtime optimizer can work with.

There is an open research question on the design of the optimizer along at least two dimensions. First, as with the previous implementation, the optimization algorithms can borrow from prior art [52], but need to be extended for serverless-specific behaviors and hybrid platform properties. Second, the optimizer is a runtime component and should itself be architected with serverless best-practices in mind.

4.2 Discussion

The sequence of implementations for a hybrid serverless application above required a combination of custom implementations by the developer, support from a compiler, and the presence of standards that vendors adopt. The Table 3 summarizes these capabilities. The choice of implementation will be based on the maturity of standards available and adopted by vendors.

We believe that new capabilities will typically follow a pipeline, first being implemented manually, such as the eventing bridging architecture in Implementation Step 2. As design patterns form around common architectural designs, they will get adopted by compilers, and as these patterns mature, the community will settle on ad-hoc standards or define official ones to support these architectural patterns.

Table 4 illustrates this progression of features. The names of the features are not important as we want to illustrate the pattern of development over time periods. Notice that some features, such as "*feature_4*", may never get promoted to compiler

Table 3 A summary of the capabilities that were implemented manually or with the help of compilers and standards

Implementation	Custom code	Compiler support	Required standards
1: Blocking API call	API call		
2: Event bridge	Eventing		
3: Event standard			Eventing
4: Event bridge with compiler		Eventing	
5: Offline service selection		Service selection	+ Service, application
6: Runtime optimizer		Online replanning	+ Monitoring, deployment

support, and others, such as "*feature_2*", may never get adopted as standards. This progression is by no means unique to hybrid serverless applications, but provides a useful framework to drive towards the hybrid serverless computing vision despite community inertia or vendor opposition.

Where standards don't exist, the compiler-based approaches will help application developers build hybrid applications without having to bother with the particulars of bridging across vendor platforms. Implementation 4 is an example of this approach.

When standards are more mature, we can consider other approaches. For example, having vendors agree on only an eventing standard leads to Implementation 3, whereas a full set of standards on the services and platform supports the full hybrid serverless vision in Implementation 6.

We believe that the landscape of serverless platforms is evolving too quickly now to expect standards to cover all the functional and non-functional specifications of services and applications requirements. For example, vendors may want to offer platforms that are HIPPA-certified, run on green-energy, offer specialized hardware for ML training workloads, or provide quantum compute instances. All these new innovations will be supported first by manual implementations by developers, then by the compiler-based approach where compilers have baked-in knowledge of

Table 4 An illustration of how capabilities exercised by hybrid serverless application architectures will get promoted from manually being implemented by application developers, to being supported by compilers, and finally getting adopted as ad-hoc or official standards

	Implemented manually	Supported by compilers	Adopted as standards
Time 1	feature_1		
Time 2	feature_2	feature_1	
Time 3	feature_3	feature_2	feature_1
Time 4	feature_4	feature_2, feature_3	feature_1
Time 5	feature_4, feature_5	feature_2	feature_1, feature_3
Time 6	feature_4	feature_2, feature_5	feature_1, feature_3

these vendor-specific capabilities. As these innovations commoditize, standards will emerge and developers can transition to the standards-based approach. There will be an ongoing transition from compiler-based to standards-based.

4.3 Vision

As different approaches mature and scope of Hybrid Serverless Computing will grow to include Edge Computing, IoT and emerging new computing paradigms.

We expect that over a longer period of time different types of computing (Serverless, Cloud, IoT, etc.) will become more and more similar eventually creating a Hybrid Computing paradigm (see Fig. 13) that will be as easy to use as Serverless Computing is today leading to Hybrid Serverless Computing.

5 Opportunities and Challenges

A hybrid serverless model brings with it a number of challenges across the stack that span multiple disciplines—from universal identity management to task scheduling and data placement, as well as code and data migration issues. Furthermore, there is a need to address the impedance mismatch when bridging across serverless platforms from multiple providers, including the non-functional properties such as latency, scalability, availability, and cost. For example, while optimizing for cold-start behavior

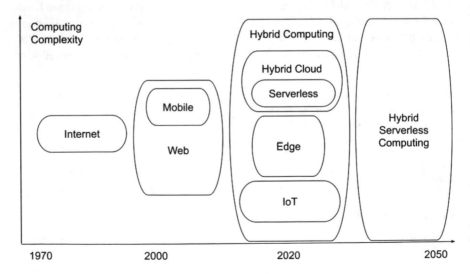

Fig. 13 Evolution of computing towards hybrid serverless computing

has been studied within the context of serverless functions deployed on a single plat-form [71, 72], scheduling and data placement at scale will greatly impact the emergent cold-start behavior when serverless functions operate across providers. There are also functional mismatches, such as security policies, and messaging semantics that need to be reconciled. Several opportunities and challenges for serverless computing have been identified [5, 28]. In this section, we review them within the context of Hybrid Serverless Computing (HSC) and identify new challenges and open research prob-lems. We start with a particular challenge that we believe is critical to adoption of HSC.

5.1 Standards

Currently, lack of serverless standards and lack of interoperability across providers has led to a chaotic landscape for application developers. Surveys indicate that devel-opers view complexities such as vendor lock-in as one of the major pain points [73, 74], and developers use dozens of disconnected tools, spending a significant amount of time managing their development environments. Serverless development lacks many of the abstraction layers available in other, more mature platforms.

In particular stickiness and vendor lock-in is a growing problem [74], for example there may be large amounts of data stored in one cloud provider and it is very expensive to move them out and some features are specific to one provider and not supported in the same way in other cloud providers.

While creating a universally accepted standard is a process that will evolve over time, the standards need to grow organically and should follow a bottom-up approach similar to IETF Internet Standards that standardize best practices and depend on multiple implementations before standards are endorsed.

The history of Web Services standards as created by IBM and Microsoft and other large enterprises show how that evolution works in practice. Although there was expectation that tooling would hide complexity of SOAP, XML and WS-* standards eventually RESTful Web services won by building on simplicity and adding bottom-up missing features to already working HTTP-related standards [75].

We recommend similar simplicity and building standards on existing implementa-tions as the only way to create successfully adopted standards for Hybrid Serverless Computing.

Below are some of the main elements that we think should be the building blocks of a new standard on Hybrid Serverless Computing.

- **Basic code packaging standards**: One packaging format is required and CNCF Open Container Initiative (OCI) provides an open standard. Similarly, container orchestration is required. Kubernetes has emerged as the most popular open plat-form and is becoming the de-facto standard for code packing. The main advantage of using containers is it not only provides flexibility for the developer to package the code and libraries in a container but can also seamlessly run in Kubernetes cluster environments [76]. In fact, many if not all cloud providers have out of the box Kubernetes services available to run their customer containers.

- **Service Specification Standards**: It is critical not only to have a standard for code packaging and execution, we assume that these containers are going to be working on different computing platforms e.g. Cloud, Edge, IoT, and even consumer devices such as smartphones and PCs. We should avoid creating separate and incompatible standards for each of these platforms. We hope that emerging standards such OpenAPI [77] and AsyncAPI [69] will gain in popularity and will be used in HSC.
- **Open Source APIs as evolving standards**: both the breadth of services and the fast paced nature of change in the cloud native development landscape makes developing industry-wide standards challenging. Sometimes, de facto standards arise from the popularity of a service—for example, many platforms now support the S3 API for object storage. Researchers are proposing the use of open source APIs as a foundation for establishing practical standards that can evolve over time as the popularity of a service waxes and wanes [16, 50]. This community consensus approach does not lock a domain to a specification, but rather allows the specification to arise (and retire) organically from actual usage.

In addition to the above-mentioned building blocks, there may be a need to standardize other aspects of the hybrid serverless application such as eventing (e.g. Cloud Events [78]), runtime monitoring, application requirement specification, and vendor deployments.

Both academia and industry can work together to create best practices based on successful open source projects (for example by creating compiler-based approaches we described) and make informal standards into successful open international standards of the same level of impact as the Internet interoperable standards had in the past.

5.2   Foundational Challenges in Hybrid Serverless Computing

While many of these challenges have been outlined and partly addressed in a single cloud server computing provider, below we outline how they change when adapted to Hybrid Serverless Computing (HSC) settings.

There has been much progress in serverless computing research since 2017 [66, 71, 79]. Generous free tiers from cloud providers, as well as the availability of open source platforms allows for larger scale experimentation. As an example, Apache OpenWhisk is a popular open source FaaS platform used in commercial offerings such as Adobe I/O Runtime [80], IBM Cloud Functions [81], and Nimbella [82] as well as various academic research projects [83, 84, 85]. Researchers have characterized real workloads presented to serverless platforms, as well as the behavior of real production platforms with respect to QoS [86, 87]. There is a better understanding of the potential attack surface of serverless [88]. Researchers have also worked to optimize the configuration of serverless jobs [81, 82]. Multiple drawbacks of serverless platforms were identified, for example [89] listed limitations such as short lifetimes of serverless functions, I/O bottlenecks, communication through slow storage, and lack of specialized hardware. Future improvements have also been identified, such as fluid code and data placement, heterogeneous hardware support, supporting long-running code, SLAs, and security concerns.

More recent work investigates the merits of hybrid serverless [63], for example scheduling stateless functions across heterogenous (e.g. edge) target platforms [55, 84, 90] and work on data aware scheduling across different serverless target nodes [91]. Mampage et al., presents a comprehensive model of serverless applied to the classic problem of resource management to include fog and edge nodes [92]. In their view, serverless is an ideal for resource management in a distributed cloud environment but comes with three key areas of challenges—workload modeling, resource scheduling, and resource scaling. In this chapter, we consider the influence a serverless perspective provides to the many challenges (and opportunities) encountered in any hybrid cloud setting.

Cost: Takes on a considerably new dimension in Hybrid Serverless settings, which needs to be optimized across multiple cloud providers instead of one and take into consideration on-prem resources if available. What makes this challenging is the availability of different pricing models across each provider, and also different tuning knobs across serverless providers, to minimize the resource usage both when it is executing and when idle.

Cold start, resource limits, and scaling: challenges for an application in an HSC environment are different and more complex than for a regular serverless environment. Cold starts may be very different across multiple environments which adds complexity but also an opportunity for dynamic routing that takes into consideration cost and resource limits. Developers need to deal with multiple scaling policies across different providers. While this approach may be considered a step back from the serverless promise of relieving the developer from scalability concerns, development of automated tools for observability and scalability can help bridge that gap.

Security: It is a complex issue for serverless computing and becomes even more complex in a HSC setting where you have different levels of function isolation across providers, and the need to have accounts with multiple providers and different access control mechanisms. The attack surface of a typical FaaS platform is potentially large given the larger number of components [88], and gets even larger in a hybrid setting. One solution is to handle mission critical functions that require privacy and security through on-prem resources.

Heterogeneity: How to identify a single API abstraction, security, privacy, access control, QoS, and deployment mechanisms between edge, on-prem, and cloud.

Legacy systems: How existing systems, including existing serverless computing applications, can interoperate and be integrated into HSC. It should be easy to access legacy Cloud, and non-cloud systems from serverless code running in HSC platforms but how does one migrate existing applications to take full advantage of HSC?

Service Level Agreements (SLAs): providing QoS guarantees still remains difficult in a serverless environment. Providing QoS guarantees across different serverless platforms presents an even bigger challenge for a developer: First, some providers may have SLAs on a particular set of serverless services as opposed to other sets of services. Second, an SLA provided by one provider may not be the same as that of another provider (e.g. 99% uptime versus 99.99% uptime). Third, the dimensions considered for QoS, may or may not be available in specific regions (e.g. GDPR). Providing an SLA across the end2end flow of the application through different HSC platforms may need an SLA formalism and set of SLA transformations that is easy to understand and modify by the application developer.

Serverless at the Edge: There have been recent advancements in the use of serverless platforms at the edge [90]. However, we consider this issue challenging because HSC platforms may have more diversity in computing resources and how they are connected. Consequently, the code running at one edge device, and in the cloud may not just be embedded but virtualized to allow movement between devices and cloud. That may lead to specific requirements that redefine cost. For example, energy usage may be more important than speed.

Orchestration: As serverless applications are gaining in complexity when moving to HSC environments they will need better tooling for orchestration of services and resources. There is also an opportunity to optimize for performance, reliability, minimize costs and cold starts, select for resource availability and take advantage of available scaling while meeting SLAs for performance and cost.

Developer experience with tools and integrated development environments (IDEs): While there have been some advancements in capabilities that enable developers to build, monitor, test, and debug serverless applications [93, 94], it is far from complete [95]. In HSC, there is the need to raise the level of abstraction and develop new approaches to enable developers to debug across platforms and services locally.

Observability: In an HSC environment, as functions keep running for shorter amounts of time, there will be many orders of magnitude more of them running making it harder to identify problems and bottlenecks. Recent work on call and causal graphs [96, 97] helped identify dependencies among functions and their executions. In an HSC, event correlation across this inherently distributed environment is needed to understand the behavior of the application across environments and dependency among components.

State: Real applications often require state. It is not clear how to manage state in HSC stateless serverless functions, particularly because a state generated in one platform may need to be accessed in another platform. Serverless compute is potentially portable, but state is more difficult due to data gravity and or regulatory constraints that dictate where state must reside. The performance and cost of an application may greatly depend on the ability to migrate state, which in turn requires optimizations on how current stateless functions can communicate state efficiently. Today, serverless functions make use of a proxy store which is a bottleneck. Stateful programming models, like actors, may offer some solutions in easing the burden of creating stateful serverless applications.

Recovery Semantics: Identifying recovery semantics, such as exactly once, at most once, and at least once semantics, in an HSC setting is critical to ensure the successful execution of application and enabling recovery from failures.

5.3 High-Impact Challenges in Academia and Industry

In this section we highlight high-impact opportunities and challenges for research in academia and industry that may be critical to evolving Serverless Computing and Hybrid Cloud into Hybrid Serverless Computing (HSC):

Interface description: What should the interface to custom services look like? Should it include functional specifications, such as the input and output parameters, and non-functional ones, such as the cost per invocation and response time estimates? What is common across multiple services and vendors? What could be standardized?

Application and solution descriptions: How could application developers define the requirements and composition of their applications as they solve business problems? What is common across vendors that could be standardized for vendor runtime monitoring and vendor deployment interfaces?

Continuous Integration/Continuous Deployments (CI/CD) pipelines: As diversity of HSC deployments increases it becomes very important to automate them with CI/CD pipelines. This is a key challenge identified for the future of cloud engineering [98], and is gaining some attention in serverless settings [99]. With HSC, the complexity increases as it includes deployments to multiple clouds, edges, and other upcoming Hybrid Computing environments.

Container adaptation: Can containers be adapted to run in all possible computing environments, from public cloud to edge to IoT devices? How to manage tradeoffs in resources (CPU, memory, network) versus cost and energy usage. Can security be maintained across multiple environments in one simple consistent way? Can new upcoming packaging formats like WebAssembly (Wasm) with common system-level interface (such as WASI and related proposed APIs) [100, 101] be used?

Container orchestration: Can cold-start and other related challenges (such as container security when compared to virtual machines) be addressed to make container-based and container orchestration successful long-term?

Compiler optimizations: How can HSC compilers optimize the selection of services based on application requirements? What would be the designs and implementations of such compilers? What optimizations could be applied? To what degree can we optimize applications to take full advantage of diverse distributed resources? How transparent are such compiling and optimizations for developers and to what degree can they support observability and developer productivity? For example, could monitoring be done in a standard way and still support multiple platforms?

Sustainable and energy-aware environment-friendly computing: as HSC environments are growing in complexity and diversity there is great opportunity to not only optimize for cost but also for environmental impact. The challenges in this area are already identified for serverless computing [45, 102] providers must make it easier to discover energy usage of energy by computing resources and schedule to use green resources.

Everything-as-a-Software: As "Software is Eating World" [103] does that mean that HSC hardware is abstracted and treated as a software resource that can be managed with uniform HSC software tooling and treated as software code? Can existing approaches such as Infrastructure as Code (IaC) be extended to contain all of HSC and go beyond infrastructure to represent any relevant resources as software code?

Ease-of-use: Simplicity of serverless computing is the main reason why it is popular. To what degree can we retain serverless simplicity in HSC? We do not only have multiple cloud and hybrid clouds environments but also edge and other emerging computing paradigms. One particular vision worth outlining as a grand challenge for HSC programming model outlined in [104] is identifying that what users want is the simplicity of being able to work locally and use a simple button to deploy remotely: "... *students simply wish they could easily "push a button" and have their code—existing, optimized, single-machine code—running on the cloud.*". The diversity of environments comprising HSC make transparently providing tools and runtime support to improve developer productivity [105] even more challenging.

Next Generation Tooling, Productivity and Developer Experience: How does the decreasing cost of running containers impact developer experience? The cost for 1 GB of memory per hour is estimated to be less than $0.01 in the near future (about $0.06 in 2022) [106]. However, one cost is unlikely to change, which is developer time per hour. It is crucial for HSC to support consistent experience for developers regardless of where containers are running: local, remote, cloud or edge. Ideally, developers can have a consistent local and remote development experience (aka transparency [105]) where they use identical code and tooling across these different estates. Is it possible to create common ground based on compilers and standards that allow interoperability between providers in a serverless cloud, similar to what Internet Protocol did for the Internet [21]?

Artificial Intelligence (AI) and Machine Learning (ML): as complexity of HSC environments is growing it becomes a great target to apply AI/ML techniques to not only do compiler-based or dynamic optimizations but also to better understand how developers and customers are building applications and solutions. For example, recent trends that focus on building Intelligent assistants that can help developers [107] use the best practices for building and evaluating code [108], help customers discover vulnerabilities [109], repair bugs [110], understand choices in deploying solutions and optimizing configurations [111], and provide easy-to-understand recommendations based on monitoring data available. These techniques can serve as building blocks to address the challenges in a HSC setting. Another opportunity is to use AI/ML to gain better understanding of HSC applications operations by extending existing serverless observability solutions in Application Performance Management (APM) [112], artificial intelligence for IT operations (AIOps) [113, 114, 115], and related areas.

6 Conclusion

Serverless today provides us with guidance on how to build the cloud platform of the future. In our view, Hybrid Serverless Compute provides a generalized view of how we can simplify cloud programming with a combination of higher-level abstractions coupled to a platform geared to optimize resource usage for Hybrid Cloud and beyond. We presented opportunities and challenges for future research. These are applicable in more specific domains for future directions—for example, extending research efforts on FaaS platforms running on hybrid platforms that span multiple clouds and Edge, more general approaches to serverless resource management [92], and serverless as a facilitator for integrating traditional enterprise compute stacks with a diverse set of distributed compute platforms like HPC, quantum compute, and complex event processing.

There exists a huge opportunity to build one open-standards and open-source based Hybrid Serverless Computing platform that will have as large an impact as the Internet by providing computing standards to allow building hybrid serverless computing solutions anywhere. Parts of Hybrid Serverless Computing may not be based on open standards when there are computing niches dominated by existing providers. Still Hybrid Serverless Computing may provide one set of abstractions and standards that provide alternatives that may win in the longer term. It is unlikely that one provider can build Hybrid Serverless Computing solutions. It is most likely that the best solutions will require a diverse set of computing resources and will need standardization to support Hybrid Serverless Computing.

References

1. Cisco 2022 Global hybrid cloud trends report. [Online]. Available: https://www.cisco.com/c/en/us/solutions/hybrid-cloud/2022-trends.html
2. Baldini I et al. (2017) Serverless computing: current trends and open problems. In: Research advances in cloud computing, Chaudhary S, Somani G, Buyya R (eds). Singapore: Springer, pp 1–20. https://doi.org/10.1007/978-981-10-5026-8_1
3. Kaviani N, Kalinin D, Maximilien M (2019) Towards serverless as commodity: a case of Knative. In: Proceedings of the 5th international workshop on serverless computing—WOSC '19. Davis, CA, USA, pp 13–18. https://doi.org/10.1145/3366623.3368135
4. Kehoe B Serverless is a State of Mind.https://ben11kehoe.medium.com/serverless-is-a-state-of-mind-717ef2088b42
5. Wade Tyler Millward, Red Hat CEO Matt Hicks: Partners Key To $1 Trillion Open Hybrid Cloud Market (2022). https://www.crn.com/news/cloud/red-hat-ceo-matt-hicks-partners-key-to-1-trillion-open-hybrid-cloud-market
6. Parkhill DF (1966) The challenge of the computer utility. Addison-Wesley Publishing Company
7. The grid: blueprint for a new computing infrastructure (The Elsevier Series in Grid Computing): Foster, Ian, Kesselman, Carl: 9781558604759: Amazon.com: Books. https://www.amazon.com/Grid-Blueprint-Computing-Infrastructure-Elsevier/dp/1558604758. Accessed 29 Jul 2022
8. Armbrust M et al (2010) A view of cloud computing. Commun ACM 53(4):50–58. https://doi.org/10.1145/1721654.1721672
9. Cloud-computing definition | Oxford Advanced Learner's Dictionary. https://www.oxfordlearnersdictionaries.com/definition/english/cloud-computing. Accessed 29 Jul 2022
10. Buyya R, Yeo CS, Venugopal S, Broberg J, Brandic I (2009) Cloud computing and emerging IT platforms: vision, hype, and reality for delivering computing as the 5th utility. Future Gener Comput Syst 25(6):599–616. https://doi.org/10.1016/j.future.2008.12.001
11. OpEx versus CapEx: the real cloud computing cost advantage. https://www.10thmagnitude.com/opex-vs-capex-the-real-cloud-computing-cost-advantage/
12. Yussupov V, Breitenbücher U, Leymann F, Müller C (2019) Facing the unplanned migration of serverless applications: a study on portability problems, solutions, and dead ends. In: Proceedings of the 12th IEEE/ACM international conference on utility and cloud computing. Auckland New Zealand, pp 273–283. https://doi.org/10.1145/3344341.3368813
13. Linthicum D (2019) Serverless computing's dark side: less portability for your apps. https://www.infoworld.com/article/3336498/serverless-computings-dark-side-less-portability-for-your-apps.html
14. Trends in cloud computing: 2022 State of the cloud report. Flexera Blog. 21 Mar 2022. https://www.flexera.com/blog/cloud/cloud-computing-trends-2022-state-of-the-cloud-report/. Accessed 29 Jul 2022
15. Multicloud: everything you need to know about the biggest trend in cloud computing. ZDNet. https://www.zdnet.com/article/multicloud-everything-you-need-to-know-about-the-biggest-trend-in-cloud-computing/. Accessed 29 Jul 2022
16. Multicloud strategy by design or by accident? Forrester. https://www.forrester.com/what-it-means/ep269-multicloud-strategy/. Accessed 29 Jul 2022
17. GAIA-X. https://www.data-infrastructure.eu/GAIAX/Navigation/EN/Home/home.html. Accessed 29 Jul 2022
18. Kaur K, Sharma S, Kahlon KS (2017) Interoperability and portability approaches in interconnected clouds: a review. ACM Comput Surv 50(4):49:1–49:40. https://doi.org/10.1145/3092698
19. What is hybrid cloud? | IBM. 02 May 2022. https://www.ibm.com/cloud/learn/hybrid-cloud. Accessed 29 Jul 2022

20. What is Hybrid Cloud Computing—Definition | Microsoft Azure. https://azure.micros oft.com/en-us/resources/cloud-computing-dictionary/what-is-hybrid-cloud-computing/. Accessed 29 Jul 2022

21. Stoica I, Shenker S (2021) From cloud computing to sky computing. In: Proceedings of the workshop on hot topics in operating systems. NY, USA,pp 26–32. https://doi.org/10.1145/ 3458336.3465301

22. Pemberton N, Schleier-Smith J, Gonzalez JE (2021) The RESTless cloud. In: Proceedings of the workshop on hot topics in operating systems. Ann Arbor Michigan, pp 49–57. https://doi. org/10.1145/3458336.3465280

23. Zhang Y, Sani AA, Xu GH (2021) User-defined cloud. In: Proceedings of the workshop on hot topics in operating systems. Ann Arbor Michigan, pp 33–40. https://doi.org/10.1145/345 8336.3465304

24. Dell Technologies APEX. https://www.dell.com/en-us/dt/apex/index.htm. Accessed 29 Jul 2022

25. HPE GreenLake—Edge-to-cloud platform | HPE. https://www.hpe.com/us/en/greenlake. html. Accessed 29 Jul 2022

26. Crossplane. https://crossplane.io/. Accessed 29 Jul 2022

27. Mytton D (2020) How much energy do data centers use? David Mytton. David Mytton. 10 Feb 2020. https://davidmytton.blog/how-much-energy-do-data-centers-use/. Accessed 29 2022

28. Computers sold in the world this year—Worldometer. https://www.worldometers.info/com puters/. Accessed 29 Jul 2022

29. IDC—Personal Computing Devices—Market Share, IDC: The premier global market intelligence company. https://www.idc.com/promo/pcdforecast. Accessed 29 Jul 2022

30. Turner A (2018) How many people have smartphones worldwide (July 2022). 10 Jul2018. https://www.bankmycell.com/blog/how-many-phones-are-in-the-world (accessed Jul. 29, 2022).

31. 21+ Internet of Things Statistics, Facts & Trends for 2022. 15 Feb 2022. https://findstack. com/internet-of-things-statistics/. Accessed 29 Jul2022

32. Worldwide Spending on Edge Computing Will Reach $250 Billion in 2024, According to a New IDC Spending Guide—Bloomberg. https://www.bloomberg.com/press-releases/2020-09-23/worldwide-spending-on-edge-computing-will-reach-250-billion-in-2024-according-to-a-new-idc-spending-guide. Accessed 29 Jul 2022

33. Google Trends Hybrid Cloud and Serverless. Google Trends. https://trends.google.com/tre nds/explore?date=all&q=%2Fg%2F11c0q_754d,Hybrid%20Cloud. Accessed 29 Jul 2022

34. Schleier-Smith J et al. (2021) What serverless computing is and should become: the next phase of cloud computing. Commun. ACM 64(5):76–84. 1145/3406011

35. Episode #52: The past, present, and future of serverless with Tim Wagner—serverless chats. https://www.serverlesschats.com/52/. Accessed 29 Jul 2022

36. IBM Code Engine. [Online]. Available: https://www.ibm.com/cloud/code-engine

37. Google Cloud Run. [Online]. Available: https://cloud.google.com/run

38. AWS Aurora Serverless. [Online]. Available: https://aws.amazon.com/rds/aurora/serverless/

39. Azure SQL database serverless. [Online]. Available: https://docs.microsoft.com/en-us/azure/ azure-sql/database/serverless-tier-overview?view=azuresql

40. Apache OpenWhisk. [Online]. Available: https://openwhisk.apache.org/

41. OpenFaaS. [Online]. Available: https://www.openfaas.com/

42. Knative. [Online]. Available: https://github.com/knative

43. CloudFlare Workers. [Online]. Available: https://workers.cloudflare.com/

44. Snowflake. https://www.snowflake.com/

45. Patros P, Spillner J, Papadopoulos AV, Varghese B, Rana O, Dustdar S (2021) Toward sustain-able serverless computing. IEEE Internet Comput 25(6):42–50. https://doi.org/10.1109/MIC. 2021.3093105

46. Raza A, Zhang Z, Akhtar N, Isahagian V, Matta I (2021) LIBRA: an economical hybrid approach for cloud applications with strict SLAs. In: 2021 IEEE International conference on cloud engineering (IC2E). San Francisco, CA, USA, pp 136–146. https://doi.org/10.1109/IC2 E52221.2021.00028

47. Grossi M et al. (2021) A serverless cloud integration for quantum computing. ArXiv Prepr. ArXiv210702007
48. Luthra M, Hennig S, Razavi K, Wang L, Koldehofe B (2020) Operator as a service: stateful serverless complex event processing. IEEE international conference on big data (Big Data) 2020:1964–1973
49. Pérez A, Risco S, Naranjo DM, Caballer M, Moltó G (2019) On-premises serverless computing for event-driven data processing applications. In: 2019 IEEE 12th International conference on cloud computing (CLOUD), pp 414–421. https://doi.org/10.1109/CLOUD.2019.00073
50. Multicloud is the New Normal. [Online]. Available: https://www.cisco.com/c/dam/global/en_uk/solutions/cloud/overview/cloud_business_cloud_advisor_infobrief_eng_FY18Q3.pdf
51. Lenarduzzi V, Daly J, Martini A, Panichella S, Tamburri DA (2021) Toward a technical debt conceptualization for serverless computing. IEEE Softw 38(1):40–47. https://doi.org/10.1109/MS.2020.3030786
52. Taibi D, Kehoe B, Poccia D (2020) Serverless: from bad practices to good solutions
53. SUSE Rancher Kubernetes Cluster Management Platform I SUSE. https://www.suse.com/products/suse-rancher/. Accessed 29 Jul. 2022
54. Funqy. https://quarkus.io/guides/funqy. Accessed 29 Jul 2022
55. Takabi H, Joshi JBD, Ahn G-J (2010) Security and privacy challenges in cloud computing environments. IEEE Secur Priv Mag 8(6):24–31. https://doi.org/10.1109/MSP.2010.186
56. Raza A, Matta I, Akhtar N, Kalavri V, Isahagian V (2021) SoK: function-as-a-service: from an application developer's perspective. J Syst Res 1(1). https://doi.org/10.5070/SR31154815
57. Opara-Martins J, Sahandi R, Tian F (2014) Critical review of vendor lock-in and its impact on adoption of cloud computing. In: International conference on information society (i-Society 2014), pp 92–97. https://doi.org/10.1109/i-Society.2014.7009018
58. Building the hybrid serverless multiclouds of the future. Wikibon Res. 08 Nov 2018. https://wikibon.com/building-hybrid-serverless-multiclouds-future/. Accessed 29 Jul 2022
59. Balakrishnan H et al (2021) Revitalizing the public internet by making it extensible. SIGCOMM Comput Commun Rev 51(2):18–24. https://doi.org/10.1145/3464994.3464998
60. Bernhardsson E (2022) Storm in the stratosphere: how the cloud will be reshuffled. Erik Bernhardsson. https://erikbern.com/2021/11/30/storm-in-the-stratosphere-how-the-cloud-will-be-reshuffled.html. Accessed 29 Jul 2022
61. How snap made its old stack disappear. Protocol. 26 Aug 2022. [Online]. Available: https://www.protocol.com/enterprise/snap-microservices-aws-google-cloud
62. AWS RedShift Serverless. [Online]. Available: https://aws.amazon.com/redshift/redshift-serverless/
63. Baarzi AF, Kesidis G, Joe-Wong C, Shahrad M (2021) On merits and viability of multi-cloud serverless. In: Proceedings of the ACM symposium on cloud computing. Seattle WA USA, pp 600–608. https://doi.org/10.1145/3472883.3487002
64. Chard R et al. (2020) funcX: a federated function serving fabric for science. In: Proceedings of the 29th international symposium on high-performance parallel and distributed computing, pp 65–76. https://doi.org/10.1145/3369583.3392683
65. Defining supercloud—Wikibon research. https://wikibon.com/defining-supercloud/. Accessed 29 Jul 2022
66. Baldini I et al. (2017) The serverless trilemma: function composition for serverless computing. In: Proceedings of the 2017 ACM SIGPLAN international symposium on new ideas, new paradigms, and reflections on programming and software. NY, USA, pp 89–103. https://doi.org/10.1145/3133850.3133855
67. Knative Eventing. Knative Eventing. https://knative.dev/docs/eventing/. Accessed 06 Aug 2022
68. Web Services Description Language (WSDL) Version 2.0 Part 1: Core Language. https://www.w3.org/TR/wsdl20/. Accessed 29 Jul 2022
69. AsyncAPI Initiative for event-driven APIs. https://www.asyncapi.com/. Accessed 29 Jul 2022

70. Sun L, Dong H, Hussain FK, Hussain OK, Chang E (2014) Cloud service selection: state-of-the-art and future research directions. J Netw Comput Appl 45:134–150. https://doi.org/10.1016/j.jnca.2014.07.019
71. Ishakian V, Muthusamy V, Slominski A (2018) Serving deep learning models in a serverless platform. In: 2018 IEEE International conference on cloud engineering (IC2E). Orlando, FL, pp 257–262. https://doi.org/10.1109/IC2E.2018.00052
72. Castro P, Ishakian V, Muthusamy V, Slominski A (2019) The rise of serverless computing. Commun ACM 62(12):44–54. https://doi.org/10.1145/3368454
73. Hassan HB, Barakat SA, Sarhan QI (2021) Survey on serverless computing. J Cloud Comput 10(1):39. https://doi.org/10.1186/s13677-021-00253-7
74. O'Reilly serverless survey (2019): Concerns, what works, and what to expect—O'Reilly. https://www.oreilly.com/radar/oreilly-serverless-survey-2019-concerns-what-works-and-what-to-expect/. Accessed 29 Jul 2022
75. REST and web services: In theory and in practice | SpringerLink. https://link.springer.com/chapter/https://doi.org/10.1007/978-1-4419-8303-9_2. Accessed 15 Aug 2022
76. Shepherd D (2022) Introducing Acorn | Acorn labs. https://acorn.io/introducing-acorn/. Accessed 22 Aug 2022
77. OpenAPI specification—Version 3.0.3 | Swagger. https://swagger.io/specification/. Accessed 15 Aug 2022
78. CloudEvents. https://cloudevents.io/. Accessed 29 Jul 2022
79. Singhvi A, Balasubramanian A, Houck K, Shaikh MD, Venkataraman S, Akella A (2021) Atoll: a scalable low-latency serverless platform. In: Proceedings of the ACM symposium on cloud computing. NY, USA, pp 138–152. https://doi.org/10.1145/3472883.3486981
80. https://developer.adobe.com/runtime/. Accessed 15 Aug 2022
81. IBM cloud functions—overview. 20 Jul 2021. https://www.ibm.com/cloud/functions. Accessed 15 Aug 2022
82. Serverless Cloud Platform for Developers | Nimbella.com®. https://nimbella.com/opensource. Accessed 15 Aug 2022
83. Kuntsevich A, Nasirifard P, Jacobsen H-A (2018) A distributed analysis and benchmarking framework for apache openwhisk serverless platform. In: Proceedings of the 19th international middleware conference (Posters). NY, USA, pp 3–4. https://doi.org/10.1145/3284014.3284016
84. Quevedo S, Merchán F, Rivadeneira R, Dominguez FX (2019) Evaluating apache open-Whisk—FaaS. In: 2019 IEEE fourth ecuador technical chapters meeting (ETCM), pp 1–5. https://doi.org/10.1109/ETCM48019.2019.9014867
85. Djemame K, Parker M, Datsev D (2020) Open-source serverless architectures: an evaluation of apache openwhisk. In: 2020 IEEE/ACM 13th international conference on utility and cloud computing (UCC), pp 329–335. https://doi.org/10.1109/UCC48980.2020.00052
86. Shahrad M et al. (2020) Serverless in the wild: characterizing and optimizing the serverless workload at a large cloud provider. In: 2020 USENIX annual technical conference (USENIX ATC 20), pp 205–218. [Online]. Available: https://www.usenix.org/conference/atc20/presentation/shahrad
87. Tariq A, Pahl A, Nimmagadda S, Rozner E, Lanka S (2020) Sequoia: enabling quality-of-service in serverless computing. In: Proceedings of the 11th ACM symposium on cloud computing. NY, USA, pp 311–327. https://doi.org/10.1145/3419111.3421306
88. How to Design a Secure Serverless Architecture. [Online]. Available: https://cloudsecurityalliance.org/artifacts/serverless-computing-security-in-2021/
89. Hellerstein JM et al. (2019) Serverless computing: one step forward, two steps back, CIDR 2019
90. Wang B, Ali-Eldin A, Shenoy P (2021) LaSS: running latency sensitive serverless computations at the edge. In: Proceedings of the 30th international symposium on high-performance parallel and distributed computing. NY, USA, pp 239–251. https://doi.org/10.1145/3431379.3460646

91. Smith CP, Jindal A, Chadha M, Gerndt M, Benedict S (2022) FaDO: FaaS functions and data orchestrator for multiple serverless edge-cloud clusters. In: 2022 IEEE 6th International conference on fog and edge computing (ICFEC), pp 17–25

92. Mampage A, Karunasekera S, Buyya R (2022) A holistic view on resource management in serverless computing environments: taxonomy and future directions. ACM Comput Surv CSUR

93. Klingler R, Trifunovic N, Spillner J (2021) Beyond @CloudFunction: powerful code annotations to capture serverless runtime patterns. In: Proceedings of the seventh international workshop on serverless computing (WoSC7) 2021, Virtual Event Canada, pp 23–28. https://doi.org/10.1145/3493651.3493669

94. Spillner J (2017) Practical tooling for serverless computing. In: Proceedings of the 10th international conference on utility and cloud computing. Austin Texas USA, pp 185–186. https://doi.org/10.1145/3147213.3149452

95. Lenarduzzi V, Panichella A (2021) Serverless testing: tool vendors' and experts' points of view. IEEE Softw 38(1):54–60. https://doi.org/10.1109/MS.2020.3030803

96. Lin W-T et al. (2018) Tracking causal order in AWS lambda applications. In: 2018 IEEE International conference on cloud engineering (IC2E), Orlando, FL, Apr 2018, pp 50–60. https://doi.org/10.1109/IC2E.2018.00027

97. Obetz M, Patterson S, Milanova A (2019) Static call graph construction in AWS lambda serverless applications. In: 11th USENIX workshop on hot topics in cloud computing (HotCloud 19), Renton, WA, Jul 2019. [Online]. Available: https://www.usenix.org/conference/hotcloud19/presentation/obetz

98. Bermbach D et al. (2021) On the future of cloud engineering. In: 2021 IEEE International conference on cloud engineering (IC2E), Oct 2021, pp 264–275. https://doi.org/10.1109/IC2E52221.2021.00044

99. Dalla Palma S, Catolino G, Di Nucci D, Tamburri DA, van den Heuvel W.-J (2022) Go serverless with RADON! A practical DevOps experience report. IEEE Softw 0–0. https://doi.org/10.1109/MS.2022.3170153

100. WebAssembly (abbreviated Wasm). https://webassembly.org/. Accessed 29 Jul 2022

101. The WebAssembly system interface (WASI). https://wasi.dev/. Accessed 29 Aug 2022

102. Sharma P Challenges and opportunities in sustainable serverless computing, p 7

103. Andreessen M () Why software is eating the world. Andreessen Horowitz, 20 Aug 2011. https://a16z.com/2011/08/20/why-software-is-eating-the-world/. Accessed 15 Aug 2022

104. Jonas E, Pu Q, Venkataraman S, Stoica I, Recht B (2017) Occupy the cloud: distributed computing for the 99%. In: Proceedings of the 2017 symposium on cloud computing, New York, NY, USA, Sep 2017, pp 445–451. https://doi.org/10.1145/3127479.3128601

105. García-López P, Slominski A, Shillaker S, Behrendt M, Metzler B (2020) Serverless end game: disaggregation enabling transparency. arXiv, 01 Jun 2020. Accessed 29 Jul 2022. [Online]. Available: http://arxiv.org/abs/2006.01251

106. Slominski A, Muthusamy V, Isahagian V (2019) The future of computing is boring (and that is exciting!). In: 2019 IEEE International conference on cloud engineering (IC2E), Jun 2019, pp 97–101. https://doi.org/10.1109/IC2E.2019.00023

107. Finnie-Ansley J, Denny P, Becker BA, Luxton-Reilly A, Prather J (2022) The robots are coming: exploring the implications of openAI codex on introductory programming. In: Australasian computing education conference, virtual event Australia, Feb 2022, pp 10–19. https://doi.org/10.1145/3511861.3511863

108. Chen M et al. (2021) Evaluating large language models trained on code. https://doi.org/10.48550/ARXIV.2107.03374

109. Pearce H, Tan B, Ahmad B, Karri R, Dolan-Gavitt B (2021) Examining zero-shot vulnerability repair with large language models. https://doi.org/10.48550/ARXIV.2112.02125

110. Prenner JA, Robbes R (2021) Automatic program repair with openAI's codex: evaluating quixBugs. https://doi.org/10.48550/ARXIV.2111.03922

111. Akhtar N, Raza A, Ishakian V, Matta I (2020) COSE: configuring serverless functions using statistical learning. In: IEEE INFOCOM 2020—IEEE conference on computer communications. Toronto, ON, Canada, pp 129–138. https://doi.org/10.1109/INFOCOM41043.2020. 9155363
112. Thurner P (2022) Seamless AI-powered observability for multicloud serverless applications. Dynatrace News, 09 Feb 2022. https://www.dynatrace.com/news/blog/seamless-ai-powered-observability-for-serverless/. Accessed 29 Aug 2022
113. Artificial intelligence for IT operations (AIOps), 25 Jul 2022. https://www.ibm.com/cloud/learn/aiops. Accessed 29 Aug 2022
114. Fisher T (2021) Introducing the first integration of instana's enterprise observability platform with IBM Watson AIOps. Instana, 24 May 2021. https://www.instana.com/blog/introducing-the-first-integration-of-instanas-enterprise-observability-platform-with-ibm-watson-aiops/. Accessed 29 Aug 2022
115. Datadog (2022) Machine Learning Based Monitoring|Datadog, Machine Learning Based Monitoring. https://www.datadoghq.com/solutions/machine-learning/. Accessed 29 Aug 2022

A Taxonomy of Performance Forecasting Systems in the Serverless Cloud Computing Environments

Sena Seneviratne, David C. Levy, and Liyanage C. De Silva

Abstract The Serverless Clouds Computing environment (or platform) manages the resource management of its respective clients who generally submit their respective applications as sets of functions (tasks). A client may submit his application as a set of tasks (functions) or as a monolithic task (single function). Each set of functions (tasks) compiled in the form of Directed Acyclic Graph (DAG), where each node is a function representing a fine-grained task and each edge represents a dependency among two functions. The decisions made through performance forecasting systems (PFS) or resource forecasting engines are of immense importance to such resource management systems. However, the forecasting of future resources is a complex problem. Several of PFS projects span over several computer resources in several dimensions. The most of the PFS projects have already been designed for performance forecasting of resources on the Distributed Computing Environments such as Peer-Peer, Queue systems, Clusters, Grids, Virtual machine organizations and Cloud systems and therefore in software engineering point of view, the new code can be written to integrate their forecasting services on the Serverless (Edge) Clouds platforms. In this chapter the taxonomy for describing the PFS architecture is discussed. The taxonomy is used to classify and identify approaches which are followed in the implementation of the existing PFSs in the Distributed Computing Environments and to realise their adaptation in the Serverless (Edge) Cloud Computing.

Keywords Performance forecasting system · Data mining · Serverless clouds · Containers · MicroVM · RMS

S. Seneviratne (✉) · D. C. Levy
Computer Engineering Laboratory, School of Electrical and Information Engineering, The University of Sydney, Camperdown, Australia
e-mail: ssen2304@uni.sydney.edu.au

L. C. De Silva
School of Digital Science (SDS), University of Brunei Darussalam, Bandar Seri Begawan, Brunei

© The Author(s), under exclusive license to Springer Nature Switzerland AG 2023
R. Krishnamurthi et al. (eds.), *Serverless Computing: Principles and Paradigms*,
Lecture Notes on Data Engineering and Communications Technologies 162,
https://doi.org/10.1007/978-3-031-26633-1_4

1 Introduction

Serverless Computing Platforms (SCPs) (or Serverless clouds) have emerged as a popular method for deploying applications on the cloud [6, 36, 88]. The internal server of SCP can do the Resource Management better with the help of PFS and can make autonomous fine grained scaling of computational resources; high availability of resources, fault tolerance, and billing only for actual computing time. To make such capabilities a reality, SCPs leverage ephemeral infrastructure such as MicroVMs or application containers of Containers as a Service (CaaS). The SCP architecture promises better server utilization as cloud providers can effectively deploy client workloads to occupy available capacity thereby saving both cost of the clients and energy of the venders [44, 83]. The Serverless architecture guarantees fair and generous hosting costs as fine-grained resources are provisioned on demand and charges reflect only actual computing time.

The required middleware platforms such as CaaS are a form of container-based virtualization in which containers are given to the users as a service from a Serverless cloud provider. It is used to leverage SCP to deploy, host, and scale resources on demand for individual functions known as "micro-services". On CaaS platforms, temporary infrastructure (container) containing user code plus dependent libraries are created and managed to provide granular infrastructure for each micro-service [53]. The SCP must create, destroy, and load balance micro-service requests across available server resources. The total number of micro-service invocations, runtime, and memory utilization are metered (charged) to the nearest second for the sake of providing a fair service to the clients.

The internals of the working of a Serverless provider's Resource Management System (RMS) can be generally described as follows. When an event occurs, in order to invoke the relevant function(s) as per the defined set of rules, a request(s) is sent to the API gateway. Before scheduling function(s), it is imperative to perform the Performance Forecasting for all the available worker-nodes, and then select the best set of performing worker node(s) which the function(s) (or job task(s)) should be submitted to. This is where; a provider has the opportunity of applying the available Performance Forecasting Algorithms to predict the future performance of the function(s) or job task(s). If he is keen to use a Machine learning forecasting algorithm, then it is required to archive all the historical load data Independent Data Tuples (IDPs) of all the worker nodes, in a relational database because the training of a certain Machine Learning (ML) predictor needs to be done just before the usage of that predictor. This is quite possible in the Serverless cloud environment because the provider has a server (though we call this system Serverless incorrectly) to manage all the nodes of the Cloud. However, if the Serverless provider uses an Analytical (Mathematical) forecasting model, he does not have to archive a big historical collection of load data from the worker-nodes.

The SCPs have been introduced to support highly scalable, event-driven applications consisting of short-running, stateless functions triggered by events generated from middleware, sensors, micro-services, or users [6]. Its use cases include: Multimedia processing, data processing pipelines, IoT data collection, Chat bots, Backend APIs, Continuous integration and continuous deployment services (CI/CD) etc.

2 Background and Related Work

The analytical modelling and Machine Learning techniques have been used in abundance in developing workload characterization techniques and forecasting models in SCPs. The difficult nature of performance forecasting on SCPs, including the need to address performance variance resulting from hardware heterogeneity is identified [38]. The several of researchers recognise how pay-as-you-go pricing models, as the sophistications of Serverless job deployments, leads to the key pitfall which is unpredictable makes spans and costs. In contrary to application hosting with VMs, SCPs complicate budgeting as organizations must predict service utilization to estimate hosting costs. Performance variance of Serverless workloads and accuracy of runtime forecasting are invariably connected scenarios.

The key factors that are responsible for producing performance variance are resource contention, provisioning variation and the hardware heterogeneity. Ou and Farley identified the existence of heterogeneous CPUs that host identically labelled VM types on Amazon EC2, leading to IaaS cloud performance variance [28, 54]. Rehman et al. identified the problem of "provisioning variation" in IaaS clouds in [57]. Provisioning variation is the random nature of VM placements that generates varying multi-tenancy across physical servers producing performance variance from resource contention. Schad et al. showed the unpredictability of Amazon EC2 VM performance resulting from provisioning variation and resource contention from VM multitenancy in [61]. Ayodele et al. and Lloyd et al. demonstrated how resource contention from multi-tenant VMs can be identified using the cpuSteal metric [4, 44].

Evaluating the performance of FaaS platforms for hosting a variety of workloads has been conducted as a part of early investigations on Serverless Cloud computing. The performance implications for hosting scientific computing workflows have been looked into by several scientists [47, 74]. Others have investigated FaaS performance for Machine Learning inference [9], NLP inference [31], and even neural network training [30]. Boza et al. designed CloudCal, to succour cost comparison of SCP versus IaaS cloud. It is a tool to estimate hosting costs for service-oriented workloads on IaaS (reserved), IaaS (On Demand), and FaaS platforms [10]. The minimum number of VMs to maintain specified average request latency to compare hosting costs to FaaS deployments can be determined by the CloudCal.

In other efforts, the case studies have been conducted to compare costs for running specific application workloads on IaaS versus FaaS [82], and FaaS versus PaaS [62]. The characterization of the performance variance of workloads across FaaS platforms was done by Cordingly et al. He had used their approach which was Linux time accounting method to predict FaaS workload runtime and cost [13].

There are many forecasting Engines which have already been designed to predict performance on the DCEs except Serverless Computing Platforms. With certain required alterations such forecasting engines can be used appropriately by the providers to forecast the computing resources of SCPs. Therefore in this survey, we optimistically count on and consider any forecasting engines which are already used in the DCEs because in software point of view, any such forecasting engine can be adapted in any DCE. In Fig. 3, the Resource Type Taxonomy describes the adaptability of PPEs in the higher levels of the hierarchy.

The plain truth is that the demand of Forecasting Engines (FEs) for the Serverless clouds overarches all of the cloud resources. Therefore the Performance Prediction needs to consider different kinds of FEs. Thus, if the internal server of SCP needs to have forecasts of several different resources, that require to be conducted using a middleware platform such as CaaS which provide access to different FEs. The FE developers need to think about new performance metrics which are appropriate for the Serverless cloud environments.

In the current study, not only do we look into a large number of FEs which have already been used in the DCEs to forecast the computer resources, but also categorise them using several taxonomies which are developed, making an effort to define and categorise the FEs with respect to different levels of resources and applications and to motivate the researchers to soil their hands on new forecasting methods.

The rest of the chapter has been organised as follows. Section 3 provides the basic difficulties and challenges in finding the resource forecasting solutions in SCPs. Section 4 presents our threefold taxonomy specifically for the SCPs and in general for all DCEs. Section 5 provides a brief description of each of the prominent PFSs which are specifically used in SCPs or could be adaptable to use in SCPs after required software adjustments. Section 6 analyses the survey in terms of the aforementioned taxonomy. Section 7 provides the conclusion to the book chapter.

3   Challenges

A fundamental difficulty in finding forecasting solutions is due to the heterogeneous and distributed nature presented in the Serverless (edge) cloud (or SCP). This causes as a result of underlying differences in the different types of jobs and resources of Serverless (edge) cloud platforms (or SCP). The archived information profiles of a particular Serverless (edge) cloud (or any DCE) can easily be accessed and would be of great assistance for the performance prediction systems. However, past experience indicates their effectiveness depends on the skilful usage of them. The reality is none of the forecasting algorithm is entirely accurate. It is also well known that some algorithms can be more efficiently and effectively used to meet a given particular specialised objective than others. For instance, forecasting of runtimes of parallel batch job tasks in a homogeneous cluster by Modelling Workloads for the Grids [73]. Thus, it is fairly prudent to think over joining multiple algorithms to obtain the best outcome.

In the past, a single reading of point valued variables might have been acceptable for forecasting of runtime of a short job task. However, the DCEs such as clusters, grids and Serverless (edge) cloud (or SCP) application consists of long job tasks and therefore such variables may produce wrong results, since they can only represent a certain point of time. Thus, it is suitable to design relevant Serverless cloud performance metrics to represent a certain duration of time [66].

For the proper designing of the Forecasting Engines (FEs), it is imperative to address the following list of problems which require the solutions in terms of the nature of the Serverless (edge) cloud environments [14].

1. Forecasting of different variables such as latency (queue waiting time, resource set up time, runtime and communication time), throughput, event arrival rate, average throttle, energy consumption, resource efficiency (utilization), communication time of MPIs of DAGs, data transfer time (remote input devices), user cost.
2. Forecasting of errors of the predicted resources in the system. If we can collect information on possible forecasting errors, then it is possible to make statistical corrections for the forecasting errors of resources.
3. If possible, novel metrics should be designed to suit the different specialised SCP. In conventional parallel computing, function's runtime and system utilization are appropriate. However in the SCP, we may require to update such traditional metrics, because the Serverless cloud is a dynamic environment.
4. The FEs of Serverless cloud environments must forecast the quality of service.
5. The FEs of Serverless cloud environments must forecast the overheads of the system.
6. The FEs of Serverless cloud environments must forecast the availability of the required data storage.

4 The Taxonomy

The FEs have been analysed under three different taxonomies, namely the Design methodology of forecasting engines, Forecasting algorithm, Resource hierarchy tree.

4.1 Classification of PFS by the Design Methodology of Forecasting Algorithms

Our objective is to forecast the DCE (Peer-peer, Queue system, Cluster, Grid, VM, Clouds, including Serverless Clouds) job task's (function's) performance on each of the prospective hosts. The Explicit Forecasting Method and Implicit Forecasting Method [21] are the two major approaches to attain the design objective.

4.1.1 Explicit Forecasting Method

In the Explicit method some performance metric of the job task (or function) for each host in the Distributed Computing Environment is directly forecast. For instance, they include end-to-end latency, runtime, cost etc. Thereafter a host whose forecasting performance is appropriate is selected.

4.1.2 Implicit Forecasting Method

In the Implicit method, it is assumed that the job task's (or function's) performance on each of the prospective hosts could be ordered according to some task-independent metric on the host. For instance the current resultant background load which is measured at this point of time can be thought of as a task-independent metric as suggested by Wolsi's NWS for the submission of short job tasks [86]. However the computer-node's background load can vary unexpectedly and therefore, this method is somewhat unreliable for the long job tasks.

The Explicit Method has two branches. They are the application-oriented method and resource-oriented method.

4.1.3 Application-Oriented Versus Resource-Oriented Method

Figure 1 shows the relationship between the dependencies involved in choosing an appropriate host to run a task. The question is whether to perform forecasting at the runtime node-level or resource availability node-level. This also determines whether to follow application-oriented or resource-oriented forecasting. In a time shared computer-node where the background load varies with time, a job task's runtime depends on the variable background load. Therefore it is required to collect and store in a database the average background load of the job task together with other related characteristics. In the resource oriented method the prospective job task's background load is forecast. This is directly related to forecasting the available CPU capacity or available CPU resource level. In the resource oriented method forecasting the resource level can be done independent of the number of job tasks which are running during certain period of time. Once the resource level (e.g. host load) is predicted the runtime can be estimated.

In an application oriented method such as Smith's where the previous job tasks which are similar to the prospective job task needs to be identified using advance search techniques for the estimation of the prospective job task's runtime [69].

The forecasting algorithms are only to be effectively used in the computer nodes of grid, cluster, peer-peer or Serverless (edge) cloud computing environment and our requirements are limited to it [66, 67]. Analysing the problem at the resource level has become clearer and easier as the CPU loads and Disk IO loads are periodically collected for each DCE user and the system administrator (or root user) [78]. This

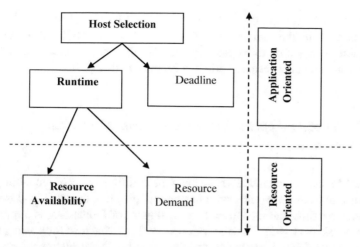

Fig. 1 Dependencies of forecasting [21]

means the quality of resource collection has improved and therefore we are encouraged to consider the available forecasting opportunities under a resource-oriented method.

Resource-oriented forecasting is related to forecasting of individual resources (e.g. host load) which have high epochal behaviour and therefore statistical techniques may be used in the forecasting of the host load [18, 67]. In computer-nodes, regardless of whether they belong to any DCE, the load profiles of CPU bound job tasks with minor disk r/w often show consistent behaviour which allows the use of a simpler linear mathematical/statistical forecasting model on each sampling unit (time step) and then aggregate the results with required adjustments to reflect the behaviour and existence of disk r/w segments.

The application-oriented method has the advantage of operating directly on the performance metric (deadlines, runtimes etc.) with which the scheduler is ultimately concerned. The measurements and forecasting made using this method are involved with the specifics of the job task (function) and therefore are not useful to other jobs tasks (functions). This historical data for each job task running on a particular computer-node is collected separately. Borrowing of measurement information is always possible as the periodic measurement of the system is collected on each node. This means that the measurement information can be shared with, or is available to, other forecasting advisors.

In an application-oriented forecasting, metrics are forecast that are closer to the actual parameters which can be used for the scheduling decisions (e.g. end-to-end latency, runtime, CPU utilization) [69]. In a resource-oriented forecasting the resource availability (e.g. host load) must first be predicted and then converted into runtime of a particular job task. As a result of this conversion, the accuracy of forecasting may suffer. Please see Fig. 1 for more details. In the computer-nodes of DCEs, however, a better model which accurately forecasts the runtime/latency can be achieved by improving the underlying CPU and Disk load (resource) measurement system. Thus the introduction of Division of Load (DOL) in the OS kernel allows more information to be gathered, specifically about user loads, making the development of a better resource oriented model a reality [63].

4.2 Forecasting Approaches Trained from Historical Execution

Analytical/Mathematical models have been the traditional way of modelling prediction algorithms. However, early on it has been shown that the CPU load is strongly correlated over time and modelling historical data for forecasting is a great possibility by Wolski et al. [85]. Dinda's work further shows that load forecasting which is based on historical data is practicable and the linear time series such as autocorrelation models may be used in forecasting [20].

The statistical analysis can be used on past archived data to realize their characteristics. For instance, Modelling workloads for grid systems [72] and Queue wait time forecasting in space shared environments [22] are based on the statistical analysis of historical data. The workload modelling is introduced to make use of the collected historical data for analysis and simulation in an analytical and manageable way [23, 29].

There are many streams to do modelling. For instance AI methods can be used. The Grid Performance Forecasting System (GPRES), which is based on the architecture of the expert systems has been developed by Kurowski et al. [41].

It is quite possible to find similar datasets using the Data mining rules. The Distance Function has been used to categorize similar jobs and resources by Li et al. [42]. The Genetic Search Algorithm is used to search for certain weights of the nearest neighbours from the historical data. The details of the algorithm are found in the relevant literature [42, 43].

The ML forecasting models branches off to the following major types. They are the Spatial Temporal Correlation Models (STC) and the models that analyse data as independent data tuples (datasets) (IDT).

4.3 Resource Type Taxonomy (Resource Type Hierarchy Tree)

There are different types of fundamental resources that are utilized by the users of the Distributed Computer Environments including the Serverless Clouds. They are (1) CPU (Processor time), (2) Memory, (3) Disk (Access cost), and (4) Network bandwidth per CPU. The performance forecasting of each fundamental resource type can be done separately as each one has different characteristics and behaviour and each resource serves a different purpose.

There are two ways to share the resources and they are time-shared method and space-shared method. An example of a time shared resource is the manner in which the CPU in a desktop PC shares its job tasks (functions). In this case, a few similar priority job tasks are running in round-robin fashion during their allocated time slots or time slices. In contrast, the space-shared CPU can only be allocated to a single job task at a time. The next job task may be allocated to the next available CPU. A good example of this is a Cluster computing system where a number of CPUs is managed in a space-shared manner. The Network bandwidth and Disk storage are space-shared resources.

The Computing Cloud may contain well connected homogenous resources such as homogenous clusters. Further, the resource can be shared or dedicated. For instance, a cluster of nodes, which is permanently available for HPC tasks, can be considered as a dedicated resource. For instance, in the Google Serverless Cloud, there can be a collection of homogenous and scalable worker nodes or a cluster (Cloud Bigtable cluster) [25, 55].

For a particular resource, its usage can be defined in several metrics. Therefore, it is important to measure these resources with relevant metrics that are easy to forecast.

Figure 3 shows the vivid levels of resource forecasts on the Distributed Computer System or DCE hierarchy. The most basic resources such as CPU, Memory, Disk space and I/O bandwidth per computer-node are on the lowest level (L-0). Level-1 (L-1) contains computer node, Network bandwidth and a Disk storage unit. Level-2 (L-2) contains the Data storage facility. Level-3 (L-3) contains the Grids and the Clusters. Level-4 (L-4) contains virtual organizations. Above them, at the very top level stay Serverless clouds. Some of the Cluster and Grid PFSs which are developed to forecast the basic resources of Level-0 and Level-1 can be easily modified to forecast the same basic resources of the Serverless clouds. Please refer to the hierarchical diagram of Fig. 3 to understand the relationship between the various levels of resources.

Level-0: This level consists of the basic elements such as CPU, memory etc.

CPU: CPU, which is available for a new job task (of a parallel application), can be predicted on a single node by using Dinda [19], Smith et al. [69], ASKALON [27], OpenSeries and StreamMiner [1], DIMEMAS [5], eNANOS [58], MWGS [72], GPRES [41], Li et al. [42], PACE [51], PPSKel [71], FREERIDE-G [32], Minh and Wolters [49], FAST [17], Forecasting of Variance [89], AWP [87], FVMRRP [40], WP_ARIMA [11], UPVM_Energy Efficient [37], DRDPred [81] or TPM [66].

Some of the above mentioned forecasting methodologies, mathematical/analytical algorithms are used. However in most of the methodologies involve the analysis of time series (STC) workload flow data. In Serverless clouds too, the forecasting of request (event) rate and end to end latency can be conducted using time series (STC) workload data sets. Therefore in both of the above situations, the Box Jenkin's models are used with certain ML and heuristic techniques. For instance, Barister [9], SPOCK [34], STOIC [91], Atomic forecasting suite for cloud resource provisioning [50] etc. However, in the Serverless clouds, end-to-end latency of a function (s) is made up of a combination of cold start time, computing time, communication time and storing time and therefore its relationship with the CPU load is a complex phenomenon. Therefore, the Serverless cloud resources are listed on the highest level of the tree.

Memory: This basic metric can be forecast on a single node by using PACE [51], FAST [17] or OpenSeries and StreamMiner [1].

Disk: In FREERIDE-G [32] predict the disk space access cost (time) or the data retrieval time.

I/O Bandwidth per Node: Both Network I/Os and Disk I/Os inherit I/O Bandwidth per node (please see next level).

Level-1: At this level, there are 3 main resource components that can be predicted either using the parameters of the level-0 or directly.

Machine/Node: The availability of a computer-node can be forecast directly using historical data as the service provider expresses the time intervals in the day that the machine is available for Grid/Cloud users. NWS [86] or OpenSeries and StreamMiner [1] predicts the availability of computer-nodes. The availability of computer-node can be forecast after performing L-0 level forecasting on CPU, memory, or disk (access cost) resources.

Network Bandwidth: This metric can be forecast by using NWS [84], Faerman et al. [26], PACE [51], EDG ROS [8], FREERIDE-G, FAST, Vazhkudai and Schopf [79] or PDTT [90].

Data storage unit: A large amount of replica data is stored in different Hierarchical Storage Management (HSM) systems. The access latency consists of two major components and they are network access cost and storage access cost. In EDG ROS [8], the forecasting of storage access cost is conducted by CrossGrid data access estimator. FREERIDE-G can forecast the data storage access time.

Level-2: At this level, there is a single resource component that can be forecast either using the parameters of the level-0 and level-1 or directly.

Data storage facility: The forecasting of the access cost of the Data storage facility is done through the forecasting of individual data storage units at level-1. If a data storage unit consists of several individual computer-nodes, after predicting the disk space access cost of each machine at level-0, the total data storage access time can be calculated.

Level-3: The resources at this level can be forecast either using the forecast information of level-0, level-1 and level-2 or directly. The Grid systems belong to this level.

Cluster (Parallel application's total runtime): DIMEMAS [5] can forecast the communication and computational times of a MPI parallel application. Smith et al., eNANOS, Li et al., Minh and Wolters or PQR2 [48] predict the parallel job's runtime. MWGS [72] or GPRES [41] forecast the parallel job's runtime.

Cluster (Parallel application's required number of nodes): The MWGS [72] or RBSP [7] forecasts a parallel application's required number of computer-nodes. The suitability of a DAG parallel application to a particular cluster of nodes (of a loud) can be forecast using the GAMMA Model [33] therefore, it also belongs to level-3.

Cluster (Available memory): PQR2 [48] or eNANOS [58] forecast the available memory.

Cluster (Queue wait time): In the available PFSs, the queue waiting time is defined for a space-shared cluster of computer-nodes and therefore it belongs to level-3. Downey, Smith, ASKALON [27], Li et al., QBETS [52] or eNANOS forecasts the queue waiting time. Also, MWGS or GPRES forecast the queue wait time.

Grids: The resource prediction is conducted at level-0, level-1 and level-2.

Sanjay and Vadhiya [59] or HIPM [24] forecasts the MPI parallel job's runtime on a Grid. GIPSY [80] forecasts the parameter sweep applications runtime on a Grid. LaPIe [76] can forecast the total communication time of a MPI parallel application on a Grid or a Cloud.

Level-4: The suitability of a particular VO requirement to be forecast using the forecast information of the levels below them (level-0, level-1 and level-2).

Level-5: Serverless (Edge) Cloud Level: The most suitably the Cloud resources can be predicted using information at level-0. Besides the Cloud resources can be forecast using the forecast information of level-1, level-2, level-3 and level-4.

Archipelago [68] forecasts the latency/deadline of a DAG of functions of serverless clouds. The SPOCK's predictive scheme [34] which is based on moving window Linear Regression (LR) can closely forecast the average request rate in the 500s window on serverless clouds. STOIC [91] accurately predict total response time or deployment time from edge to the cloud. Das et al. [15] have developed models (POEC) for accurately forecasting end-to-end latencies (Computing time on the edge device, communication time and cloud storage time) and cost for functions running from edge to cloud. The SONIC and many other performance prediction systems are available [45] in our survey.

5 Survey

Our objective is to focus on PFS for Serverless (edge) clouds. Nevertheless it is imperative to consider the PFSs which are related to the lower branches of the Resource type hierarchy tree of Fig. 3, because in basic software engineering point of view, the resources of an upper level (branch) generally inherits the resources of the lower level. Therefore with certain code changes the PFSs which are designed to forecast the resources of the lower levels can be used to forecast the resources of the helm of the tree. The Serverless (edge) cloud stays at the very helm of the tree (Fig. 3) and therefore it inherits all the resources at lower levels. Therefore all the PFSs which are designed to predict the resources at the lower levels are included in our survey. In other words our survey includes all the PFSs which have been developed for all the DCEs.

The summary of the survey which is conducted using the Taxonomy is included in Tables 1 and 2.

5.1 The Surveyed Performance Forecasting Systems

The following systems have been investigated in our survey. They have been categorised into 3 sections namely Analytical/Mathematical Models, ML Models: Spatial Temporal Correlation (STC) and ML Models: Usage of Independent Data Tuples (IDT).

5.1.1 Analytical/Mathematical Models

Archipelago (Serverless clouds) [68]
CPU Cap Control in Serverless Comp. Platform [39] (Serverless clouds)
Performance Optimization for Edge-Cloud Serverless Platforms (POEC) [15] (Edge-Cloud)
SPOCK [34] (Serverless clouds)
SONIC [45] (Serverless clouds)
PACE [51] (Clusters and Grids)
The Task Profiling Model (**TPM**) [66] (Clusters and Grids)
DIMEMAS [5] (Single Machines/Clusters)
LaPIe [75, 76] (Clusters and Grids)
ASKALON [27] (Single Machines/Clusters)
GAMMA [27] (Clusters)
Performance Forecasting with Skeletons [71] (**PPSkel**) (Clusters and Grids)
Performance Forecasting Model for FREERIDE-G [32] (Grids)
Fast Agent's System Timer [17] (**FAST**) (Grids)
Forecasting of the QoS [12] (Grids)
EDG Replica Optimization Service [8] (**ROS**) (Grids).

5.1.2 Machine Learning Models: STC

STOIC [91] (Edge-cloud)
Barista [9] (Serverless clouds)
Mahmoudi and Khazaei [46] (Serverless clouds)
Atomic forecasting suite for cloud resource provisioning [50] (Clouds)
Wolski [86] (DCE)

Table 1 The classification of performance forecasting approaches

Performance forecasting model	Category of the predict model					Input metrics/training data	Res. type	Class of job (time/space shared)
	Machine learning model				Analytical mo			
	STC	IDT	Selection of data (Man./Auto.)					
			Man	Auto				
Archipelago [68]					Y	Given the execution time of a function and the SLA, they model how requests of the function arrive to determine the minimum number of sandboxes needed. They make an assumption that request inter-arrival times follow an exponential distribution and model the number of requests expected in a given time interval T as a Poisson distribution	Worker node availability	Functions of DAG (space)
CPU cap control in serverless computing [39]					Y	(1) the "throttled time" of each worker node obtained from the underlying operating system, and (2) the number of "unprocessed functions" in the queue of each worker	Latency (CPU, memory)	Functions of DAG (space)

(continued)

Table 1 (continued)

Performance forecasting model	Category of the predict model					Input metrics/training data	Res. type	Class of job (time/space shared)
	Machine learning model				Analytical mo			
	STC	IDT	Selection of data (Man./Auto.)					
			Man	Auto				
Performa optimization for serverless edge-cloud [15]					Y	Input data size. For training: upload time, cold start time, warm start time, compute time, storage time and the time to send the results from the edge device to the cloud IoT Core, including the network transfer time and the framework-induced overhead	Latency (CPU, memory)	Functions on the smart edge devices (space)
SPOCK [34]					Y	Rate of requests	Deadline (CPU, memory)	ML based web services and applications with multiple request types
STOIC [91]	Y			Y		Batch size, historical runtimes of functions	CPU, bw memory	Functions of DAG (space)
Barista [9]	Y			Y		Work loads Number of requests	CPU, memory	Functions (space)
Mahmoudi and Khazaei [9]	Y			Y		Task arrival rate, response time, maximum concurrency level, initialization time	CPU, bw memory	Functions (space)

(continued)

Table 1 (continued)

Performance forecasting model	Category of the predict model					Input metrics/training data	Res. type	Class of job (time/space shared)
	Machine learning model				Analytical mo			
	STC	IDT	Selection of data (Man./Auto.)					
			Man	Auto				
Skedulix [16]		Y		Y		Function compute time Framework overhead, file size	CPU, bw, memory	Functions of DAG (space)
Fifer [35]		Y		Y		Past arrival rates	CPU, memory	Functions of DAG (space)
Sonic [45]					Y	Each function's memory foot print, compute time, intermediate data size, fan-out type/degree etc.	Latency Memory	Functions of DAG (space)
Downey [22]		Y		Y		Jobs' historical runtimes, queue times, cluster size and details of their processors are collected from the similar sites to plot the distribution of total allocation time of jobs in log space	Queue	Parallel jobs (space)
Dinda [21]	Y			Y		(1) Historical host load data is automatically stored. (2) Free load runtime of the new job task needs to be recorded	CPU	short j. tasks (100 ms to 10 s) (time)

(continued)

Table 1 (continued)

Performance forecasting model	Category of the predict model					Input metrics/training data	Res. type	Class of job (time/space shared)
	Machine learning model				Analytical mo			
	STC	IDT	Selection of data (Man./Auto.)					
			Man	Auto				
NWS [86]	Y			Y		Historical values of CPU usage, memory usage, TCP end-to-end bandwidth and latency, and connection time are stored automatically	CPU availability, BW	parallel j tasks (time)
Faerman [26]	Y			Y		Historical information of data transfer NWS measurements such as TCP end-to-end bandwidth and latency and the connection time are automatically stored	BW	Appli. with data transfer infor. (time)
Vazhkudai and Schopf [79]	Y			Y		GridFTP logs, disk throughput observations and network throughput data are automatically stored	BW	Appli. with data transfer infor. (time)
Smith [70]		Y		Y		Sets of template attributes and their profiles of historical workload are automatically stored	CPU Queue	Parallel job (space)
DIMEMAS [5]					Y	Sets of computation bursts and calls to MPI primitives. Descript. of application architecture	CPU and BW	MPI parallel jobs (space)

(continued)

Table 1 (continued)

Performance forecasting model	Category of the predict model					Input metrics/training data	Res. type	Class of job (time/space shared)
	Machine learning model				Analytical mo			
	STC	IDT	Selection of data (Man./Auto.)					
			Man	Auto				
Predict. of variance [89]	Y			Y		CPU load time series is automatically and online stored on each node	CPU	Parallel jobs (time)
LaPIe [75]					Y	Communication latency, message gap according to message size and number of processors	BW	MPI parallel jobs (time)
ASKALON [56]		Y		Y		Job tasks' historical information such as, job task names, runtimes, input parameter sizes, processor speeds are automatically stored from previous actual runs. If the historical information is not available then it is manually supplied from one of the identical machines	CPU, CPU (total) and queue	Parallel jobs (time)
Li et al.[42]		Y		Y		The profiles of historical workloads are automatically stored	CPU Queue	Parallel job (space)

(continued)

Table 1 (continued)

Performance forecasting model	Category of the predict model				Input metrics/training data	Res. type	Class of job (time/space shared)
	Machine learning model			Analytical mo			
	STC	IDT	Selection of data (Man./Auto.)				
			Man	Auto			
PDTT [90]	Y			Y	Network bandwidth time series is automatically and online recorded at constant width time intervals	BW	Data intensive (time)
eNANOS [58]		Y		Y	Statistical and data mining predictors need the same inputs: job name, user name, group name, no. of processors, job and script names automatically stored as load profiles	CPU Memory, queue	MPI parallel jobs (space)
OpenSeries and StreamMiner		Y		Y	Historical values of workload such as CPU idleness, percentage of free virtual memory, machine availability (switched on or off) and user presence indicator (logged on or off). The attribute selection process which has 3 phases can be semi-automated	CPU, memory and CPU availability	Parallel jobs (time)

(continued)

Table 1 (continued)

Performance forecasting model	Category of the predict model					Input metrics/training data	Res. type	Class of job (time/space shared)
	Machine learning model				Analytical mo			
	STC	IDT	Selection of data (Man./Auto.)					
			Man	Auto				
GPRES [41]		Y		Y		Historic jobs are categorised according to static or dynamic template attributes. Then the mean of values estimated parameters are calculated for each category. Categories with specific set of values fed to knowledge database as rules and this can be automated	CPU Queue	Parallel jobs (space)
MWGS [72]	Y			Y		User-name, submission time, job ID, number of nodes requested, user-predicted job runtime, actual job runtime are the input to generate two Markov chains for the runtimes and num. of nodes. Data collection and generation process can be automated	Queue No. of nodes CPU (total)	Parallel jobs (space)
GAMMA [33]					Y	Computational and network traffic information of the application and the cluster and costing parameters	Cluster	Parallel job (space)

(continued)

Table 1 (continued)

Performance forecasting model	Category of the predict model					Input metrics/training data	Res. type	Class of job (time/space shared)
	Machine learning model				Analytical mo			
	STC	IDT	Selection of data (Man./Auto.)					
			Man	Auto				
PACE [51]					Y	Software code of the applications and machine and environmental details	CPU, bw queue Mem	MPI parallel J. (time)
TPM [66]					Y	FLPs of the job tasks and machine environmental details. Disk IO maps of the job tasks	CPU, Disk	Parallel job (space/time)
PPSkel [71]					Y	The records of execution activities of the CPU usage, memory consumption and MPI message exchanges are taken from the same program	CPU BW Mem	MPI parallel jobs (time)
EDG ROS [8]					Y	Characteristic details of CPU, network and storage	Disk BW	Remote data processing appli. (time)

(continued)

Table 1 (continued)

Performance forecasting model	Category of the predict model						Input metrics/training data	Res. type	Class of job (time/space shared)
	Machine learning model					Analytical mo			
	STC	IDT	Selection of data (Man./Auto.)						
			Man	Auto					
Sanjay and Vadhiyar [59]		Y		Y			Available CPUs and available BWs are automatically measured for all processors and links at periodic intervals. The calculated coefficients are used in the total runtime equation	CPU and BW	MPI parallel jobs (time)
PQR2 [48]		Y		Y			Application and system-specific attributes such as cluster name, CPU clock, amount of memory, location of data, CPU speed, memory speed, disk speed, number of threads. The data sets can be collected automatically	CPU Mem, disk	MPI parallel jobs (space)
QBETS [52]	Y						Historical data profiles of similar jobs. The data is collected online: collection can be automated	Queue	Parallel jobs (space)

(continued)

Table 1 (continued)

Performance forecasting model	Category of the predict model					Input metrics/training data	Res. type	Class of job (time/space shared)
	Machine learning model				Analytical mo			
	STC	IDT	Selection of data (Man./Auto.)					
			Man	Auto				
FREERIDE-G [32]					Y	No. of storage nodes, dataset size, network BW, execution speed, disk speed, no. of computing nodes, and the corresponding values of the outputs	Disk BW CPU	Remote data processing appli. (space)
Minh and Wolters [49]		Y	Y			Original inputs: user_name, group_name, queue_name, job_name, Intermediate parameters: historical database size (N), no of nearest neighbour jobs K, the factor α and β Training parameters: user_name, group_name, queue_name, job_name, point_of _separate, N, K, α, β The traces are collected automatically and training parameters are calculated automatically	CPU	Parallel jobs (space)

(continued)

Table 1 (continued)

Performance forecasting model	Category of the predict model					Input metrics/training data	Res. type	Class of job (time/space shared)
	Machine learning model				Analytical mo			
	STC	IDT	Selection of data (Man./Auto.)					
			Man	Auto				
HIPM [24]		Y		Y		Data/activities of the workflow application, such as type (e.g. metric multiplication), name, arguments, problem size, preparation time, user name, grid site, submission time, queue time, external load, processors, execution time can be automatically collected and the predictor can be trained fast using Bayesian network	CPU and BW	Workflow job (time)
RBSP [7]		Y		Y		Job's input variables, number of machines under consideration are inputs to the regression equation. The collection of the training data can be automated	Cluster	MPI parallel jobs (time)

(continued)

Table 1 (continued)

Performance forecasting model	Category of the predict model					Input metrics/training data	Res. type	Class of job (time/space shared)
	Machine learning model				Analytical mo			
	STC	IDT	Selection of data (Man./Auto.)					
			Man	Auto				
AWP [87]	Y			Y		Historical workload points are collected online	CPU	Parallel jobs (time)
FAST [17]					Y	Dynamically collected data such as CPU speed, workload, BW, available memory, batch system	CPU Memory BW	Parallel jobs (space)
GIPSY [80]		Y		Y		(a) Initial training sample selection can be automated (b) Selection of the model may be automated subjected to the condition. The selected models can be run until the predicted runtime error converges, and then the most suitable model will be selected.	CPU	Parameter sweep jobs (space)
Forecasting of the QoS [12]					Y	The current measurement of the balance of resources of all the nodes (CPU, mem etc.)	CPU, memory	Parallel jobs (space)

Table 2 The classification of the resource types

Forecasting model	Res. type	Res. level	Predicted metrics	Centralized/decentralised	Homogeneous/heterogeneous	Dedicated/shared
Archipelago [68]	Worker node availability	L-5	Worker node availability	Decentralise	Homogeneous	Dedicated
CPU cap control in serverless computing [39]	CPU, memory	L-5	Latency	Decentralise	Homogeneous	Dedicated
Performa optimization for edge-cloud serverless [15]	CPU, memory	L-5	Latency	Decentralise	Homogeneous	Dedicated
SPOCK [34]	CPU, memory	L-5	Deadline	Decentralise	Homogeneous	Dedicated
STOIC [91]	CPU, memory, BW	L-5	Overall execution time	Centralized	Homogeneous	Dedicated
Barista [9]	CPU, memory, BW	L-5	Execution time	Centralized	Homogeneous	Dedicated
Mahmoudi and Khazaei [46]	CPU, memory, BW	L-5	Average response time, probability of cold start, and the average number of function instances in the steady-state	Centralized	Homogeneous	Dedicated
Skedulix [16]	CPU, memory, BW	L-5	Function execution time, network latency	Centralized	Homogeneous	Dedicated
Fifer [35]	CPU, memory	L-5	Required number of containers	Centralized	Homogeneous	Dedicated

(continued)

Table 2 (continued)

Forecasting model	Res. type	Res. level	Predicted metrics	Centralized/decentralised	Homogeneous/heterogeneous	Dedicated/shared
Sonic [45]	BW, CPU	L-5	Optimal performance/cost Low latency	Centralized	Both	Dedicated
Downey [22]	Queue	L-3	Queue time	Centralized	Homo	Dedicated
Dinda [19]	CPU	L-0 L-0	Host load Job task's runtime	Both	Both	Both
NWS [86]	CPU availability BW	L-1, L-1, L-1	CPU availability, TCP end-to-end throughput, TCP end-to-end latency	Decentralised	Hetero	Shared
Faerman [26]	BW	L-1	Data transfer rate	Decentralised	Hetero	Shared
Vazhkudai and Schopf [79]	BW	L-1	Data transfer rate	Decentralised	Hetero	Shared
Smith [70]	CPU and Queue	L-3, L-3	Job's runtime, queue time	Centralized	Homo	Dedicated
DIMEMAS [5]	CPU and BW	L-3	Job's runtime	Centralized	Homo	Dedicated
Forecasting of variance [89]	CPU	L-0	CPU load mean and variance over a time	Decentralised	Hetero	Shared
LaPle [75]	BW	L-4	MPI job's communication makes span	Decentralised	Hetero	Shared
ASKALON [56]	CPU CPU (total) Queue	L-0 L-3 L-3	Job task's runtime Job's runtime Queue time	Centralized	Hetero	Dedicated

(continued)

Table 2 (continued)

Forecasting model	Res. type	Res. level	Predicted metrics	Centralized/decentralised	Homogeneous/heterogeneous	Dedicated/shared
Li [42]	CPU, queue	L-3 L-3	Job's runtime Queue time	Centralized	Homo	Dedicated
PDTT [90]	BW	L-1	Data transfer time between 2 nodes	Decentralised	Hetero	Shared
eNANOS [58]	CPU, memory, queue	L-3, L-3, L-3	Job's runtime Memory Queue time	Centralized	Homo	Dedicated
OpenSeries and StreamMiner [2]	CPU, memory, CPU availability	L-0 L-0 L-1	Idle % of CPU Memory Availability of PCs	Decentralised	Hetero	Shared
GPRES [41]	CPU, queue	L-3 L-3	Job's runtime Queue time	Centralized	Homo	Dedicated
MWGS [73]	Queue, no. of nodes, CPU (total)	L-3 L-3 L-3	The arrival time of job No. of nodes Job's runtime	Centralized	H omo-	Dedicated
GAMMA Model [33]	Cluster	L-3	1. For each cluster Γ (γ_a/γ_m) 2. Total usage cost	Centralized	Homo-	Dedicated
PACE [51]	CPU, memory, BW, queue	L-0, L-0, L-1, L-3	Job task's runtime Memory Communication time Queue time	Decentralised	Hetero	Shared
TPM [66]	CPU, disk	L-0 L-0	Load profiles of future Job tasks Disk access time	Decentralised	Hetero	Shared

(continued)

Table 2 (continued)

Forecasting model	Res. type	Res. level	Predicted metrics	Centralized/decentralised	Homogeneous/heterogeneous	Dedicated/shared
PPSke [71]	CPU, BW, memory,	L-0, L-0, L-0	MPI job task's runtime (CPU, communication and memory)	Decentralised	Hetero	Shared
EDG ROS [8]	Disk, BW	L-1 L-1	Data retrieval time and communi. time	Decentralised	Hetero	Shared
Sanjay and Vadhiyar [59]	CPU and BW	L-4	Job's runtime	Decentralised	Hetero	Shared
PQR2 [8]	CPU, memory, disk	L-3, L-3, L-3,	Job's runtime, memory Disk space	Decentralised	Hetero	Shared
QBETS [52]	Queue	L-3	Probability of past queue wait times reaching the confidence level (95%) of the predicted queue wait times and the RMS error of job tasks that delays less than the predicted value	Centralized	Homo	Dedicated
FREERIDE-G [32]	Disk, BW, CPU	L-0, L-1, L-0,	Data retrieval time, commun. time, and data processing time	Decentralised	Hetero	Shared
Minh and Wolters [49]	CPU	L-3	Job's runtime	Centralized	Homo	Dedicated
HIPM [24]	CPU and BW	L-4	Job's runtime	Decentralised	Hetero	Shared
RBSP [7]	Cluster	L-3	No of machines	Centralized	Homo	Dedicated

(continued)

Table 2 (continued)

Forecasting model	Res. type	Res. level	Predicted metrics	Centralized/decentralised	Homogeneous/heterogeneous	Dedicated/shared
AWP [87]	CPU	L-0	Load profile	Decentralised	Hetero	Shared
FAST [17]	CPU, memory, BW	L-0, L-0, L-1	Processing runtime, memory, Communication time	Decentralised	Hetero	Shared
GIPSY [80]	CPU	L-4	Job's runtime	Decentralised	Hetero	Shared
Forecasting. Of QoS [12]	CPU, Mem etc.	L-0, L-0	Future available balance of resources (CPU, mem etc.)	Decentralised	Hetero	Shared

Dinda [21] (Clusters and Grids)
Modelling Workloads for Grid Systems (MWGS) [73] (Parallel Computers with Batch Queue Systems)
Forecasting of Variance [89] (Grids)
Forecasting of Data Transfer Time (FDTT) [90] (Data Grids)
Adaptive Workload Forecasting in Confidence Window (AWP) [87] (Grids)
QBETS [52] (for Parallel Computers with Batch Queue Systems).

5.1.3   Machine Learning Models: (IDT)

Skedulix [16] (for Serverless clouds)
Downey [22] (Clusters)
Fifer [35] (Serverless clouds)
Smith [69, 70] (Single Machine/Cluster)
Li's data mining method [42] (Clusters)
eNANOS [58] (Clusters)
OpenSeries and StreamMiner [2] (Desktop Grids)
GPRES [41] (Clusters)
Faerman [26] (Collection of Distributed Resources)
Vazhkudai and Schopf [79] (Data Grids)
PQR2 [48] (Clusters)
Sanjay and Vadhiyar [59, 60] (Grids)
Minh and Wolters [49] (Backfilling parallel systems)
Hybrid Intelligent Forecasting Model (HIPM) [24] (Grids)
Regression-Based Scalability Forecasting [7] **(RBSP)** (Clusters)
Grid Information Forecasting System [80] **(GIPSY)** (Grids).

6   Analysis of the Survey

6.1   Introduction

Since our intent is to analyse the outcome of the survey, we consider the existing PFS that are already used in the RMS of the Serverless Cloud Platforms. Since it is possible to upgrade PFSs that are used on DCEs, to be compatible with the Serverless Cloud Environments, we consider such PFSs also in our survey. After analysing over 50 different PFSs, we found that the accurate estimation of the forecasting information in the serverless cloud environment is a complex task. The results of our analysis have been tabulated in Tables 1 and 2.

In our study, the most of the PFSs (26) forecast the runtime of the job task (in the case of serverless clouds, a function's end to end latency). The Parallel job's queue waiting time has been forecast by 8 PFSs on the clusters. The Parallel job's runtime has been forecast by 7 PFSs on the clusters. MPI parallel job's combined computational and communicational time has been forecast by a single PFS [5] on the clusters. The available memory has been forecast by 2 PFSs on the clusters. MPI parallel job's combined computational and communicational time has been forecast by 2 PFSs on a Grid. The communication time between two points has been forecast by 8 PFSs. The total communication time has been forecast by a single PFS on the Grid (LaPIe). The computer-node memory has been forecast by 3 PFSs. The available number of PCs has been forecast by 4 PFSs. The suitability of a parallel application to a cluster (GAMMA model) has been forecast by a single PFS [33].

There is also a forecasting effort on QoS of the resources [12].

6.2   Meeting the Challenges

6.2.1   Appropriate Performance Metrics for the Serverless Computing Environments

It is vital to introduce the metrics that are compatible with Serverless cloud environment. In this regard, it is important to study the forecasting mechanisms of other DCEs because we can adapt them to perform forecasting on the Serverless clouds. In the Resource level taxonomy (Fig. 3), for making forecasts of resources of a certain level, there is a possibility of adapting the forecasting algorithms which stay below that level of resource type hierarchy of the DCS tree.

We found that there are novel metrics with a few PFSs. For instance, Yang et al. [89] have forecast the average CPU load over a certain time interval and variation of CPU load over some future time interval. Yang et al. [90] have developed a forecasting technique to predict the effective bandwidth over the time interval of the data transfer using predicted means and variances in the shared networks. This forecasting technique can be easily adapted to forecast the data transfer time from edge devices to the cloud.

Wu et al. [87] forecast the CPU load across the n time steps within a certain confidence window. Their AHModel calculates the Mean Square Error of forecasting over n and w steps, and therefore this forecasting model is much relevant to the Clusters and Serverless Clouds where long job tasks are usually executed.

The new metric namely "individual Free Load Profile or FLP" of future job tasks has been an essential foundation of the TPM which has been developed by Seneviratne and Levy [64, 66]. This forecasting technique can be easily adapted to forecast in the serverless cloud platforms.

The GAMA model [33] helps forecast the most suitable cluster for a particular MPI parallel application. A set of computer nodes in a serverless cloud can be forecast using the GAMMA Model [33].

The new metric namely total communication time of a set of MPI tasks or workflow can be forecast by LaPIe model. This method can also be adapted to forecast the communication time of a DAG of a Serverless cloud [75, 76].

QBETS [52] has used a new metric to predict the delay times on the queue using the probability of past queue wait times reaching the confidence level (95%) of the forecasted queue wait time.

6.2.2 Forecasting of the Bandwidth (Communication Time)

The forecasting of network bandwidth has been achieved by fewer PFSs than that of job task's (function's) run time. For edge cloud computing based smart (mobile) devices it is essential to have the most optimal upstream network time (latency) from the devices to the Serverless cloud server. The following network bandwidth forecasts which have been conducted at the lower levels of the Resource type hierarchy tree of Fig. 3 can be adapted to meet such an objective.

The regression based techniques can be used for the forecasting of data transfer times. Yang et al. [90] is based on the NWS [86] forecasting engines, making forecasting over certain time duration, making it more relevant than others [79] for the Serverless edge clouds. The forecasting of data transfer using the linear regression has been conducted at level L-1 by Faerman et al. [26]. The TCP end-to-end bandwidth, latency and connection time have been predicted by NWS [86] at resource level L-1. The multivariate regression technique has been used by Vazhkudai and Schopf [79] in their predictors.

After the development of NWS by Wolski [86], there have been several efforts to forecast the communication time or bandwidth. The communication time of a job task can be predicted by the PACE toolkit [51] at L-1 (Fig. 3). The total communication time of a set of MPI tasks or a workflow can be predicted by LaPIe [75] at L-4. FAST [17] predicts communication time without considering the background congestion.

6.2.3 Forecasting of Multiple Metrics

The GPRES, eNANOS, and PACE predict several different parameters at a time.

When different classes of various applications run in the same DCE, prediction of several parameters may be required. For example, in Clusters, Grids, Clouds and Serverless Clouds, the forecasting of multiple metrics may be required. The disadvantage of such prediction systems is that they do not specialise in a particular resource type and therefore their forecasting accuracy may be low as fine tuning them is difficult.

6.2.4 Forecasting of Data Access

The data access time is the time required to access replicated data from any location of the serverless cloud (or Grid, or Cloud). This could be a certain bottleneck in the field of high performance computing [77]. Therefore it is imperative to forecast the access times for the Hierarchical Storage Management (HSM) systems for reading data efficiently. EDG Replica Optimization Service [8] and Forecasting Model for FREERID-G [32] are two PFSs which have been included in this study.

6.3 Taxonomy

6.3.1 Resource Type Hierarchy Tree (Resource Type Taxonomy)

There are several resource types which are required to be forecast for efficient and effective scheduling in the Serverless clouds. According to the Resource type hierarchy tree on Fig. 3, the forecasting of resources of certain level can be done using the forecasting methods belong to that level and all the levels below that level. Therefore, it is imperative to consider forecasting methods which have been successfully used to forecast the computer resources of Grids, clusters, peer-peer engines, virtual machines, queue systems etc. because they all stay below the top level which contains the Serverless clouds or Serverless edge clouds.

In the tree of Fig. 3, the metrics on a leaf of a higher level depends on the metrics on a leaf of the lower level therefore the forecasting which is done on a lower leaf can be transferred to a higher leaf without loss of information. For instance, the availability of a particular computer-node (which belongs to Level-1) can be done by forecasting the CPU load, Memory usage and Disk usage which belong to the Level-0. However, the forecasting information such as CPU capacity or Memory capacity may not be derived from the value forecast at Level-1. Similarly, if the most suitable future cluster is chosen using the Gamma model (at Level-3), the predicted CPU capacity, Memory capacity or Disk access cost of a particular node (of that cluster) cannot be extracted from this information. In the upper levels of the tree of the Fig. 3, the information is tightly entangled and therefore accurately separating them is either extremely difficult or impossible [21, 64].

A metric of the lowest level (Level-0) of the tree is a fundamental or basic measurement (CPU load, Disk load, Memory capacity etc.) of the system and therefore it can be measured directly and easily. Its characteristics can be completely understood too. Therefore the accuracy of their predicted value is better than that of the higher levels. For instance, Dinda [19], PACE [51], PPSKel [71], TPM [64]. For the Serverless clouds (Level-5), it is most appropriate to forecast at the lowest level of the tree and transfer the forecast results to the top level without any loss of accuracy. Therefore, the PFSs (Level-0) which are developed to predict the basic metrics at the lowest level (Level-0) can easily be adapted for the performance prediction of applications on the Serverless clouds.

A resource at the higher level depends on the many types of metrics that belong to the lowest level, and therefore its behaviour becomes complex. This makes it difficult to analyse or forecast its behaviour at the higher level.

Similarly, the forecasting at the highest level which Serverless clouds belong to, can be done at basic lower levels and transfer the results meaningfully to the highest level. Therefore the forecasting methods which have been already designed to perform the prediction of fundamental elements of resources such as usage of CPU (of computer-node), memory (of computer-node) and Disk (of computer-node) can be easily adapted to forecast the resources for Serverless clouds.

In our study, it has been revealed that only few PFSs are available to do forecasting at the higher levels of the tree (Fig. 3) because it is rather difficult to do forecasting accurately at the higher levels. For instance, at level-4, to forecast the computational and communication times of a parallel application on the Grid only a single PFS namely is available [60]. There is only a single PFS to predict the execution time of a workflow on the Grid [24].

At the lowest level of the tree which is Level-0, 18 applications predict the CPU resource or job task's runtime. At Level-1, there are at least 9 PFSs to forecast the communication time between two points. At Level-1, there are 2 PFSs to forecast the data storage access time. At the Level-0, 2 PFSs can predict the available memory on a computer-node. Please refer to the descriptions of the Fig. 3 for the details of these prediction algorithms.

Therefore, in the Serverless cloud platforms (SCP) it is better to conduct the forecasting at the lowest level because the prediction of independent resources is easier and more accurate than their higher level counterparts. For the heterogeneous collection of resources (a grid of heterogeneous computer-nodes), the prediction of resources at the lowest level is more relevant than for the homogeneous set of resources because they are going to behave in more sophisticated manner with several diverse dependencies therefore the forecasting accurately at the higher level is a difficult task.

6.3.2 Forecasting Approach Taxonomy

There are 2 major types of forecasting approaches. They are Mathematical or Analytical approaches and ML approaches. They are STC family and the group of forecasting engines that analyses individual data sets (tuples). Figure 2 depicts the types in details. We have tabulated them according to their very characteristics in the Table 1.

Nudd et al. [51] had developed analytical model namely PACE toolkit which uses the job tasks' source code and machine's hardware configuration to simulate the forecasting. Since the processing of source code and hardware configuration to simulate prediction is not popular options among the clients, Seneviratne et al. [64] developed Task Profiling Model which uses Free Load Profiles instead of source code to reflect the behaviour of both CPU and disk loads, could be a worthy effort. The other interesting analytical model is GAMMA model (Gruber and Tran, 2004) which chooses the most suitable cluster for a particular parallel application. The LaPIe [75] forecasts network BW metric of a MPI parallel application.

In the Machine Learning prediction systems, it is imperative to automate the collection of the training data to make the forecasting systems efficient. The training of a ML PFS takes a few minutes, however once they are trained and ready to go, ML models can be quite competitive in the industry.

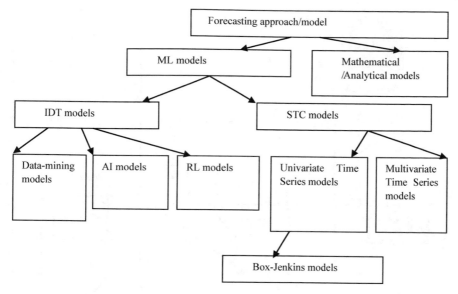

Fig. 2 Forecasting approach taxonomy [65]

When you apply the forecasting algorithm, it is vital to apply that to the most appropriate level of the tree (Fig. 3). For instance, job total runtime and queue time can be forecast for a cluster at level-3 [69, 70]. However for a Grid at level-4 Smith's prediction algorithm cannot be applied because the Grid is a network of heterogeneous resources. Sanjay and Vadhiyar [59] forecasts job runtime on a Grid at Level-4. However for a Serverless Cloud we preferably need to apply the prediction algorithms at level-0 to separately forecast job task's (function's) runtime on each worker node.

The other important fact is that unlike the Grid which is dynamic and heterogeneous collection of resources, a Serverless (edge) cloud which consists of collections of homogeneous clusters, the Level-3 prediction algorithms could well be adapted for the forecasting of SCP resources.

7 Conclusions

In this research survey, we have proposed the taxonomy for the classification of the PFSs which could support the preparation of efficient and effective application schedules for the SCPs.

The taxonomy is threefold: (a) the Design methodology of forecasting algorithms (b) the Forecasting approach and (c) the Resource type hierarchical tree. The relevant

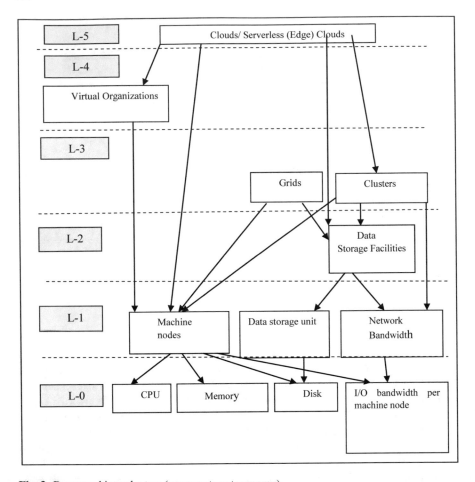

Fig. 3 Resource hierarchy tree (resource type taxonomy)

PFSs have been listed in Taxonomy Tables 1 and 2. The selected PFSs have originally been designed for the DCEs such as serverless clouds, serverless edge-clouds, clusters, grids, peer-peers systems, virtual machines etc.

The design methodology of forecasting algorithms taxonomy is presented to differentiate the forecasting methods by analysing their fundamental building blocks from ground up (Fig. 1). The forecasting approach taxonomy classifies different types of forecasting algorithms, for example Analytical/Mathematical and ML (Fig. 2). In Fig. 3, the Resource hierarchy tree taxonomy shows the various levels of resources in the DCEs. The survey is helpful us to understand what PFSs are capable of accurate predictions of resources and at what levels they are most effective and efficient. The three fold taxonomy using the Resource hierarchy tree in Fig. 3 describes, how best the PFSs which are already designed for the DCEs such as clusters, queue systems, peer-to-peer systems, grids, clouds can be adapted to use in the Serverless cloud platforms.

The prediction capabilities of the most of PFSs are still very limited. However, the PFSs which are developed to forecast the resources at the Level-0 (Fig. 3) are considered to be most reliable of them all. The reason for this is at the lowest level; the most fundamental resources are forecast which can be easily measurable/observable with the first principles of science [3]. Thus the analytical/mathematical prediction algorithms can be designed from their basics fundamentals of applied mathematics. However as one goes above the trees, the ability to observe from the first principles of science gradually diminishes and therefore the designers would have to resort to the ML prediction techniques.

The several PFSs can predict parallel application's runtime and queue waiting time (level-3) on a cluster but these PFSs are useless when it comes to the Grids. However PFSs which forecast the fundamental resources on the individual computer-nodes at Level-0 can be used to predict the resources on the Grids. Similarly such PFSs can be used to predict the basic (Level-0) resources on the Serverless (edge) clouds.

Finally we find that none of the PFSs in our study predict the forecasting errors. There is no predictor to forecast when its required input information is incomplete. Further, none of the PFSs could forecast the overheads of the DCE.

Acknowledgements We thank Prof David Abramson of the Queensland University (who was the former head of the school of Computer Science & Engineering, Monash University) and Redmond Barry Distinguished Prof Rajkumar Buyya of the Melbourne University for their invaluable guidance to Dr Sena Seneviratne in the fields of the Distributed Computer Systems.

We acknowledge that some of the classification methods which are applied on the prediction algorithms of the Serverless Cloud Computing in this chapter had been previously used by Dr Sena Seneviratne in the field of Grid computing in the book chapter namely "Taxonomy of Performance Prediction Systems for Parallel and Distributed Computing Systems" which was published in 2015, *In:* BARBOSA, J. G. (ed.) *Grid Computing Techniques and Future Prospects.* New York: Nova Science Publishers. We acknowledge that he had submitted to ARXIV repository a similar article namely "A Taxonomy of Performance Prediction Systems in the Parallel and Distributed Computing Grids". Further, we have used the same classification techniques in the paper namely "Taxonomy & Survey of Performance Prediction Systems For the Distributed Systems Including the Clouds" which was presented by him at *2021 IEEE International Conference* CPSCom which was held in Melbourne, Australia. Further, we acknowledge that some of the Machine Learning Algorithms which are contained in this chapter had been discussed by him in the paper namely "Adapting the Machine Learning Grid Prediction Models for Forecasting of Resources on the Clouds" and presented at the 2019 IEEE conference namely *"Advances in Science and Engineering Technology"* which was held in Dubai, United Arab Emirates.

References

1. Andrzejak A, Graupner S, Plantikow S (2006) Predicting resource demand in dynamic utility computing environments. In: International conference on autonomic and autonomous systems (ICAS'06), Santa Clara, USA, July 2006

 2. Andrzejak A, Domingues P, Silva L (2006) Predicting machine availabilities in desktop pools. In: IEEE/IFIP network operations & management symposium (NOMS 2006), Vancouver, Canada, April 2006
 3. Aristotle 323 BC. Metaphysics
 4. Ayodele AO, Rao J, Boult TE (2015) Performance measurement and interference profiling in multi-tenant clouds. In: 2015 IEEE 8th international conference on cloud computing, pp 941–949
 5. Badia RM, Labarta J, Gimenez J, Escalé AF (2003) DIMEMAS: predicting MPI applications behavior in grid environments. In: Workshop on grid applications and programming tools (GGF8), June 2003
 6. Baldini I, Chang K, Chang P, Flink S, Ishakian V, Michell N, Muthusamy V, Rabbah R, Slominsky A, Sutter P (2017) Serverless computing: current trends and open problems. Res Adv Cloud Comput
 7. Barnes BJ, Rountree B, Lowenthal DK, Reeves J, Supinski B, Schulz M (2008) A regression-based approach to scalability prediction. In: 22nd international conference on supercomputing (ICS '08). ACM, Kos, Greece
 8. Bell WH, Cameron DG, Capozza L, Millar P, Stockinger K, Zini F (2002) Design of a replica optimization service. EU Data-Grid Project, Geneva, Switzerland
 9. Bhattacharjee A, Chhokra AD, Kang Z, Sun H, Gokhale A, Karsai G (2019) BARISTA: efficient and scalable serverless serving system for deep learning prediction services
10. Boza EF, Abad CL, Villavicencio M, Quimba S, Plaza JA (2017) Reserved, on demand or serverless: model-based simulations for cloud budget planning. In: 2017 IEEE second Ecuador technical chapters meeting (ETCM), pp 1–6
11. Calheiros RN, Masoumi E, Ranjan R, Buyya R (2015) Workload prediction using ARIMA model and its impact on cloud applications' QoS. IEEE Trans Cloud Comput 3:449–458
12. Carvalho M, Miceli R, Maciel Jr PD, Brasileiro F, Lopes R (2010) Predicting the quality of service of a peer-to-peer desktop grid. In: 10th IEEE/ACM international conference on cluster, cloud and grid computing (CCGrid). IEEE Comput. Soc., Melbourne, Australia
13. Cordingly R, Shu W, Lloid W (2020) Predicting performance and cost of serverless computing functions with SAAF. In: IEEE 6th international conference on cloud and big data computing
14. Coregrid 2006. Review of performance prediction models and solutions. Institute on Resource Management and Scheduling
15. Das A, Imai S, Patterson S, Wittie MP (2020) Performance optimization for edge-cloud serverless platforms via dynamic task placement. In: 2020 20th IEEE/ACM international symposium on cluster, cloud and internet computing (CCGRID), 11–14 May 2020, pp 41–50
16. Das A, Leaf A, Varela CA, Patterson S (2020) Skedulix: hybrid cloud scheduling for cost-efficient execution of serverless applications. In: 2020 IEEE 13th international conference on cloud computing (CLOUD), 19–23 Oct 2020, pp 609–618
17. Desprez F, Quinson M, Suter F (2002) Dynamic performance forecasting for network-enabled servers in a heterogeneous environment. In: International conference on parallel & distributed processing techniques & applications (PDPTA), Las Vegas, USA
18. Dinda PA (1999) The statistical properties of host load. Sci Program 7:211–229
19. Dinda PA (2002) Online prediction of running time of tasks. Clust Comput 5:225–236
20. Dinda PA, O'Hallaron DR (2000) Host load prediction using linear models. Clust Comput 3:265–280
21. Dinda PA (2000) Resource signal prediction and its application to real time scheduling advisors. PhD, Carnegie Mellon University, USA
22. Downey AB (1997) Predicting queue times on space-sharing parallel computers. In: 11th international symposium on parallel processing, Geneva, Switzerland, 1997, pp 209–218
23. Downey AB, Feitelson DG (1999) The elusive goal of workload characterisation. Perform Eval Rev 26:14–29
24. Duan R, Nadeem F, Wang J, Zhang Y, Prodan R, Fahringer T (2009) A hybrid intelligent method for performance modeling and prediction of workflow activities in grids. In: 9th IEEE/ACM international symposium on cluster computing and the grid (CCGRID '09), Shanghai, China

25. Eyupoglu C (2019) Big data in cloud computing and Internet of Things. In: 2019 3rd international symposium on multidisciplinary studies and innovative technologies (ISMSIT), 11–13 Oct 2019, pp 1–5
26. Faerman M, Su A, Wolski R, Berman F (1999) Adaptive performance prediction for distributed data-intensive applications. In: ACM/IEEE international conference on super computing, Portland, OR, USA, 1999 (CDROM), Article No. 36
27. Fahringer T, Jugravu A, Pllana S, Prodan R, Seragiotto C Jr, Truong HL (2005) ASKALON: a tool set for cluster and grid computing. Concurr Comput: Pract Exp 17:143–169
28. Farley B, Juels A, Varadarajan V, Ristenpart T, Bowers KD, Swift MM (2012) More for your money: exploiting performance heterogeneity in public clouds. In: Proceedings of the third ACM symposium on cloud computing. Association for Computing Machinery, San Jose, CA
29. Feitelson DG (2002) Workload modelling for performance evaluation. School of Computer Science & Engineering, Hebrew University, Jerusalem, Israel
30. Feng L, Kudva P, Silva DD, Hu J (2018) Exploring serverless computing for neural network training. In: 2018 IEEE 11th international conference on cloud computing (CLOUD), 2–7 July 2018, pp 334–341
31. Fotouhi M, Chen D, Lloyd WJ (2019) Function-as-a-Service application service composition: implications for a natural language processing application. In: Proceedings of the 5th international workshop on serverless computing. Association for Computing Machinery, Davis, CA, USA
32. Glimcher L, Agrawal G (2007) A performance prediction framework for grid-based data mining application. In: International parallel and disributed processing symposium (IPDPS)
33. Gruber R, Tran TM (2004) Parameterisation to Tailor commodity clusters to applications. EPFL Supercomput Rev 14:12–17
34. Gunasekaran JR, Thinakaran P, Kandemir MT, Urgaonkar B, Kesidis G, Das C (2019) Spock: exploiting serverless functions for SLO and cost aware resource procurement in public cloud. In: 2019 IEEE 12th international conference on cloud computing (CLOUD), 8–13 July 2019, pp 199–208
35. Gunasekaran JR, Thinakaran P, Nachiappan NC, Kandemir MT, Das CR (2020) Fifer: tackling resource underutilization in the serverless era. In: Proceedings of the 21st international middleware conference. Association for Computing Machinery, Delft, Netherlands
36. Hendrickson S, Sturdevant S, Harter T, Venkataramani V, Arpaci-Dusseau AC, Arpaci-Dusseau RH (2016) Serverless computation with openLambda. In: Proceedings of the 8th USENIX conference on hot topics in cloud computing. USENIX Association, Denver, CO
37. Hsieh S-Y, Liu C-S, Buyya R, Zomaya AY (2020) Utilization-prediction-aware virtual machine consolidation approach for energy-efficient cloud data centers. J Parallel Distrib Comput 139:99–109
38. Jonas E, Schleier-Smith J, Sreekanti V, Tsai C-C, Khandelwal A, Pu Q, Shankar V, Carreira J, Krauth K, Yadwadkar NJ, Gonzalez J, Popa RA, Stoica I, Patterson DA (2019) Cloud programming simplified: a Berkeley view on serverless computing. arXiv:1902.03383
39. Kim YK, Hoseinyfarahabady MR, Lee YC, Zomaya AY (2020) Automated fine-grained CPU cap control in serverless computing platform. IEEE Trans Parallel Distrib Syst 31:2289–2301
40. Kumar J, Singh AK, Buyya R (2020) Ensemble learning based predictive framework for virtual machine resource request prediction. Neurocomputing 397:20–30
41. Kurowski K, Oleksiak A, Nabrzyski J, Kwiecien A, Wojtkiewicz M, Dyczkowski M, Guim F, Corbalan J, Labarta J (2005) Multi-criteria grid resource management using performance prediction techniques. In: CoreGrid integration workshop, Pisa, Italy, Nov 2005
42. Li H, Groep D, Wolters L (2007) Mining performance data for metascheduling decision support in the grid. Futur Gener Comput Syst 23:92–99
43. Li H, Sun J, Sun BL (2009) Financial distress prediction based on OR-CBR in the principle of K-nearest neighbors. Expert Syst Appl 36:643–659
44. Lloyd W, Ramesh S, Chinthalapati S, Ly L, Pallickara S (2018) Serverless computing: an investigation of factors influencing microservice performance. In: IEEE international conference on cloud engineering (IC2E 2018)

45. Mahgoub AY, Shankar K, Mitra S, Klimovic A, Chaterji S, Bagchi S (2021) SONIC: application-aware data passing for chained serverless applications. In: USENIX annual technical conference, 2021
46. Mahmoudi N, Khazaei H (2020) Performance modeling of serverless computing platforms. IEEE Trans Cloud Comput 10:2834–2847
47. Malawski M, Gajek A, Zima A, Balis B, Figiela K (2020) Serverless execution of scientific workflows: experiments with HyperFlow, AWS Lambda and Google Cloud Functions. Future Gener Comput Syst 110:502–514
48. Matsunaga A, Fortes JAB (2010) On the use of machine learning to predict the time and resources consumed by applications. In: 10th IEEE/ACM international conference on cluster, cloud and grid computing (CCGRID). IEEE Comput. Soc., Melbourne, VIC, Australia
49. Minh TN, Wolters L (2010) Using historical data to predict application runtimes on backfilling parallel systems. In: 18th Euromicro conference on parallel, distributed and network-based processing (PDP '10). IEEE Comp. Soc., Pisa, Italy
50. Nikravesh AY, Ajila SA, Lung C-H (2017) An autonomic prediction suite for cloud resource provisioning. J Cloud Comput 6:3
51. Nudd GR, Kerbyson DJ, Panaefstathiou E, Perry SC, Harper JS, Ewilcox DV (2000) Pace— a toolset for the performance prediction of parallel and distributed systems. High Perform Comput Appl 14:228–252
52. Nurmi DC, Brevik J, Wolski R (2007) QBETS: queue bounds estimation from time series. In: SIGMETRICS '07. International conference on measurement & modeling of computer systems. Springer, San Diego, CA, USA
53. Oakes E, Yang L, Houck K, Harter T, Arpaci-Dusseau AC, Arpaci-Dusseau RH (2017) Pipsqueak: lean lambdas with large libraries. In: IEEE 37th international conference on distributed computing systems workshops (ICDCSW 2017)
54. Ou Z, Zhuang H, Lukyanenko A, Nurminen JK, Hui P, Mazalov V, Ylä-Jääski A (2013) Is the same instance type created equal? Exploiting heterogeneity of public clouds. IEEE Trans Cloud Comput 1:201–214
55. Palmer N, Sherman M, Wang Y, Just S (2015) Scaling to build the consolidated audit trail: a financial services application of Google Cloud Bigtable
56. Prodan R (2007) Specification and runtime workflow support in the ASKALON grid environment. Sci Program 15:193–211
57. Rehman MS, Sakr MF (2010) Initial findings for provisioning variation in cloud computing. In: 2010 IEEE second international conference on cloud computing technology and science, 30 Nov–3 Dec 2010, pp 473–479
58. Rodero I, Guim F, Corbalán J, Labarta J (2005) eNANOS: coordinated scheduling in grid environments. In: Parallel computing: current & future issues of high-end computing, Parco, 2005
59. Sanjay HA, Vadhiyar S (2008) Performance modeling of parallel applications for grid scheduling. Parallel Distrib Comput 68:1135–1145
60. Sanjay HA, Vadhiyar S (2009) A strategy for scheduling tightly-coupled parallel applications on clusters. Concurr Comput: Pract Exp 21:2491–2517
61. Schad J, Dittrich J, Quiané-Ruiz J-A (2010) Runtime measurements in the cloud: observing, analyzing, and reducing variance. Proc VLDB Endow 3:460–471
62. Scheuner J, Leitner P (2020) Function-as-a-Service performance evaluation: a multivocal literature review. J Syst Softw 170:110708
63. Seneviratne S, Levy DC, Hong W, De Silva LC, Hu J (2021) Introduction of the new Os kernel internals for the new metrics for the performance prediction on the distributed computing environments. In: Petrova VM (ed) Advances in engineering research. Nova Science and Technology, New York
64. Seneviratne S (2009) A framework for load profile prediction for grid computing. PhD, Sydney University
65. Seneviratne S, Levy D, Rajkumar B (2015) Taxonomy of performance prediction systems for parallel and distributed computing systems. In: Barbosa JG (ed) Grid computing techniques and future prospects. Nova Science Publishers, New York

66. Seneviratne S, Levy DC (2011) Task profiling model for load profile prediction. Futur Gener Comput Syst 27:245–255
67. Seneviratne S, Witharana S, Toosi AN (2019) Adapting the machine learning grid prediction models for forecasting of resources on the clouds. In: 2019 advances in science and engineering technology international conferences (ASET), Dubai, United Arab Emirates
68. Singhvi A, Houck K, Balasubramanian A, Shaikh MD, Venkataraman S, Akella A (2019) Archipelago: a scalable low-latency serverless platform. arXiv:1911.09849
69. Smith W, Foster I, Taylor V (2004) Predicting application run times with historical information. Parallel Distrib Comput 64:1007–1016
70. Smith W, Taylor V, Foster I (1999) Using runtime predictions to estimate queue wait times and improve scheduler performance. In: International workshop on job scheduling strategies for parallel processing, San Juan, Puerto Rico, 1999. Springer, pp 202–219
71. Sodhi S, Subhlok J, Xu Q (2008) Performance prediction with skeletons. Clust Comput 11:151–165
72. Song B (2005) Workload modelling for parallel computers. PhD, University of Dortmund
73. Song B, Ernemann C, Yahyapour R (2004) Parallel computer workload modelling with Markov chains. In: International conference on job schdeduling strategies for parallel processing. Springer, NY, USA
74. Spillner J, Mateos C, Monge DA (2018) FaaSter, Better, Cheaper: the prospect of serverless scientific computing and HPC. In: High performance computing. Springer, Cham, pp 154–168
75. Steffenel LA (2005) LaPIe: Communications Collectives Adaptées aux Grilles de Calcul. PhD, INPG
76. Steffenel LA, Mounie G (2008) A framework for adaptive collective communications for heterogeneous hierarchical computing systems. Comput Syst Sci 74:1082–1093
77. Stockinger K, Stockinger H, Dutka L, Slota R, Nikolow D, Kitowski J (2003) Access cost estimation for unified grid storage systems. In: Fourth international workshop on grid computing (GRID03), Phoenix, AZ, USA
78. Ullah QZ, Shahzad H, Khan GM (2017) Adaptive resource utilization prediction system for infrastructure as a service cloud. Comput Intell Neurosci 2017
79. Vazhkudai S, Schopf JM (2003) Using regression techniques to predict large data transfers. High Perform Comput Appl 17:249–268
80. Verboven S, Hellinckx P, Arickx F, Broeckhove J (2008) Runtime prediction based grid scheduling of parameter sweep jobs. In: IEEE international conference of Asia-Pacific services computing (APSCC), Yilan, Taiwan
81. Verma M, Gangadharan GR, Narendra NC, Ravi V, Inamdar V, Ramachandran L, Calheiros RN, Buyya R (2016) Dynamic resource demand prediction and allocation in multi-tenant service clouds. Concurr Comput: Pract Exp 28:4429–4442
82. Villamizar M, Garcés O, Ochoa L, Castro H, Salamanca L, Verano M, Casallas R, Gil S, Valencia C, Zambrano A, Lang M (2016) Infrastructure cost comparison of running web applications in the cloud using AWS lambda and monolithic and microservice architectures. In: 2016 16th IEEE/ACM international symposium on cluster, cloud and grid computing (CCGrid), 16–19 May 2016, pp 179–182
83. Wang L, Zhang M, Li Y, Ristenpart T, Swift M (2018) Peeking behind the curtains of serverless platforms. In: USENIX annual technical confeerence
84. Wolski R (1998) Dynamically forecasting network performance using the network weather service. Clust Comput 1:119–132
85. Wolski R, Spring N, Hayes J (2000) Predicting the CPU availability of time-shared Unix systems on the computational grid. Clust Comput 3:293–301
86. Wolski R, Spring N, Hayes J (1999) The network weather service: a distributed resource performance forecasting service for metacomputing. Future Gener Comput Syst 15:757–768
87. Wu Y, Hwang K, Yuan Y, Zheng W (2010) Adaptive workload prediction of grid performance in confidence windows. IEEE Trans Parallel Distrib Syst 21:925–938
88. Yan M, Castro P, Cheng P, Ishakian V (2016) Building a chatbot with serverless computing. In: Proceedings of the 1st international workshop on mashups of things and APIs

89. Yang L, Schopf JM, Foster I (2003) Homeostatic and tendency-based CPU load predictions. In: 17th international parallel and distributed processing symposium (IPDPS 2003), Los Alamitos, CA, USA, 2003. IEEE CD-ROM, p 9
90. Yang L, Schopf JM, Foster I (2005) Improving parallel data transfer times using predicted variances in shared networks. In: Fifth IEEE international symposium on cluster computing and the grid (CCGRID05). IEEE Computer Society, Washington, DC, USA
91. Zhang M, Krintz C, Wolski R (2021) Edge-adaptable serverless acceleration for machine learning Internet of Things applications. Softw: Pract Exp 51:1852–1867

Open-Source Serverless for Edge Computing: A Tutorial

Priscilla Benedetti, Luca Gattobigio, Kris Steenhaut, Mauro Femminella, Gianluca Reali, and An Braeken

Abstract Serverless computing is a recent deployment model for cloud, edge and fog computing platforms, which ultimate goal is to provide cost reduction and scalability enhancement with no additional deployment overhead. The main implementation of this model is Functions-as-a-Service (FaaS): Developers deploy modular functions, which are typically event-driven, on the platform without the need to manage the underlying infrastructure. Moreover, using the so called warm start mode, the FaaS containers hosting the application are kept up and running after initialization, granting the user the impression of high availability. Conversely, in a cold start mode scenario, those containers are deleted when no application requests are received within a certain time window, to save resources. This focus on resources efficiency and flexibility could make the serverless approach significantly convenient

P. Benedetti (✉) · M. Femminella · G. Reali
Department of Engineering, University of Perugia, via G.Duranti 93, 06125 Perugia, Italy
e-mail: priscilla.benedetti@studenti.unipg.it; priscilla.benedetti@vub.be

M. Femminella
e-mail: mauro.femminella@unipg.it

G. Reali
e-mail: gianluca.reali@unipg.it

P. Benedetti · K. Steenhaut
Department of Electronics and Informatics (ETRO), Vrije Universiteit Brussel, Pleinlaan 2, 1050 Brussels, Belgium
e-mail: kris.steenhaut@vub.be

L. Gattobigio · A. Braeken
Department of Engineering Technology (INDI), Vrije Universiteit Brussel, Pleinlaan 2, 1050 Brussels, Belgium
e-mail: luca.gattobigio@vub.be

A. Braeken
e-mail: an.braeken@vub.be

M. Femminella · G. Reali
Consorzio Nazionale Interuniversitario per le Telecomunicazioni (CNIT), 43124 Parma, Italy

for edge computing based applications, in which the hosting nodes consist of devices and machines with limited resources, geographically distributed in proximity to the users. In this paper, we explore the available solutions to deploy a serverless application in an edge computing scenario, with a focus on open-source tools and IoT data.

Keywords Serverless computing · FaaS · Edge · IoT · Open-source

1 Introduction

Driven by Internet-of-Things (IoT) devices and new technologies such as the ones used in self driving cars, the exponentially increasing volume of real-time and data-intensive services has led to the shift from a centralized cloud computing model to a more distributed service model such as edge computing. In fact, to avoid potential bottlenecks and latency caused by transport and computation of large data volumes, edge computing attempts to move data processing closer to the source, relying on resource-constrained nodes instead of a centralized cloud data center approach. Moreover, IoT data typically lead to the implementation of various applications, deployed in various network located in different locations, with varying computational demands, to be often satisfied by the same computing platform. Edge computing provides local computation and storage of data, with the opportunity to send the collected data to a centralized cloud computing platform if needed, hence granting the flexibility required by the IoT domain [1, 2].

The term edge computing has been introduced with many definitions. For the Linux Foundation, such as reported by the 5G Infrastructure Public Private Partnership (5G PPP) Technology Board [3], edge computing can be defined as:

> The delivery of computing capabilities to the logical extremes of a network in order to improve the performance, operating cost and reliability of applications and services. By shortening the distance between devices and the cloud resources that serve them, and also reducing network hops, edge computing mitigates the latency and bandwidth constraints of today's Internet, ushering in new classes of applications. In practical terms, this means distributing new resources and software stacks along the path between today's centralized data centers and the increasingly large number of devices in the field, concentrated, in particular, but not exclusively, in close proximity to the last mile network, on both the infrastructure and device sides.

Edge computing can provide a variety of functions and network locations, hosting multiple applications simultaneously. In fact, there is no unique location where edge computing must be deployed. For example, edge nodes can be included in network routers, radio towers, WiFi hotspots and local data centers. Any device with computational power that is near the user's location can act as an edge node, as long as it can process a computational workload. As shown in Fig. 1, the edge computing layer is typically placed between user devices (more than tens of millions of points of presence) and centralized computing datacenters whether they are public clouds or telecom operator (Telco) cloud facilities. Device computing resources are hard to leverage and control because of their heterogeneity and the type of network environment they are connected to (typically Local Area Network (LAN) environments) [3]. Providing various available deployment configurations, edge-computing based applications can be designed with temporal and spatial closeness to final users, real-time and interactive features, to be compliant to the requirements imposed by use cases such as video analytics, healthcare, industrial control and multimedia services [4]. The standardization process of this computing paradigm started in the European Telecommunications Standards Institute (ETSI) in late 2014 with the creation of the Industry Specification Group (ISG) on Multi-Access Edge Computing (MEC). Standing at the ETSI ISG MEC definition, MEC grants computing power and an Information Technology (IT) service environment to application developers and content providers as in a cloud-computing scenario, but being located at the edge of the network. Typically, MEC is characterized by [5]:

- **On-Premises**: While having access to local resources, edge application can be isolated from the rest of the network. These isolation boundaries can be a convenient feature for Machine-to-Machine communication requirements. This could be the case of security or safety systems in which high levels of resilience and privacy are needed.
- **Proximity**: Granting spatial closeness to data sources, edge computing is particularly efficient at processing information for big data analytics, providing as well direct access to the devices when needed by the specific business logic.
- **Lower latency**: The aforementioned proximity feature provides a beneficial latency reduction as well. This can sensibly improve service reactiveness and user experience, reducing the chance of traffic bottlenecks in the platform.
- **Resource efficiency**: Being edge devices usually equipped with limited hardware, the optimization of resource usage is fundamental.

Resource efficiency is a fundamental feature of edge computing. However, the traditional approach of service provisioning via virtual machines (VMs) significantly limits the number of concurrent applications and users in an edge system. In fact, virtual machines are inefficient because they are often overprovisioned by design and can incur into large computing overhead and long boot time. To overcome these issues, a more and more popular approach is to deploy applications in containerized packages: the so called containers are processes that encapsulate applications with their runtimes, directly executing them on the host system. Therefore, containers are sharing the same host system kernel while keeping isolated filesystems. This property enables containers to boot up to an order of magnitude faster than virtual machines and to do so with increased resource efficiency. A new deployment paradigm that makes a prominent use of containers is serverless computing.

When using a serverless computing approach [6], computing resources, as well as configuration and management services, are dynamically provisioned by the platform at runtime. This is made possible by relying on an event-driven allocation model, called *function as a service* (FaaS). This new model could be considered somehow similar to the well-known *software as a service* (SaaS) paradigm, but it puts more emphasis on event-based architecture and autoscalability, leading to a platform comparable to a back-box independent from the underlying servers (i.e. serverless), at least from the developers' and final users' point of view. In fact, developers of serverless applications can focus only on the services logic, typically split into atomic modules and wrapped into containers, differently than previous methods in which resources configuration and provisioning must be cautiously planned during application design. Infrastructure provisioning and maintenance operations are managed by the platform itself, in a transparent way to final users and developers [7]. Hence, when relying on the FaaS paradigm, even complex application are designed and released as a set of functions hosted in containers, which can be dynamically loaded on demand, often with parallel execution, without any additional overhead due to the control and management of the application deployment at operating-system level, because the hardware infrastructure maintenance operations are handled by cloud providers. Hence, serverless computing facilitates the exposure of edge services to application developers, offering to developers and IT/operations teams a fine-grained deployment model in which the overhead involved in the provisioning, update and management of the hardware infrastructure is eliminated for the application developer.

The FaaS approach, focused on resource usage efficiency and flexible provisioning, makes serverless computing a promising deployment paradigm for edge scenarios [8, 9]. FaaS can tailor the allocation of resources to the actual service needs, taking advantage of containerization highly flexible configuration and speed. This feature is convenient specially when dealing with short and event-triggered jobs such as in the IoT scenario. In fact, serverless computing allows dynamic processing of data streams from IoT devices, that can be event-triggered and inactive most of the time, transmitting messages in a periodic or random fashion. The reception of these messages can then trigger the activation of the needed serverless modules, considering two different operational modes, warm and cold start. When the warm start mode is selected, the containers hosting the application are kept up and running after initialization, granting the user the impression of high availability. Conversely, with a cold start configuration, those containers are deleted when no application requests are received within a certain time window, to save resources. However, a cold start approach may cause some latency issues. In fact, typical latencies of edge computing systems range from $\sim$20 ms in the telecom infrastructure to $\sim$5 ms in customer premises, as shown in Fig. 1. In this scenario, a service deployed using a serverless approach with "warm" containers would provide an appropriate response time. However, the same serverless service deployed using a cold start mode could lead to severe latency penalties, providing a response time that does not comply with the required quality of service, as it is up to three orders of magnitudes larger than the edge typical latencies shown in Fig. 1 [10]. This happens because, in cold start, each serverless function Pod[1] needs to be redeployed to serve a new request after a period of inactivity. However, alternatives to the warm or cold start modes such as

---

[1] A Pod is a single running entity containing one or more containers.

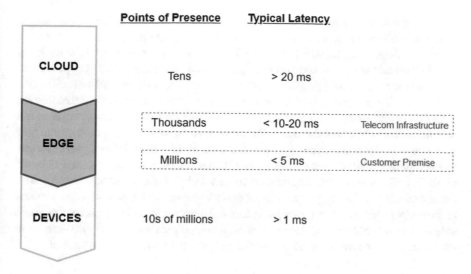

Fig. 1 Edge computing location [3]

caching and reusing previously employed containers tie up memory and can lead to information leakage. Hence, cold start issues and possible solutions are receiving an increasing interest from the academic community [11, 12].

The aim of this tutorial is to provide an exhaustive overview of the open-source serverless technologies state-of-art for edge computing. Our objective is to delve into the possible solutions to deploy serverless application on edge nodes, with a special focus on IoT data, a major use-case for edge computing. Moreover, we provide some insights on new lightweight virtualization technologies for serverless, and the analysis of the benefits and the limitations of serverless computing in term of security. Finally, we provide an overview of the current state-of-art for edge and cloud interworking in multiclouds, a potential new environment for serverless technologies.

Hence, the contribution of this work consists of the following points:

- The analysis of the currently available open-source serverless platforms, providing a comparison of their features, benefits and disadvantages.
- A detailed description of OpenFaaS, the most used open-source server platform, based on Kubernetes.
- An overview of new lightweight technologies as potential alternatives to containers for serverless application on the edge.
- The analysis of serverless computing on edge nodes' benefits and drawbacks from a security perspective.
- The description of a state-of-art serverless application for IoT data.
- A discussion on the current ways to interconnect edge and cloud platforms and the role of serverless computing in this scenario.

In the remainder of this tutorial, in Sect. 2 we illustrate the available open-source solutions to deploy a serverless application in an edge computing scenario, with a particular focus on OpenFaaS [13], a flexible and intuitive framework with the highest adoption rate among all the existing open-source serverless computing platforms. Moreover, we show how to deploy a serverless function in OpenFaaS, describing the different available templates of this serverless platform. In Sect. 3, we delve into the deployment alternatives to containers for a serverless application, analyzing features, benefits and drawbacks of these new lightweight virtualization technologies and how to use them in OpenFaaS. Subsequently, Sect. 4 presents a security analysis of serverless computing related to an edge scenario. In Sect. 5, we present a detailed example of an OpenFaaS serverless application to handle IoT data in an edge cluster with limited resources. Finally, for the sake of completeness, Sect. 6 provides an overview of the cooperation between the edge layer and the centralized cloud, presenting some solutions to interconnect them in order to migrate computational-heavy services from the serverless-empowered edge to the cloud. Section 7 concludes this tutorial.

2 Open-Source Serverless Platforms

Serverless technologies have been introduced with Amazon's product called Amazon Web Services (AWS) Lambda, which was followed by similar products from other Cloud providers, such as Google Cloud Functions, Microsoft Azure Functions or IBM OpenWhisk. All of these technologies share the concept of deploying and executing small code snippets without any control over the compute resources on which the software is running. These technologies inspired the introduction of open-source projects such as *OpenLambda* [14], an open-source serverless computing platform for building next-generation web services and applications, creating an ecosystem based on the concept of the Lambda model, which is the equivalent to the idea of serverless computation itself.

However, the aforementioned serverless platforms are offering extra functionalities tied the specific product in use, as somehow providers tend to lock-in their customers with proprietary services. For this reason, the adoption of open-source serverless frameworks is growing, having the advantage of simplifying the design, development and deployment of serverless application, abstracting the process from the specific platform and infrastructure in use [15].

In literature, some studies can be found that compare some of these frameworks from different points of view. As an example, the study presented in [16] assesses serverless frameworks on top of a Kubernetes cluster, taking into account the support offered by the framework on each software's lifecycle phase. The study of [17] performs evaluations and measures various metrics such as the service time, it also analyzes the quality of service and auto-scaling's influence on some of these metrics.

Table 1 Overview of highly adopted open-source serverless frameworks

Features	OpenFaaS	Knative	OpenWhisk	Fission	Nuclio
Adoption rate (stars)	20.7k	4.2k	5.5k	6.6k	4.2k
Container orchestrator	Kubernetes	Kubernetes	No orchestrator required, Kubernetes supported [17]	Kubernetes	Standalone Docker, or Kubernetes
Framework development language	Go	Go	Scala	Go	Go
Programming languages supported	Node.js, C#, Go, Java, Python, Ruby, PHP [18]	C#, Go, Java, Kotlin, Shell, Node.js, PHP, Python, Ruby, Scala, Rust [19]	Javascript, Swift, Python, PHP, Java, Go [17]	Python, Node.js, Ruby, Perl, Go, Bash, .NET, PHP [17]	Go, Python, Shell, Node.js, .NET, Java [20]
Message queue	NATS, Kafka, MQTT	Kafka, RabbitMQ	Kafka	NATS, Azure storage queue	Kafka, NATS, Kinesis, RabbitMQ, MQTT
Autoscaling metrics	CPU, RPS, custom metrics	CPU, RPS, concurrency	RPS	CPU	RPS

2.1 A Taxonomy of Open-Source Serverless Platforms

Examples of these frameworks, in terms of open-source solutions, are OpenFaaS, Knative, Fission, OpenWhisk or Nuclio among others. An overview of the features of these frameworks is shown in Table 1. As shown by the table, all the considered open-source serverless frameworks rely, by default or optionally, on Kubernetes [21] for containers orchestration. In fact, it is the standard de-facto to manage containers lifecycle and deployment on distributed platforms in a reliable way. Each framework offers various available languages to develop serverless functions, making use of different metrics to manage containers autoscaling. Autoscaling can be implemented in two ways: workload-based, i.e., providing additional resources following increasing traffic volume as offered by AWS Lambda, or resource-based [22], which is often relying on the Kubernetes Horizontal Pod Autoscaler (HPA) [23] to implement condition-based scaling via the definition of per-instance CPU or memory utilization thresholds.

Knative [24] offers various Kubernetes-based modules to deploy and expose serverless functions, along with a set of autoscaling mechanisms, considering CPU usage, requests per second (RPS) and the concurrency limit, which is the number of parallel processed requests. It provides a specific Command Line Interface (CLI), *kn*, and supports various tools, i.e., eventing components, to use an event-driven architecture, such as Apache Kafka and RabbitMQ brokers.

Apache OpenWhisk [25], such as Nuclio [26], is a standalone serverless platform developed by IBM. OpenWhisk is independent from Kubernetes, but it is capable of integrating it for load balancing and scaling. OpenWhisk's logic architecture is based on three basic concepts: actions, triggers and feeds. In this platform, a module executing a code snipped is defined an action. Following the FaaS concepts, actions can be dynamically scheduled, deployed and requested relying on various triggers types. In fact, triggers can be caused by events from external sources, i.e., feeds, such as IoT sensors, message queues, web applications and more. OpenWhisk also supports Hypertext Transfer Protocol (HTTP) requests to invoke actions. Kafka is used as a communication middleware between invokers and controller. The controller is responsible for authentication and authorization of every incoming request. Invokers handle container deployment and resource allocation to create a runtime for function execution [27].

Fission [28] is an open source project from Platform9 Systems. It orchestrated by Kubernetes and it allows the configuration of a set of pre-instantiated (warm) containers to provide low service time [17]. Fission optimizes the performance of cold-start latency overhead of functions by using a pool of pre-warmed containers running on the cluster. Functions are terminated when idle, ensuring there is no compute resource consumption when code is not running. Fission is based on functions, environments and triggers. Functions are single-entry point modules, written in a supported language and executed by the platform. These modules are compliant with the interface defined for the specific programming language in use, as expected by the FaaS paradigm [7]. Environments are the language-specific components that contain a pool of available resources for functions deployment. Each Environment relies on a container equipped with an HTTP server and a function loader, called the runtime. Fission invokes functions with HTTP calls. The triggers bind events to function invocations. The most common triggering method used in Fission is via HTTP. Nevertheless, Fission supports additional triggering methods such as Timers, and triggers based on open-source messaging system NATS and Azure Message Queue [28].

In the Fission framework, the component called executor controls function Pods lifecycle and the available resources. The available executors in Fission are Poolmgr and Newdeploy, as seen in Fig. 2. Poolmgr is a pool-based executor which interacts with Kubernetes and instantiates a pool of three Pods per environment, by default [28]. As previously mentioned, each of these pods includes a loader to deploy the function in a container. When the function is invoked, one Pod is removed from the pool and used for execution. The same Pod is kept in a running state to serve the incoming requests, to be eventually cleaned up when there is no traffic. Nevertheless, Poolmgr does not support autoscaling features. The new-deployment executor (Newdeploy) is significantly depending on Kubernetes functionalities. In fact, Newdeploy relies on Kubernetes Deployments, Services and Horizontal Pod Autoscaler (HPA) to deploy, execute and scale functions, with the opportunity to include load balancing capabilities. This latter executor type is not recommended when dealing with low-latency services. Newdeploy uses a minimum number of replicas set to zero by default, i.e., relying on a cold start mode for the deployed functions, but for low-latency applications this value should be set to greater values, in order to avoid latency issues, as mentioned in Sect. 1.

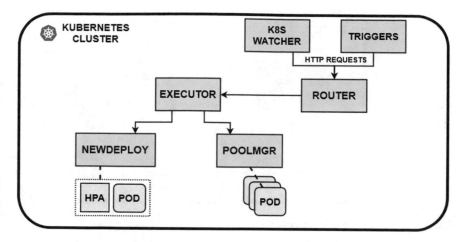

Fig. 2 Fission executor architecture [28]. In purple, Fission's native components

Nuclio [26] is a project derived from Iguazio,[2] a Platform as a Service for data science. Nuclio functions can be executed either in Docker containers or with additional Kubernetes orchestration. Nuclio's main feature is the Processor architecture, which leverages multiple parallel worker processes that can be hosted in each container. Nuclio supports direct function invocation from an external client, without requiring any Application Programming Interface (API) gateway or ingress controller, in contrast to the majority of the currently available serverless platforms. In Nuclio, the function Pod contains two kinds of processes: the previously mentioned worker processes (the actual deployed serverless function) and the event-listener. The degree of parallelism, i.e., the number of worker processes, can be configured as a static parameter. This enables the opportunity to leverage parallel execution when using multi-core nodes [22]. Like most of the other serverless frameworks, Nuclio offers various triggers such as Event Listeners, HTTP, and support for various message queues (RabbitMQ, MQTT, NATS) and data streams (Kinesis, Kafka). Nuclio is very performant, it can run up to 400,000 function invocations per second with a simple Go[3] function. However, the open-source version comes with limitations, lacking shared volume support between functions and autoscaling [26].

Among these solutions, OpenFaaS shows an adoption rate higher than the others by an order of magnitude, as shown in Table 1, where the number of GitHub stars for each framework is considered. Moreover, as already highlighted in [17], this framework offers great flexibility and good performance across different scenarios. For this reason, it represents a good choice to deploy a serverless application in an edge computing platform, taking advantage of the underlying Kubernetes functionalities to manage containers in such a distributed scenario.

---

[2] https://www.iguazio.com/.

[3] Go is a statically typed, compiled programming language with memory safety, garbage collection and structural typing features.

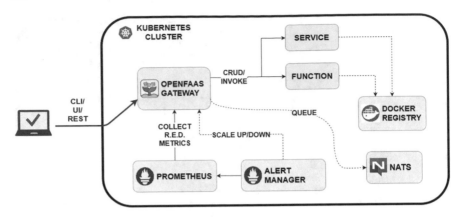

Fig. 3 OpenFaas framework's stack [29]

2.2 Deploying a Serverless Application with OpenFaaS

OpenFaaS [13] is an open-source serverless framework from cloud native computing foundation (CNCF). With OpenFaaS, developers can use templates of different languages or create their own templates to define, build and deploy serverless functions. As well as the majority of the open-source serverless frameworks, it relies on Docker images and Kubernetes control plane to run applications, with additional secret management, autoscaling and high availability features, mainly inherited from the underlying Kubernetes architecture. As shown in Fig. 3, in the OpenFaaS framework, the OpenFaaS Gateway component, similarly to a reverse proxy, is in charge of managing the function Pods lifecycle, exposing a Representational State Transfer (REST) API for all interactions. Prometheus [30] collects Rate, Errors and Duration (R.E.D.) metrics from the OpenFaaS Gateway to monitor the serverless applications in the cluster and, eventually, trigger replicas auto-scaling. In fact, triggered by the alerts sent by Prometheus and the Alert Manager, the gateway can scale functions according to the load. This scaling is regulated by modifying the replica count parameter in the dedicated Kubernetes resource. It is possible to leverage resource-based autoscaling as well, relying on Kubernetes Horizontal Pod Autoscaler [31]. OpenFaaS functions are triggered using the OpenFaaS Gateway API. For each serverless function created, there is an associated Kubernetes service. Kubernetes uses a service as an abstraction to define a policy by which accesses a set of Pods, in this case, the serverless function's Pods. Each function is deployed using a number of Pod replicas defined by the developer with the minimum and maximum scaling parameters. At present, functions can also scale to zero replicas and back again through use of the *faas-idler*[4] or the REST API, as shown in the snippet in Code 1. OpenFaaS Gateway REST API is fully documented with Swagger.[5]

---

[4] Available only in OpenFaaS PRO.

[5] https://raw.githubusercontent.com/openfaas/faas/master/api-docs/swagger.yml.

```
1 curl -X POST -d '{"service": "function-name", "replicas":
    0}' -u admin:$PASSWORD $OPENFAAS_URL/system/scale-
    function/{function-name}
```

Code 1 REST API call to scale a serverless function (function-name) to zero.

Docker images associated with serverless functions must be saved in a Docker registry. This can be either Docker Hub, the official cloud-based Docker's Container Registry, or a local registry in the cluster. The registry of choice is configurable via Kubernetes ImagePullPolicy. The most common way to trigger a function in OpenFaaS is by using a HTTP request. Moreover, functions can be invoked in an asynchronous way on OpenFaaS. In fact, long-running tasks can be invoked on a specific asynchronous endpoint. An invocation can be run in a queue using NATS Streaming [32]. This component splits the HTTP transaction between the caller and the requested service. The HTTP request is serialized to NATS Streaming through the gateway as a "producer". The queue-worker acts as a subscriber and, at a later time, it deserializes the HTTP request to use it to directly invoke the function. An optional callback Uniform Resource Locator (URL) can be provided as well [29].

Developers can create and deploy serverless functions using OpenFaaS' command line interface (*faas-cli*). Typically, FaaS modules must be atomic, stateless and compliant to the specific FaaS platform's interface [7]. The function watchdog is implemented as an HTTP server that enables requests concurrency. Moreover, for each function the corresponding watchdog handles its timeouts and checks the health of the associated pods. Every OpenFaaS function can use a watchdog as an entry point which is the init process for the Docker container hosting the function. In OpenFaaS, two different categories of templates are available, based on the type of watchdog in use:

- *Classic*: These templates make use of the classic watchdog, which uses STDIO to communicate with the developed serverless functions. Classic templates pair one process per request, providing high flexibility. This setup is mainly intended for didactic use [33].
- *Of-watchdog*: These templates rely on the associated of-watchdog, which uses HTTP to communicate with functions. It is optimized for high-throughput and reuse of expensive resources, such as database connection pools or machine-learning models. To grant high availability, it is configured to keep function processes warm between invocations [33].

For example, when choosing to deploy a Python 3 serverless application based on Flask HTTP framework [34], OpenFaaS of-watchdog template provides the following items:

- A YAML stack file, containing metadata such as the OpenFaaS Gateway's address, function name and specific Kubernetes secrets.[6]

---

[6] Kubernetes Secrets can be passwords, tokens or keys. Secrets are stored in the API server's underlying data store (etcd).

- The function's folder, containing a file to handle dependencies (*requirements.txt*) and the function itself (*handler.py*).

```
1 def handle(req):
2     """handle a request to the function
3     Args:
4         req (str): request body
5     """
6     return "Hello," + str(req) + "!"
```

Code 2 A simple Python function in OpenFaaS to print the name received in the request body. The following file must be saved as handler.py to be compliant with OpenFaaS Python 3 template.

A simple example of a Python function in OpenFaaS, relying on the of-watchdog, is shown in Code 2. It is important to note that more serverless functions in Open-FaaS can share the same stack file, to define shared metadata and interactions for a serverless modular application. The primary Python template uses Alpine Linux as a runtime environment because of its minimal size, but other environments such as Debian can be configured. Once a Docker image is built for the serverless function, one or more Pods are deployed in the cluster.

3 Beyond Containers: Lightweight Virtualization Techniques

Containers are the standard de facto virtualization approach for deploying serverless functions, being faster and lighter than virtual machines. Nevertheless, various reports [35, 36] have shown that the usage of a shared kernel can entail some security and isolation issues, as well as increased latency when deploying the containers from scratch (cold start). Hence, to find alternative virtualization solutions for fast resource and provisioning efficiency, along with the need of system security, new lightweight virtualization technologies have been proposed. These alternatives include unikernels, micro virtual machines (microVMs) and minimal hypervisors [37].

3.1 Unikernels

Unikernels, defined as runtime environments with minimal surface, have been presented as a more secure and faster virtualization solution than containers. However, with the design choice of pruning the support for dynamic forking, which is a basis for the common multi-process abstraction of conventional UNIX applications, unikernels lose flexibility and applicability. Moreover, some recent research has highlighted that, by running directly in the kernel ring, unikernels could also incur into some security flaws, since dangerous kernel functions could be invoked from the virtualized applications and privilege separation is not feasible [38, 39]. Unikernels technology inspired the IBM Nabla [40] project, which has improved isolation by using a specific, unikernel-optimized, hypervisor, Nabla Tender, in place of a general-purpose hypervisor (e.g., QEMU). Google gVisor as well [41] runs a specialized guest kernel, Sentry, in user space to support sandboxed applications. In fact, it fetches all the system calls from the sandboxed applications to the host kernel, handling them within the safe boundaries of gVisor's custom kernel [37].

3.2 Kata Containers

Kata Containers [42] are a promising lightweight virtualization technology that leverages microVMs to isolate containers processes and namespaces. This solution aims at merging the speed of containerization and the isolation granted by virtual machines into a single sandbox. In fact, Kata Containers' runtime (*kata-runtime*) deploys a Kernel-based Virtual Machine (KVM) [43] for each instantiated Pod, offering support for multiple hypervisors, including minimal ones specifically designed for serverless deployments, as shown in Fig. 4. At present, the main serverless-specific hypervisor is Firecracker [44], a minimalist, open-source Virtual Machine Monitor that relies on KVM virtualization to deploy and manage microVMs, developed in Rust[7] [45]. Similarly to unikernels, Firecracker is designed with minimal device emulation, to provide reduced memory consumption and faster startup time for each microVM [46].

3.3 Integration in Serverless Platforms

It is already possible to integrate these new virtualization technologies in serverless platforms. For example, both Knative and OpenFaaS support the configuration of various runtimes, relying on the use of Kubernetes Runtime Class feature. In Knative, it is sufficient to edit the feature flags ConfigMap to enable the *runtimeClassName* feature flag and then use that flag in the desired service [47].

```
1 kind: Profile
2 apiVersion: openfaas.com/v1
3 metadata:
4     name: gvisor
5     namespace: openfaas
6 spec:
7     runtimeClassName: gvisor
8 EOF
```

Code 3 OpenFaaS gVisor Profile

---

[7] Rust is a multi-paradigm, general-purpose programming language designed for performance and security, especially safe concurrency. Syntactically similar to C++, it can also guarantee memory safety by using a borrow checker to validate references.

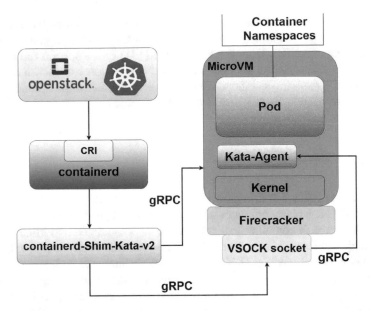

Fig. 4 Deployment of a Kata container with Firecracker hypervisor, relying on containerd, an industry-standard container runtime interface, for integration with Kubernetes or Openstack containers service

```
1 provider:
2   name: openfaas
3   gateway: \url{http://127.0.0.1:8080}
4
5 functions:
6   gvisorfunction:
7     image: functions/alpine:latest
8     annotations:
9       com.openfaas.profile: gvisor
```

Code 4 OpenFaaS stack file

In OpenFaaS, as shown in Code 3 [48], a Profile must be enabled and created to apply the desired runtime class. Then, referring the defined Profile by its *runtime-ClassName* in the serverless function YAML stack file, as shown in Code 4, OpenFaaS will deploy the serverless function using the selected lightweight technology.

Care should be taken when using lightweight virtualization technologies in a production environment. In fact, there are still some limitations and lack of maturity [37, 46]. At present, it must be noted that, when using Kata Containers with Firecracker hypervisor in Kubernetes, many issues have been encountered because of serious compatibility limitations [49–52].

4   Serverless Security for Edge Computing

In an edge computing environment, the data generated by end devices is processed and stored closer to these devices, providing fast responsiveness, fast processing and lower network traffic. However, the distribution of data in vast networks such as for edge scenarios composed by multiple interconnected devices is a challenging task that could lead to security issues. The complexity of these networks makes security control and data protection very hard, being each of the various devices present in the cluster a potential attack vector. Moreover, considering the typical multi-tenancy of the edge computing paradigm, security and isolation become an essential requirement when deploying services on the edge. Some of the security vulnerabilities of edge computing are [53, 54]:

- Denial of Service (DoS) attacks
- Data tampering
- Insecure communication between devices
- Updates' late reception and implementation
- Weak credentials for protection.

4.1   Serverless Security Advantages for Edge Computing

The usage of serverless computing to deploy services in the edge layer can lead to security advantages. Serverless functions are typically atomic, ephemeral and stateless, therefore the adoption of the serverless paradigm can reduce significantly the surface exposed to the attackers. Indeed, even if attacks are still feasible in a serverless scenario, this paradigm sensibly reduces the time window for the adversary to fetch target data from functions. The serverless approach provides the edge layer with an increased resistance to traditional DoS attacks. Serverless autoscaling features can enhance applications robustness to DoS attacks aiming to disrupt the service by overloading the servers. Nevertheless, if serverless applications owners are billed on consumed resources, some countermeasure must be introduced to protect the platform against Denial of Wallet attacks (DoW), which aims at sensibly raising the costs for application owners [55].

Table 2 Comparison of serverless execution environments security features (Kata-fc is Kata Containers with Firecracker Hypervisor)

Features	VMs	Containers	Unikernels	g-Visor	Nabla	Kata-fc
Startup time overhead	Very high	Medium	Low	Low	Low	Medium
Isolation	Very high	Low	Low	Medium	Medium	High
Language safety	No	No	No	Yes (go)	Yes (go)	Yes (go + rust)

4.2   The Tradeoff Between Security and Performance

As previously mentioned in Sect. 3, the choice of the execution environment strongly influences the performance and the security offered by the serverless platform in use. There exist various virtualization technologies with changing level of isolation between different applications and different users. In this context, the presence of a trade-off between security of the environment and performance is manifest, as can be seen in Table 2. For example, virtual machines (VMs) offer an high level of isolation with strong boundaries, but with a great startup time overhead. On the contrary, other solutions such as containers and unikernels are faster-to-deploy alternatives that guarantee an efficient resource utilisation as well, at the cost of weaker isolation boundaries [35, 36, 39]. To address these limitations and to try to jointly provide good levels of security and performance, new lightweight virtualization technologies such as Google gVisor, IBM Nabla, Amazon Firecracker and Kata Containers have been introduced in the recent years. These new solutions, already described in the previous section, rely on micro VMs to host more containers in a single strongly isolated jailer managed by an optimized minimal hypervisor (Kata Containers with Firecracker). Otherwise, Nabla and gVisor insert a guest kernel to sandbox applications containers, intercepting and handling most of the system calls invoked by the containers. Namely, gVisor only relies on 53 of the 237 available system calls, reducing of the more than the 75% the interface interacting with the host OS kernel, along with the exposed attack surface, while providing a startup time comparable with the one obtained by using containers. Nevertheless, these solutions are still too immature to be prominently adopted in serverless deployments.

4.3   The Cold Versus Warm Start Problem

An additional security-related consideration must be made regarding serverless warm and cold start modes. Warm containers, i.e., containers that are held in a running state to keep serving requests, are being more and more adopted in serverless deployments because of the significant startup times reduction and improved efficiency. In fact, a warm start mode reuses local caches or maintains connections between invocations. However, the advantages offered by warm containers entail less security for the serverless platform. While warm containers reset the default values in the filesystem at each function completion, they keep some disk space in a /tmp/ location to share state across different function invocations. So, these shared data in the system could be exploited to keep executing long-lasting attacks [55]. To avoid such vulnerabilities, application owners can disable the possibility of recycling the execution environment to run the same function multiple times, specially when dealing with security-sensitive tasks. For example, in [56] the authors present a serverless monitoring framework aimed at blocking sensitive data usage in third-party security analytics platforms, relying on OpenFaaS. In this work, containers used by the

FaaS runtime are modified to enforce lightweight processes and storage isolation within the container to provide secure poly-instantiation in warm start settings. To conclude, it must be noted that the adoption of cold start instantiated containers in place of warm start ones may not always be feasible, specially for time-sensitive tasks, since this can introduce a significant penalty in terms of latency and instability (cf. Sect. 1).

5 A Focus on IoT Scenarios

A typical application of serverless computing on edge platforms is the management and processing of data coming from IoT devices. Serverless computing is particularly useful to deploy event-triggered functionalities, such as performing analytics on IoT sensors messages [7]. A typical way to ingest data in edge nodes makes use of the Message Queuing Telemetry Transport (MQTT) protocol. In this regard, OpenFaaS presents an *event connector-pattern*, which allows developers to instantiate a broker or separate microservice which maps functions to topics and invokes functions via the OpenFaaS Gateway. In this way, the serverless code does not need to be modified per trigger/event-source. OpenFaas event-connectors can be created from scratch to connect serverless functions to pub/sub topics as well as to message queues, via OpenFaaS connector-sdk.[8] Otherwise, connectors for many well-known messaging protocols, such as MQTT, are made available.

Figure 5 shows an example of a serverless application, deployed with OpenFaaS on top of Kubernetes, to manage and expose IoT data published on an MQTT Emit-

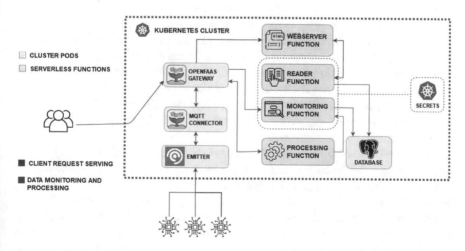

Fig. 5 Architecture of a serverless platform to monitor, process and expose via webserver external IoT event messages [10]

---

[8] https://github.com/openfaas/connector-sdk.

ter broker [57]. This system is suitable for IoT scenarios where the message rate is relatively low, such as video-surveillance applications, which typically generate sporadic transmissions, or applications for Remote Patient Monitoring (RTM) [10]. Leveraging OpenFaaS MQTT trigger feature, this serverless system is subscribed to a MQTT Emitter broker to receive and process data from the external world. To monitor, process, and expose incoming messages, four serverless Python functions have been developed, using Flask HTTP framework [34] (cfr. 2.2):

- *Monitoring Function*: This function fetch the event signalling data coming from the MQTT Emitter broker, stores them in a PostgreSQL database. If an alarm metric in the received data exceed a predefined threshold, the Monitoring Function forwards the data to the Processing Function for further processing. Once received the processed data back, it loads it in a dedicated table in the database.
- *Processing Function*: This function receives the messages with critical data from the Monitoring Function, labels them using a pre-trained machine learning classifier algorithm and sends the labeled data back.
- *Webserver Function*: This function, as the name suggests, expose a web server where registered users can monitor the stored data, both labeled and unlabeled.
- *Reader Function*: This function is called from the webserver to access and fetch stored data.

To enforce security, all external incoming requests to the Emitter broker and the OpenFaaS Gateway are handled by a firewall. Each time a message is received by the Emitter broker, OpenFaaS MQTT connector automatically triggers the invocation of the *Monitoring Function* to the OpenFaaS Gateway. It is possible to annotate a specific serverless function with the topic of interest to automatically set the event-driven invocation. Then, the serverless modules are interacting with each other via the OpenFaaS Gateway to process and expose data.

An example of a data processing chain is depicted in Fig. 6 [10].

When a new message is published in a topic (1), the Emitter broker, leveraging the OpenFaaS native MQTT connector (2), triggers the service subscribed to that topic, i.e. the monitoring function, forwarding the new data as input to the endpoint. If there are now active Pods for the function, the Gateway deploys a new replica to host the service (3) and, as a proxy, forward the received input to the instance (4). This function loads the received data in the corresponding table of the PostgreSQL database (5) and analyze the alarm metric of interest in the received message (6). If this metric does not exceed the defined threshold, the function terminates (7'). Otherwise, by calling the processing function endpoint of the OpenFaaS Gateway API (7), the monitoring function triggers its activation (8), sending the data to be processed (9). After having labelled the input using a classification pretrained algorithm, the processing function replies to the sender with the processed data (10) that it's loaded on the database by the monitoring function (11). Afterwards, the monitoring function ends its process (12). This process is described considering a warm start approach. If a cold start approach is preferred, each Pod of the serverless monitoring and processing pipeline is deleted after a predefined idle time window, to free resources with the penalty of increased latency, as shown in [10].

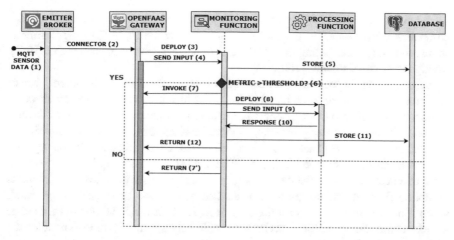

Fig. 6 Sequence diagram for the serverless processing of IoT event messages, using the MQTT protocol [10]

The read-write access to the PosgreSQL database is controlled by Kubernetes-native secrets, that can be defined by the application developers [10]. Nevertheless, more robust security mechanisms such as mutual Transport Layer Security (mTLS) can be enabled, relying on a service mesh tool like Istio,[9] to guarantee end-to-end encryption within the edge platform, from external clients all the way to the block of code that executes on a particular server and back again.

The described system is suitable for facilitating IoT data analytics and processing at the edge of the network. FaaS' dynamic provisioning of the required serverless functions promotes efficiency, a feature of paramount importance when low resource nodes are considered. It must be noted that, in case of a cold start approach, this system is convenient in terms of latency and efficiency only if the sensors message rate is considerably lower than the time needed to trigger the Pods deletion, i.e., the idle time window.

6 Handling Cloud-Edge Interworking

The adoption of edge computing keeps growing as it brings advantages for a vast range of features inherited from the architecture, from low-latency response to bandwidth savings, to a better quality of service (QoS) and quality of experience (QoE) for the user [58]. Edge environments are usually characterized by limited hardware resources, for which serverless functions can represent a perfect use case. Some efforts have been put in the standardization of edge related concepts. In Sect. 1, we referred to the European Telecommunications Standards Institute (ETSI) definition, i.e., the Multi-access Edge Computing (MEC) for Radio Access Network (RAN) [5]. There is also the OpenInfra Foundation Edge Computing Group, which released

---

[9] https://istio.io/.

a whitepaper for the definition of the architecture, the design and testing of edge computing [59]. Moreover, the Linux Foundation for the Edge (LF-Edge) aims to standardize edge computing architecture for Industrial Internet of Things (IIoT), through the EdgeX Foundry [60].

Since the edge is working with limited resources, there are more potential risks and problems that need to be handled [61]. For instance, systems can break or go offline, run out of resources, or new tasks can require more resources than the ones available. To address these problems, edge nodes often relate to a central cloud, also referred to as the edge's backend. Indeed, cloud and edge can be combined together to provide advanced service levels [58]. The edge side can offer quick response and local computation, while the cloud can perform resource-greedy tasks or jobs that do not need an immediate response. Data storage can also be shifted to the cloud, given that it can be too heavy for the edge side. As there might be one or more edge devices connected to multiple environments, in general, the cloud also handles the orchestration of the entire ecosystem.

6.1 Connecting Cloud and Edge

Multiple open-source options are available to set up the connection between cloud and edge. As there is not a preferred solution, numerous factors should be taken into account. Factors like use case, available hardware and software, knowledge of a specific technology, or specific application requirements can determine the final choice. Possible open-source solutions for setting up the connection are:

- *Software Defined Network (SDN)* or *Software Defined Wide Area Network (SD-WAN) infrastructure*. They allow to dissociate the control plane from the data plane, enabling a fine-grained control of the network flow. Open-source tools like Open Network Operating System[10] (ONOS), shown in Fig. 7, or OpenDayLight[11] (ODL) are SDN controllers that can be deployed as a standalone instance or can be integrated in cloud and edge environments like OpenStack clouds or Kubernetes clusters. An SDN approach is often combined with Network Function Virtualization (NFV) especially for 5G and mobile networks solutions [62].
- *Virtual Private Network (VPN)* or *Virtual Extensible Local Area Network (VXLAN)*. These approaches can also be adopted to set up the connection [63]. They both grant communication to devices located in different private networks as if they were directly connected, encrypting or encapsulating the packets transiting through the public internet. The most used open-source tools are Open VPN[12] (OVPN) and Open vSwitch[13] (OVS).

---

[10] https://opennetworking.org/onos/.

[11] https://www.opendaylight.org/.

[12] https://openvpn.net/.

[13] https://www.openvswitch.org/.

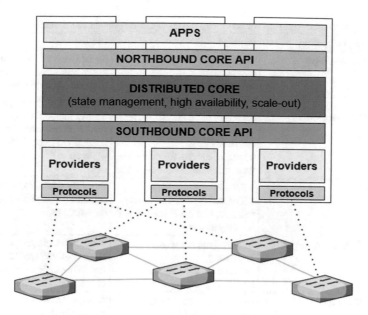

Fig. 7 An overview of ONOS SDN architecture [68]

- *Service Mesh*. Another alternative is the use of an open-source service mesh, like Istio[14] or Consul.[15] They offer VPN-like functionalities, connecting private networks while encrypting data, with some additional features such as traffic management, observability, service discovery, load balancing, etc. They have been conceived for cloud-native applications and can be perfectly integrated with public/private clouds and Kubernetes clusters.
- *Middleware*. Software solutions like middleware can be deployed to orchestrate cloud and edges intercommunications and accomplish specific use cases [64, 65]. The most common ways to set up a middleware framework are as a standalone intermediate layer or deploying it in all the entities involved. The latter usually implies end-to-end encryption for the communications. Some examples of commonly used open-source middleware are ONEedge,[16] Open IoT[17] and EdgeX Foundry.
- *Physical* or *virtual routers*. Communications can be controlled via routers, using a combination of protocols like Border Gateway Protocol (BGP) [66] and firewall rules [67]. Whereas the concept of controlling the network flow is the same as for

---

[14] https://istio.io/.

[15] https://www.consul.io/.

[16] https://oneedge.io/.

[17] https://github.com/OpenIotOrg/openiot.

Table 3 Open-source solutions for setting up the connection between cloud and edge

Networking technology	Benefits	Limitations
ONOS,OpenDayLight	Centralized controller to dissociate control plane from data plane. Fine-grained control of the network flow	Some latency is introduced. There are no specific security protocols, making the system complex to protect
OpenVPN,Open vSwitch	Encryption/encapsulation of the packets through the public internet. Grants communications for devices from different private networks	Slower connection speed and increased overhead for headers. May be difficult to scale up
Istio, Consul	VPN-like solution with additional application-level features, like observability and load balancing	Requires additional resources. Increased complexity of the environment
ONEedge,OpenIoT, EdgeX Foundry	Intermediate layer to orchestrate intercommunications. End-to-end encryption	Some products/protocols might be incompatible. Environment's adaptations could be required
OpenWRT,IPFire	No need of a centralized controller. Application at Router/Firewall level	Throughput and performances can be limited

the SDN, in this case all the operations are directly applied at router level, without the need of a centralized controller. Open-source examples are OpenWRT[18] and IPFire.[19]

A summary of the benefits and limitations of the aforementioned technologies is provided in Table 3.

6.2   Cloud and Edge Interactions

The deployment of serverless functions mostly relies on containers, where Kubernetes has a key role for their orchestration, as previously pointed out in Sect. 2. Several frameworks are available, and their deployment could differ if executed on the edge: whereas a single worker node might require less resources, an entire cluster with the relevant control plane would need much more. A worker node on the edge can join a cluster in the cloud thanks to the connection set up with one of the practical solutions listed above. In the same way a full cluster on the edge can communicate with external resources hosted in the cloud or can be joined with other clusters, producing what is defined as a cluster federation.

---

[18] https://openwrt.org/.

[19] https://www.ipfire.org/.

Interactions between cloud and edge imply data exchange, while trying to obtain bandwidth saving. In this scenario, some data transmission might be necessary for the correct functioning of the system, such as the messages from the control plane or application health checks. The interactions between cloud and edge really vary on the use case. Sensors and devices on the edge can collect and locally process information, but for long-term storage, statistics, or computationally heavy job executions, the cloud is always the best solution [69]. Backups and data replication are other examples of data transmission between edge and cloud, as well as forwarding tasks, for example when the resources available in the edge are not adequate. How and when the edge determines to interact with the cloud is another crucial aspect. It is possible that the edge waits for an idle period, or it can be based on pre-specified event-triggers, or again when the resources in use are exceeding a certain limit. Algorithms or middleware are often employed to handle the interactions. In most of the cases they work together with monitoring systems like Prometheus.[20]

In other cases, data needs to be transferred from cloud to edge, usually with the aim to improve the QoS and the QoE, for example with Content Delivery Network (CDN) or edge caching [70]. In both cases the goal is to keep content closer to the end users to make it quickly available and in the meantime offload servers in the cloud. The content stored at the edge needs to be constantly brought up to date based on pre-defined approaches, like First-In-First-Out (FIFO). In FIFO, the less recent resources are removed to make space for the latest content. Another approach is to keep the most frequently accessed logic, where the most requested elements remain available longer than the less requested ones.

It is interesting to mention also the case in which multiple edges need to communicate. A new paradigm has recently been introduced, named Collaborative Edge Computing (CEC), in which multiple edges collaborate in a distributed working model, sharing resources for tasks that cannot be managed by a single edge [71].

7 Conclusion

This work is a serverless computing for edge platforms state of the art review, which focuses on open-source solutions to process, in a serverless way, IoT data at the edge of the network. IoT devices generate data in real time, event-driven, periodically or constantly. The growing number of connected devices and sensors, along with data storage distribution on geographically sparse locations have given rise to the need of processing data closer to the source. Data processing must be efficient in order to be suitable for edge's low-resources node. This leads to the application of serverless technologies, namely FaaS frameworks, in edge computing platforms. Leveraging FaaS paradigm, the application logic is implemented by atomic, often stateless, functions, to be dynamically executed in containers deployed on-demand, without pre-allocating resources.

---

[20] https://prometheus.io/.

We have opened this tutorial with an analysis of the most popular open-source serverless frameworks existing at present, namely Knative, Apache OpenWhisk, OpenFaaS, Fission and Nuclio. The majority of these frameworks is relying on Kubernetes for containers orchestration while providing various types of triggers, i.e., eventing components, to deploy serverless functions. When considering an edge computing scenario, serverless frameworks providing IoT compliant triggers such as Apache Kafka or MQTT broker components might be useful. Moreover, traffic-based and resource-usage based autoscaling tools are an important feature in serverless frameworks, specially when dealing with workloads that are varying with time, as often happens for IoT data analytics performed on the edge. Hence, considering its adoption rate, the highest of all the considered platforms, and its great flexibility, we have chosen to focus on OpenFaaS to illustrate how to deploy a serverless application on this framework, providing insights on OpenFaaS' inner working. To get an overview of current alternatives to containers for serverless applications, we delve into lightweight virtualization techniques benefits and drawbacks, introducing Unikernels and Kata Containers. We highlight how, despite the promising features of the aforementioned technologies and the compatibility with OpenFaaS runtime classes, the actual usage of lightweight virtualization for serverless deployment is still in discussion because of various issues and lack of maturity.

Then, we provide an analysis of serverless adoption's benefits and drawbacks from a cybersecurity perspective, considering an edge computing use-case. Serverless' flexibility offers a good robustness to Denial of Service attacks, nevertheless, the usage of a warm start mode could entail some security vulnerabilities, due to shared files between invocations. This issue can be mitigated by moving to a cold start approach, but this solution is not always feasible, specially for time-sensitive tasks.

In a dedicated section, this tutorial reported a serverless architecture to process and expose IoT data on edge nodes relying on OpenFaaS, Kubernetes and the MQTT protocol. Namely, the referenced serverless functions automatically monitor the incoming messages received by the Emitter broker. This is made possible by Open-FaaS event connector for MQTT. Moreover, we reported the possibility of deploying database handlers and a webserver in a serverless way on OpenFaaS. We show the services workflow from a warm start mode perspective: a cold start mode too can be applied, but the time requirements of the services in use must be considered to evaluate possible benefits and drawbacks of this latter approach.

Finally, to provide a complete insight of the potential of the conjunction of serverless technology and edge computing, we provided a brief overview of the introduction of interworking between edge and cloud layers, to add a support for resource-greedy tasks or jobs that do not need an immediate response by the serverless-empowered edge layer.

Acknowledgements This work has been partially supported by the EU project 5G-CARMEN under grant agreement No. 825012. The views expressed are those of the authors and do not necessarily represent the project. The Commission is not responsible for any use that may be made of the information it contains. Section 6 was also supported by the Vlaio TETRA OpenCloudEdge project, project number HBC.2019.2017.

References

1. Hong C-H, Varghese B (2019) Resource management in fog/edge computing: a survey on architectures, infrastructure, and algorithms. ACM Comput Surv 52(5) [Online]. Available at https://doi.org/10.1145/3326066
2. Rimal BP, Maier M, Satyanarayanan M (2018) Experimental testbed for edge computing in fiber-wireless broadband access networks. IEEE Commun Mag 56(8):160–167
3. Artuñedo D et al (2021) Edge computing for 5g networks, v1.0. Tech. Rep., 5GPPP Technology Board Working Group [Online]. Available at https://doi.org/10.5281/zenodo.3698117
4. Sabella D, Alleman A, Liao E, Filippou M, Ding Z, Gomes Baltar L, Srikanteswara S, Bhuyan K, Oyman O, Schatzberg G, Oliver N, Smith N, Mishra SD, Thakkar P, Shailendra S (2021) Edge Computing: from standard to actual infrastructure deployment and software development—whitepaper 2021
5. Multi-access edge computing (MEC) [Online]. Available at https://www.etsi.org/technologies/multi-access-edge-computing
6. Castro P, Ishakian V, Muthusamy V, Slominski A (2019) The rise of serverless computing. Commun ACM 62(12):44–54 [Online]. Available at https://doi.org/10.1145/3368454
7. CNCF WG-Serverless Whitepaper v1.0
8. Nastic S, Rausch T, Scekic O, Dustdar S, Gusev M, Koteska B, Kostoska M, Jakimovski B, Ristov S, Prodan R (2017) A serverless real-time data analytics platform for edge computing. IEEE Internet Comput 21(4):64–71
9. Glikson A, Nastic S, Dustdar S (2017) Deviceless edge computing: extending serverless computing to the edge of the network. In: Proceedings of the 10th ACM international systems and storage conference, ser. SYSTOR '17. Association for Computing Machinery, New York, NY, USA [Online]. Available at https://doi.org/10.1145/3078468.3078497
10. Benedetti P, Femminella M, Reali G, Steenhaut K (2021) Experimental analysis of the application of serverless computing to IoT platforms. Sensors 21(3) [Online]. Available at https://www.mdpi.com/1424-8220/21/3/928
11. Mohan A, Sane H, Doshi K, Edupuganti S, Nayak N, Sukhomlinov V (2019) Agile cold starts for scalable serverless. In: 11th USENIX workshop on hot topics in cloud computing (HotCloud 19). USENIX Association, Renton, WA [Online]. Available at https://www.usenix.org/conference/hotcloud19/presentation/mohan
12. Manner J, Endreß M, Heckel T, Wirtz G (2018) Cold start influencing factors in function as a service. In: 2018 IEEE/ACM international conference on utility and cloud computing companion (UCC Companion), pp 181–188
13. OpenFaaS: open function as a service [Online]. Available at https://www.openfaas.com/. Accessed 26 Sept 2021
14. Hendrickson S, Sturdevant S, Harter T, Venkataramani V, Arpaci-Dusseau AC, Arpaci-Dusseau RH (2016) Serverless computation with OpenLambda. In: 8th USENIX workshop on hot topics in cloud computing (HotCloud 16). USENIX Association, Denver, CO
15. Kritikos K, Skrzypek P (2018) A review of serverless frameworks. In: 2018 IEEE/ACM UCC Companion, pp 161–168
16. Aggarwal V, Thangaraju B (2020) Performance analysis of virtualisation technologies in NFV and edge deployments. In: 2020 IEEE CONECCT, pp 1–5
17. Mohanty SK, Premsankar G, di Francesco M (2018) An evaluation of open source serverless computing frameworks. In: 2018 CloudCom, pp 115–120
18. OpenFaaS: templates [Online]. Available at https://www.github.com/openfaas/templates
19. Knative: code samples overview [Online]. Available at https://knative.dev/docs/serving/samples/
20. Nuclio: code examples [Online]. Available at https://nuclio.io/docs/latest/examples/
21. Kubernetes: production-grade container orchestration [Online]. Available at https://kubernetes.io/

22. Li J, Kulkarni SG, Ramakrishnan KK, Li D (2019) Understanding open source serverless platforms: Design considerations and performance. In: Proceedings of the 5th international workshop on serverless computing, ser. WOSC '19. Association for Computing Machinery, New York, NY, USA, pp 37–42 [Online]. Available at https://doi.org/10.1145/3366623.3368139
23. Kubernetes horizontal pod autoscaler [Online]. Available at https://kubernetes.io/docs/tasks/run-application/horizontal-pod-autoscale/
24. Knative [Online]. Available at https://knative.dev/docs/
25. Apache OpenWhisk is a serverless, open source cloud platform [Online]. Available at https://openwhisk.apache.org/
26. Nuclio: serverless platform for automated data science [Online]. Available at https://nuclio.io/
27. Djemame K, Parker M, Datsev D (2020) Open-source serverless architectures: an evaluation of apache OpenWhisk. In: 2020 IEEE/ACM 13th international conference on utility and cloud computing (UCC), pp 329–335
28. Triggers | Fission [Online]. Available at https://fission.io/docs/usage/triggers/
29. Stack—OpenFaaS [Online]. Available at https://docs.openfaas.com/architecture/stack/#layers-and-responsibilities
30. Prometheus—monitoring system & time series database [Online]. Available at https://prometheus.io/
31. Benedetti P, Femminella M, Reali G, Steenhaut K (2022) Reinforcement learning applicability for resource-based auto-scaling in serverless edge applications. In: 2022 IEEE international conference on pervasive computing and communications workshops and other affiliated events (PerCom Workshops), pp 674–679
32. NATS—connective technology for adaptive edge & distributed systems [Online]. Available at https://nats.io/
33. Watchdog—OpenFaaS [Online]. Available at https://docs.openfaas.com/architecture/watchdog/
34. Welcome to flask—flask documentation (1.1.x) [Online]. Available at https://flask.palletsprojects.com/en/1.1.x/
35. CVE Details: CVE-2019-5736 (2019) [Online]. Available at https://www.cvedetails.com/cve/CVE-2019-5736/
36. Exploiting CVE-2017-5123 (2017) [Online]. Available at https://reverse.put.as/2017/11/07/exploiting-cve-2017-5123/
37. Chen J (2019) Making containers more isolated: an overview of sandboxed container technologies [Online]. Available at https://bit.ly/2ZcRBOR
38. Zhang Y et al (2018) KylinX: a dynamic library operating system for simplified and efficient cloud virtualization. In: USENIX ATC 18, Boston, MA, pp 173–186
39. Talbot J et al (2019) A security perspective on Unikernels [Online]. Available at https://arxiv.org/abs/1911.06260
40. Nabla Containers [Online]. Available at https://github.com/nabla-containers
41. gvisor [Online]. Available at https://gvisor.dev/
42. Randazzo A, Tinnirello I (2019) Kata containers: an emerging architecture for enabling MEC services in fast and secure way. In: IOTSMS, pp 209–214
43. Linux KVM [Online]. Available at https://www.linux-kvm.org/page/Main_Page
44. Agache A et al (2020) Firecracker: lightweight virtualization for serverless applications. In: 17th USENIX NSDI, Santa Clara, CA, Feb 2020, pp 419–434
45. Jung R et al (2017) RustBelt: securing the foundations of the rust programming language. Proc ACM Program Lang 2(POPL)
46. Perez R, Benedetti P, Pergolesi M, Garcia-Reinoso J, Zabala A, Serrano P, Femminella M, Reali G, Banchs A (2021) A performance comparison of virtualization techniques to deploy a 5g monitoring platform. In: 2021 joint European conference on networks and communications 6g summit (EuCNC/6G Summit), pp 472–477
47. Knative Services [Online]. Available at https://gvisor.dev/docs/tutorials/knative/
48. OpenFaaS—Profiles [Online]. Available at https://docs.openfaas.com/reference/profiles/

49. FC not working with k8s in kata 2.x (2020) [Online]. Available at https://github.com/kata-containers/kata-containers/issues/1130
50. Kubernetes with containerd and kata-fc not working (2020) [Online]. Available at https://github.com/kata-containers/kata-containers/issues/1118
51. Firecracker limitation: volume support (2018) [Online]. Available at https://github.com/kata-containers/runtime/issues/1071
52. Documentation: firecracker limitations (2019) [Online]. Available at https://github.com/kata-containers/documentation/issues/351
53. Parikh S, Dave D, Patel R, Doshi N (2019) Security and privacy issues in cloud, fog and edge computing. Procedia Comput Sci 160:734–739. The 10th international conference on emerging ubiquitous systems and pervasive networks (EUSPN-2019)/The 9th international conference on current and future trends of information and communication technologies in healthcare (ICTH-2019)/affiliated workshops [Online]. Available at https://www.sciencedirect.com/science/article/pii/S1877050919317181
54. Hassan B, Askar S (2021) Survey on edge computing security
55. Marin E, Perino D, Pietro R (2021) Serverless computing: a security perspective
56. Polinsky I, Datta P, Bates A, Enck W (2021) Sciffs: enabling secure third-party security analytics using serverless computing. In: Proceedings of the 26th ACM symposium on access control models and technologies, ser. SACMAT '21. Association for Computing Machinery, New York, NY, USA, pp 175–186 [Online]. Available at https://doi.org/10.1145/3450569.3463567
57. Emitter: scalable real-time communication across devices [Online]. Available at https://emitter.io/
58. Shi W, Dustdar S (2016) The promise of edge computing. Computer 49(5):78–81
59. Edge computing—next steps in architecture, design and testing [Online]. Available at https://www.openstack.org/use-cases/edge-computing/edge-computing-next-steps-in-architecture-design-and-testing/
60. Edgex foundry [Online]. Available at https://www.edgexfoundry.org/
61. Shi W, Cao J, Zhang Q, Li Y, Xu L (2016) Edge computing: vision and challenges. IEEE Internet Things J 3(5):637–646
62. Guerzoni R, Trivisonno R, Soldani D (2014) SDN-based architecture and procedures for 5g networks. In: 1st international conference on 5g for ubiquitous connectivity, pp 209–214
63. Xiong Y, Sun Y, Xing L, Huang Y (2018) Extend cloud to edge with KubeEdge. In: 2018 IEEE/ACM symposium on edge computing (SEC), pp 373–377
64. Tang J, Yu R, Liu S, Gaudiot J-L (2020) A container based edge offloading framework for autonomous driving. IEEE Access 8:33713–33726
65. George A, Ravindran A (2019) Distributed middleware for edge vision systems. In: 2019 IEEE 16th international conference on smart cities: improving quality of life using ICT IoT and AI (HONET-ICT), pp 193–194
66. Lei B, Zhao Q, Mei J (2021) Computing power network: an interworking architecture of computing and network based on IP extension. In: 2021 IEEE 22nd international conference on high performance switching and routing (HPSR), pp 1–6
67. Singh J, Bello Y, Hussein AR, Erbad A, Mohamed A (2021) Hierarchical security paradigm for IoT multiaccess edge computing. IEEE Internet Things J 8(7):5794–5805
68. ONOS Platform: an overview [Online]. Available at https://opennetworking.org/onos/
69. Li X, Ma Z, Zheng J, Liu Y, Zhu L, Zhou N (2020) An effective edge-assisted data collection approach for critical events in the SDWSN-based agricultural internet of things. Electronics 9(6) [Online]. Available at https://www.mdpi.com/2079-9292/9/6/907
70. Baresi L, Filgueira Mendonça D (2019) Towards a serverless platform for edge computing. In: 2019 IEEE international conference on fog computing (ICFC), pp 1–10
71. Ning Z, Kong X, Xia F, Hou W, Wang X (2019) Green and sustainable cloud of things: enabling collaborative edge computing. IEEE Commun Mag 57(1):72–78

Accelerating and Scaling Data Products with Serverless

Angel Perez, Boyan Vasilev, Zeljko Agic, Christoffer Thrysøe, Viktor Hargitai, Mads Dahlgaard, and Christian Røssel

Abstract Managing a comprehensive data products portfolio with scaling capabilities is one important task for an organization-wide analytics team. Those data products can be broken down into components that use serverless offerings from cloud service providers, allowing a team of modest size to manage company-wide analytics and data science solutions while improving productivity and promoting data-driven decisions. This work describes an architecture and tools used to speed up and manage data offerings, including data visualization, pipelines, models, and APIs. Considerations of component design include re-usability, integration, and maintainability, which are discussed along with their impact to team productivity. The components described are: Data ingestion using a containerized solution as a fundamental layer for all the applications, including its execution, orchestration, and monitoring. This solution is combined with traditional pipelines in order to enrich the data available; APIs for data and model serving using containerized solutions as a building block for data products that are powered by machine learning models, and for serving a unified data ontology; Data Visualization in the form of containerized web apps that provides fast solutions for data explorations, model predictions, visualization, and

A. Perez (✉) · B. Vasilev · Z. Agic · C. Thrysøe · V. Hargitai · M. Dahlgaard · C. Røssel
Unity Technologies, Niels Hemmingsens Gade 24, 1153 Copenhagen, Denmark
e-mail: angel.perez@unity3d.com

B. Vasilev
e-mail: boyan@unity3d.com

Z. Agic
e-mail: zeljko.agic@unity3d.com

C. Thrysøe
e-mail: christoffer.thrysoe@unity3d.com

V. Hargitai
e-mail: viktorh@unity3d.com

M. Dahlgaard
e-mail: mads.dahlgaard@unity3d.com

C. Røssel
e-mail: christianr@unity3d.com

© The Author(s), under exclusive license to Springer Nature Switzerland AG 2023
R. Krishnamurthi et al. (eds.), *Serverless Computing: Principles and Paradigms*,
Lecture Notes on Data Engineering and Communications Technologies 162,
https://doi.org/10.1007/978-3-031-26633-1_6

user insights. The architectures of three data products are then described as aggregations of the distinct building blocks (components) that were developed and how those can be repurposed for different applications. This includes continuous integration and delivery as well as pairing of each solution with the corresponding products from cloud service providers (e.g. Google Cloud Platform) in order to provide some real world examples. The scaling of each solution is discussed, as well as lessons learned and pitfalls we have encountered regarding security, usability, and maintenance. The proposed components and architecture allowed a team of 7 members to cater for analytics solutions for a section (equivalent to approx. 1500 employees), which provides a clear picture of the potential of serverless in rapid prototyping and empowering effective teams.

Keywords Analytics · Architecture · Containerized · Data pipelines · Data products · Data science · Machine learning · Real-world · Serverless · Visualization · Web apps

1 Introduction

There is hardly a company out there which does not use data and analytics in some shape and form to make better and more informed decisions or to directly improve its products and user experience [1]. Based on the maturity of the data operations, data teams will vary greatly in their structure, roles, workflows, tech stack, and output. In turn, the stage of the data journey will greatly depend on the company's data needs.

The most fundamental data need is to be able to address business or product-related questions and compute metrics either ad-hoc or systematically. The data analyst is the key role associated with these tasks. The ideal analyst is characterised by great velocity [2] in navigating across the unexplored domain and data in order to deliver insights in the form of single analysis or a dashboard.

Further, the data needs will inevitably evolve in terms of:

1. Scale—supporting more teams with faster time to insights.
2. Complexity—increasing volume and complexity of data, domains and analysis.
3. Requirements—data quality, metrics standardisation, continuous tracking.

These evolved needs will lead to expanding the data team with new roles and technologies. The new roles in the data team will, most of the time, revolve around data engineering, analytics engineering, BI development, and data science [3].

Data engineers will help immensely in bringing software engineering principles in how data and related capabilities are developed, managed and served. Ideally, code will be version-controlled, and common principles and practices will be established, translating into robust workflows with fully observable and monitored data pipelines.

Analysts, analytics engineers and BI (Business Intelligence) developers will work closely to develop data models across all domains served to the company in the shape of interactive dashboards and applications.

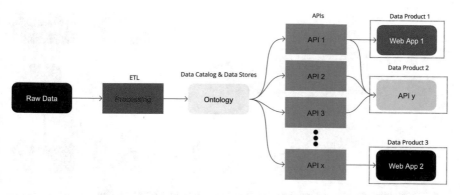

Fig. 1 Data flow used for developing data products

Data scientists can expand the scope and complexity of the generated insights by applying machine learning and advanced modelling [4, 5]. Typically, these will be served to the company in the form of interactive web applications or feeding as additional data points to existing deliveries.

As the data team and capabilities continue to mature, data and team deliveries will ultimately be treated as data products. This will result in establishing robust data, product and engineering principles which in turn will push the technology needs in the team even further.

At this state of data maturity, the need for data products [6], to inform decisions, personalize user experience and impact key business metrics inside an organization becomes one of the main responsibilities of a data science team. This responsibility is covered by a portfolio of data products that are not only to be used by the data specialist (analyst, researcher, etc.) but also by decision makers and other teams as downstream dependencies, which rely on these data products for their daily operations.

Therefore, the data science team is required to have high velocity in providing new features to be tested and deployed in order to improve business performance while keeping maintainability.

These requirements are achieved with a small team by using the serverless solutions [7–12] described in this chapter. The basic data flow used for fast prototyping and interoperability of these data products is described in Fig. 1. This modular approach in combination with serverless technology has been our proven approach for consistently delivering data products at scale with a modestly-sized team.

The flow in Fig. 1 relies heavily on a set of steps which will improve the development process, namely: obtaining a shared data model and creating a data foundation. Our first step is to build a conceptualization of the data model, which will be shared by the overall organization's data products and the data team - this is what we call our ontology [13–15].

The data ontology is crucial for building a foundation that can be reused for multiple projects and comes initially from business knowledge and conversation

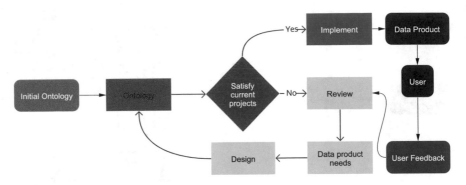

Fig. 2 Methodology for a shared data model and its continuous improvement

with key stakeholders. Its continuous update is decisive for keeping relevance and providing other teams with the right data, the process for creating and maintaining a shared data model is shown in Fig. 2. The process of iteration to improve on the shared data models is the key to building a data foundation that later will translate into standardized and commonly understood entities, attributes and relationships.

This work covers an identified gap in the literature: a comprehensive combination of serverless solutions to effectively deliver data products at an enterprise scale. In-depth books on different topics necessary to deliver data products exist individually for each cloud provider, covering data engineering [16], data science [5], or general google cloud development aspects [9, 10]. In contrast, our team can provide complete solutions that not only combine multiple google cloud offerings but also generalize the building blocks for simple migrations and applications to other cloud providers and mitigation of vendor lock-in. Additionally, the suggested architecture and solutions had been tested on a real-world application and therefore provide a production-ready solution.

The chapter is organized as follows. A general description of our methodology is described in Sect. 2, a detailed description of the tools used and the matching with a real cloud service provider is included in Sect. 3 for each one of the building blocks. A set of real-world inspired examples is provided for the sake of clearly, exposing the actual possible applications offered in Sect. 4. An analysis on the opportunity for generalization on the given architecture is discussed, with special emphasis on the ground not covered in this chapter accompanied with some final remarks can be found in Sect. 5.

2 Data Products and Their Architecture

The generalized view in which all outputs of our data teams are to be designed products, e.g. data products, allow us to decompose each task into re-usable parts as well as to maximize the maintainability of each one of our deliverables.

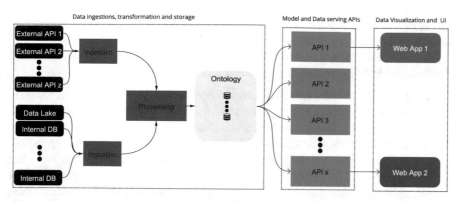

Fig. 3 Generic architecture used for our data product and fundamental building blocks

To achieve this purpose the team was required to choose a cloud architecture for the general data products that the company is expecting to maintain and produce, in such a way that all parts are reusable and require minimum extra effort for maintaining the infrastructure. In order to manage this requirement, the following paradigms were considered:

1. Infrastructure must be established as code in version-controlled repositories, that includes CI/CD capabilities. GitHub and GitHub Actions are used for this purpose.
2. Solutions based on container orchestration (e.g. Kubernetes) are preferred when implementing custom applications.
3. Serverless solutions for data processing, app serving and model training are to be preferred.

The mentioned considerations allowed the team to create a generic structure in which all our current data products can be developed, and this approach has some clear advantages regarding the team output in comparison to its size. Additionally, the serverless technology that includes auto-scaling functions allows to only use the amount of resources required at a given moment which does not only reduce the compute cost, but also the implied carbon footprint of the services that are created where much of the energy is spent by idle servers when VM's or traditional data center solution are used [17, 18].

The generic architecture used for a data product is shown in Fig. 3. The premise of the proposed architecture is that many of the APIs developed for one data product can be reused or repurposed for other data products in the future as long as the data ontology is kept relevant. The common topic with all the data products is re-usability and scalability.

The architecture in Fig. 3 allows us to decouple the API consumers from the complexities of the data, since a unified data model is exposed in the APIs to our different data products and in some cases the APIs are the product in itself. There is an extra use case of delivering dashboards and reporting which can be covered by the same model, where these deliveries are derived directly from the ontology into Looker or Data Studio—which are both managed solutions (i.e. serverless).

The team identified three main categories of components that could be treated as fundamental building blocks in the architecture in Fig. 3, namely:

Data ingestion, transformation, and storage. It is probably the largest challenge when developing data products since most of our applications are related to multiple data sources, external and internal, that are needed to integrate in order to gain insights that are actionable. It includes calls to APIs for data ingestion as well as data processing and finally storage for consumption by our data products, the details are described in Sect. 3.1. Both the code generated for ingesting the data (API calls and storage), the Docker images for the services and the orchestration of the different jobs that include ingestion and transformation are reusable. Additionally, in Sect. 4, it is presented a use case in which all solved analytics questions are version controlled allowing the team to have a repository of solutions and a historical register of changes.

Models and data serving APIs. In order to share our data, features, embeddings, and models for other applications, The solution needs to be extensible and reusable. Therefore, an API was created for each solution. Section 3 will review in detail the used solutions and Sect. 4 contains examples of applications where these blocks are vital.

Data visualization and UI. Regularly, our data, monitoring and model results need to be made available not only for other applications (e.g. personalization) but for data professionals, stakeholders, employees or users. For such scenarios small web apps with a specific purpose are created, each web app is a containerized application crafted with the intention to provide the means to get the message across with minimum effort and in a solution that traditional dashboard applications cannot provide (e.g. Looker, DataStudio, etc.).

3 Serverless Building Blocks for Data Products

In this section, the details of each of the main building blocks in our architecture are described, in order to simplify the description of the real-life study cases. The architectures are based on the solutions provided by Google Cloud Platform (GCP) [5, 10, 11, 16, 19].

The building blocks described in this section are general and expected to be transferable to other cloud services. In order to facilitate the development of such applications and the search of equivalent services, this work provides a table with alternative[1] offerings to GCP services either through other cloud providers or self-hosted options, see Table 1.

---

[1] Compare AWS and Azure services to Google Cloud in https://cloud.google.com/free/docs/aws-azure-gcp-service-comparison.

Table 1 Equivalent service offerings among Google Cloud Platform, Azure, Amazon Web Services, and self-hosted options

GCP	Azure	AWS	Self-hosted
Dataproc	Synapse	EMR[a]	Apache Spark
Composer	Data Factory	MWAA[b]/Glue	Apache Airflow
Kubernetes Engine	AKS[c]	EKS[d]	Kubernetes
Dataflow	Stream Analytics	Kinesis Data Analytics/Glue	Apache Beam
Cloud SQL	Database for PostgreSQL	RDS[e]/Aurora	PostgreSQL
App Engine	App Service	Elastic/Beanstalk	
Cloud Run	Container Instances	Fargate/Lambda/SAM[f]	

[a] Elastic MapReduce
[b] Amazon Managed Workflows for Apache Airflow
[c] Azure Kubernetes Service
[d] Elastic Kubernetes Service
[e] Relational Database Service
[f] Serverless Application Model

Since some of the services (e.g. Dataproc) offer self-managed versions, the authors would like to clarify that all solutions described are serverless, which means that the cloud provider offers managed solutions with no need for server management or resource allocation (e.g. Dataproc Serverless[2]). The word serverless is omitted on many occasions for a clear and simple exposition. The distinction is made explicit where self-managed counterparts exist, and those are not covered in this chapter.

3.1 Data Ingestion, Storage, and Transformation

The main purpose of this building block is to orchestrate a set of tasks which convert raw data into a usable ontology, that can support a wide array of varied downstream use cases. Data is being streamed from various products, within the organization, into our data lake. The data lake is maintained by a central data team, where our sources of raw data are stored.

A set of transformations and processing steps is applied to the raw storage in order to transform the data to a state that feed into our downstream data products. A non exhaustive list of transformations include:

- Imputing missing data and entity resolution.
- Formatting data and other transformations.
- Computing metrics, statistics, and features.

---

[2] https://cloud.google.com/dataproc-serverless/docs/overview.

- Joining data across product domains, financial systems, and other sources to create our ontology.
- Monitoring data quality.

The transformed data is stored using three main services:

- BigQuery [16, 19–21]: Google Cloud's petabyte scale, serverless data warehouse that powers a great deal of our processing needs by providing a highly parallel and scalable execution model, that executes processing requests across thousands of machines, through a single SQL programming interface. This simple interface allows users, across many data roles, to analyze large quantities of data, which can scale to any organization size. The data ontology can then be transformed and exposed through a simple interface, which greatly improves iteration cycles and data discovery.
- Cloud Storage [16, 22]: distributed and highly scalable object storage, that is often used to stage data before loading to BigQuery or other services.
- Cloud SQL [16, 23]: managed relational database service, which is utilized as backend, to power our in-house API's and web applications.

Each transformation, and the storage of the derived data asset, are to be scheduled at a specific cadence, e.g., weekly or daily. For this, the team needed a solution to orchestrate our workflows.

For orchestrating workflows, this work proposes the use of Airflow [24]. Airflow is a widely adopted workflow scheduling framework, among data professionals. In Airflow, workflows are expressed as DAGs (directed acyclic graphs), which are directed graphs that define the order of execution of a number of tasks in the graph, where each task is defined through some specific operator. DAGs, tasks and operators are all configured using Python. Specifically, Google Cloud's managed Airflow instance Cloud Composer [16, 25] is used, which is deployed on a Google Kubernetes Engine cluster (GKE) [16, 26]. See [27] for the Amazon Web Services (AWS) Managed Workflows for Apache Airflow. The main advantages of using a managed solution, like Google Cloud Composer, are the following:

- The managed solution requires minimal maintenance overhead, while providing much flexibility in terms of workflow execution. It allows the team to spend the majority of its time focused on building workflows and creating value for the organization. The solution also comes with community-contributed open-source execution operators, that are used within the team.
- Custom built container images can be used to execute tasks on GKE, while preserving the orchestration facilities of Airflow. These images can be contributed from the entire organization, where they can be shared in central accessible Container Registry, e.g., Google Container Registry (GCR) [28]. These images can then be re-used across multiple teams. This facilitates machine learning jobs (model evaluation and training), data integration workflows and custom processing executions. For execution, the Airflow provided `KubernetesPodOperator` [29] is used. With this operator, specific GKE Node Pools [30] can be assigned for the

workload, to support a varied set of workflows with potentially different hardware requirements. By auto-scaling the node pools, costs can be kept minimal by automatically scaling them to zero when idle. The auto-scaling nodes also allows us to run a large amount of compute intensive workflows at the same time.

- Airflow provides out of the box built-in supported operators for utilizing Big Data tooling such as Apache Spark [31] and Apache Beam [32], as well as for their serverless counterpart in CGP, namely Dataproc Serverless [33] and Dataflow [34].

Our complete solution for data processing and ingestion is implemented as follows:

- DAG workflow files are contributed to a single GitHub repository. Each DAG can potentially contain a variety of different task, for example, processing data in BigQuery, processing or moving data using Spark or Beam Jobs or custom data integration workflows executed through containerized images running some custom Python code. Workflow files, needed to execute these different workflows are also contributed to the central repository. Each team member has the freedom to choose the best set of tools, to solve the problem at hand, as long as it is possible to integrate and execute through Airflow.
- Continuous integration is achieved using GitHub Actions, which allows defining a set of custom actions that are to be executed when a pull request is merged on the repository's main branch. The main actions are:
  - copying DAGs and other workflow files to the Cloud Composer environment.
  - building and pushing custom workflow container images to the container registry [28], in order to make them available for Airflow workers, when executing some specific task.

The combined solution offers a way for data professionals to quickly update and maintain data workflows. There are other aspects that are not covered in this chapter but for the sake of completeness are stated here:

- Logging: Airflow does provide logs for all DAGs, especially important for debugging purposes. Additionally, when integrating with one of the other managed services (e.g. Dataflow) a solution for monitoring, metrics, and logging is provided by Google Cloud Operations suit [35].
- Security: In addition to implementing best practices [36], a central security team monitors the running versions of Kubernetes and suggests upgrades based on emergent security issues. Additionally, newer version of the managed services are updated constantly but it normally requires an intervention to change the version used and some tests to assure compatibility of the code with newer versions.
- Scalability: for each service the resources allocated are managed by the cloud provider and thus not usually managed by the user. However, there is a possibility to control limits in which the cloud provider allocates resources, this decision is related to budget allocation based on project importance and required pipeline updating cadence.

3.2   Models and Data Serving Using APIs

The second basic building block has as a purpose to allow the team to make available our ontology to other data teams and to applications that require personalization or machine learning models for their operation. The strategy used for this building block follows these steps:

- Create a repository that contains the API code, where Python is used alongside API frameworks such as FastAPI [37], Flask, etc. The repository contains enough information to create a Docker image for the application.
- Use GitHub Actions for continuous integration: it triggers on pull request to merge on main branch, builds of the Docker images, stores them in the registry, and deploys to Google Cloud Run [38, 39].
- GitHub offers "secrets", a way for us to store the credentials of the service account that performs the image builds using Google Cloud Build and deployment to Google Cloud run.

This solution allows us to serve data for applications at run-time (e.g. for web applications) and enables use cases such as machine learning model evaluation and recommendation engine to be served directly from our team project and with minimal maintenance overhead. Our real-life applications will be showcased in Sect. 4.

3.3   Data Visualization and User Interfaces

A big part of a data science team load is related to data visualization and storytelling. It could be used to present the results of an A/B test, evaluate model effectiveness or to demonstrate a concept by an interactive experience. Our challenge is finding the best medium to express our message clearly.

Solutions to this challenge can be tackled in multiple ways and while third-party solutions that promise just this exists, it was found that enabling maximum expressive freedom for analysts and data scientists is our best bet.

Maximum expressive freedom entails allowing the data professionals to select the tools they feel more comfortable with to produce the experiences (e.g. dashboard, web applications, etc.). This approach moves the complexity from the data professionals to the deployment process by needing to support multiple frameworks and languages.

To deal with the increased complexity on the deployment side, each data scientist/analyst will own the code that builds the custom images and CI/CD code for its deployment to App Engine. The code is based on templates developed by our data engineers and on experience with other web apps developed for the team.

In order to accommodate a dynamic mix of web frameworks, libraries, and programming languages, the following approach was developed:

- **Dashboard solutions**: Looker and Google Data Studio integrate seamlessly with BigQuery, therefore other than creating the supporting tables in BigQuery the data professional will not require any extra step in order to produce this type of visualization.
- **Web Applications**:
  - Create a repository that contains the required files for creating a docker image (e.g. Docker file, packages requirement file, etc.).
  - Use Github Actions [40] to deploy the solution to Google Cloud App Engine [41].
  - Access control to each application is done by user membership of privileged groups using Google Cloud Identity and access management [42], integration with a single sign-on solution Okta [43], and an Identity-Aware Proxy solution [44]. The latter can be configured to give access to a full company via Google Workspace and Google Cloud App Engine auto-scale capabilities.

4   Real-World Examples, Data Products Catalogue

This section is an overview of selected applications from our data products catalog. Each application is provided with a generic business context and an implementation card for easy reference. The cards follow a use case-agnostic format, namely:

Context A brief description of the business need that is addressed.
Architecture A diagram and a description of the product building blocks.
Remarks Results that were obtained with the selected architecture.

Each of the examples corresponds to a data product in our portfolio, only the most relevant study cases are portrayed and a simple structure for the ease of reproduction is displayed, more complex solutions are possible with the same architecture. The following examples are to be studied:

- Data Processing Foundation for Analytics: data product that is deployed in order to maintain our ontology.
- Ontology Visualization: a visual map of our ontology for the company to explore.
- Data Lineage Visualization: it provides a visual reference for the dependencies of each table (Lineage).
- Community Analytics: analysis of the pulse of the company's sentiment on different social networks.
- Recommendation System: powering personalization of our market place.

The main objective of this chapter is to provide the reader with a set of data products that had been tested with our architecture and that will provide a significant amount of value to the reader.

The results achieved with the presented architecture and suggested services are the following: approx. 80 pipelines (DAGs) that spawn 90 datasets totaling 2700 tables that serve the business analytics needs of an entire organization. The internal web apps see a maximum of approx. 300 active users per app and a total of approx. 100 dashboards are served. The external serving APIs (e.g. Recommender) have a volume between 100 and 3000 QPS (queries per second). All the services are provided with a monthly uptime percentage[3] of at least 99.99% with a team of only 7 data professionals.

4.1 Data Processing Foundation for Analytics

Like many data specialists working in the software industry (and beyond), our team is responsible for extracting insights from a large and growing volume of varied data. Along with the complex, event-based, non-relational usage and performance data emitted from the core software product of our company, multiple transactional systems power our business—handling, for example, accounts, licenses, e-commerce content metadata, and sales. Furthermore, thanks to the large active user community of our products, ample product feedback is provided in multiple relevant online channels (including 3rd party social media and our own forums). Additionally, new products and their instrumentation is rolled out frequently—prompt insights about their introduction to the market is very valuable for decision makers. The characteristics and requirements of the relevant data processing workflows—tackling tasks including inference-based data cleaning, data collection through APIs and storage, and feature engineering using combinations and aggregations of multiple data sources—often differ significantly, with tasks processing up to terabyte-scale datasets or mere kilobytes, and computations taking only a few seconds or up to several hours (despite parallelism).

Utilizing these heterogeneous and proliferating data sources, even from the inception of our data science and -engineering team and our data foundation, the team has enabled machine learning, data science, reporting, and interactive data exploration app use cases. The team had to address these needs despite modest engineering resources, especially initially, and support from other data teams being largely limited to ingestion and integration work (for some data sources), storage, and data governance. Thus, it was essential for us to implement an efficient solution for addressing the scaling of our data sources, in terms of both development and maintenance resources, and time to insights.

This solution was designed to support the ongoing scaling of our team itself, as well as collaboration within our team and with other data professionals and stakeholders, to enable the effective use of our growing resources, and increase our impact. Furthermore, it was designed to provide a single location where our data processing

---

[3] SLAs are given by Google services: https://cloud.google.com/appengine/sla and https://cloud.google.com/bigquery/sla.

pipelines and their infrastructure could be developed, maintained, and evolved, even by a small team, which is accessible to all interested data specialist colleagues, and easily extensible with their preferred set of tools. The reproducibility and compatibility of data pipeline development contributions to our data foundation also had to be ensured, through version control and continuous integration solutions, along with the reliable operation of the pipelines, through orchestration, monitoring, logging, and alerts. For end-to-end efficiency, this data foundation provides an extensible set of reusable data processing solutions and processed data outputs, as building blocks to power the wide range of data use cases described above. The solution has successfully addressed these needs, and our team expects it to scale further, enabling future growth and new use cases too, in light of its design.

Implementation card

Context Our team focuses on a rapidly growing volume and variety of data across several diverse data sources, and develops advanced use cases based on them—including interactive apps, machine learning models, and reporting. At the same time, the team had to consider our modest (but growing) engineering resources, and empower their efficient and effective work, as well as that of our collaborators and stakeholders. Thus, our team prioritized efficiently scalable and reusable solutions to our data processing needs, ensuring minimal time to insights and maintenance costs, and maximal developer velocity and data quality.

Architecture The full architecture of our data processing foundation can be seen in Fig. 4. The main building blocks are our data collection and ingestion, transformation, storage, workflow orchestration, version control, and continuous integration solutions, described in Sect. 3.

- GitHub [40] and GitHub Actions [45] are used for version control and continuous integration.
- Google Cloud Composer (managed Apache Airflow) [16, 24, 25] is used for orchestration, with custom operators to further increase developer velocity, and custom alerts. DAG parsing and generation are used to resolve occasional data quality incidents.
- Most internal data sources have integrations with and/or are stored in Google Big-Query [21]. Due to this, its performance, and ease of use, BigQuery is used heavily for storage and data processing in our pipelines, with custom Airflow operators to speed up DAG development. Documentation is stored in a standardized format in query files, then parsed and injected into relevant data discovery interfaces, and Python-based query conversion is used to automate the application of relevant processing for solving data quality issues efficiently.
- Kubernetes [46] is used for containerized applications within our data pipelines, with varying dependencies and hardware requirements, using the relevant Airflow operator [29], and custom, reusable utilities. Our use cases include data collection through APIs, feature engineering and other data processing (e.g. specialized NLP tasks), and machine learning batch inference.

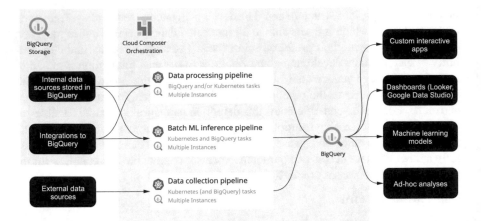

Fig. 4 General architecture for the data processing foundation and its analytics use cases, with data flow shown using black arrows

- The outputs of our data processing pipelines are generally stored in BigQuery. They are reused as building blocks of other relevant pipelines, for reporting using Looker [47] and Google Data Studio [48] dashboards, custom interactive apps, features for machine learning inference, other data science use cases, as well as ad-hoc analysis.

Remarks To address the multiple types of scaling required, and do so efficiently, serverless solutions have been a perfect fit for our data processing needs. The selected architecture has allowed a small team to rapidly produce actionable insights, with end-to-end efficiency, reusability, scalability, high data quality, and reliability, enabling use cases including custom interactive apps, machine learning, and dashboards, while empowering collaborators and other stakeholders too.

- Our team and other collaborators with diverse backgrounds have found Google Cloud Composer and the open source Apache Airflow project it is based on [24, 25] simple to learn, highly customizable, and easy to extend.
- In our experience, for our common use cases, BigQuery [21] provides competitive performance using significantly less development resources than Apache Spark [31] workflows.
- Our team has found our Kubernetes [46] orchestration with Google Cloud Composer relevant for deploying machine learning batch inference too, in addition to other types of data processing. By reusing this method and our custom utilities, our models can be deployed simply and efficiently, if the added complexity of specialized ML inference deployment solutions is not needed. Our team deployed our user segmentation models using this approach, and found the seamless integration with the feature engineering pipeline and the pipelines using the batch inference outputs of the models to be advantageous too.

4.2 Ontology Visualization and Business Logic Compliance

One of the main challenges on the data democratization process inside the enterprise is not only data literacy and universal data access across the company but the understanding of the data model. The ontology and data model are high level descriptions of the interactions between different entities inside the enterprise or software that the team is building. These entities and their interactions can be captured in a network. The network can then be displayed for qualitative analysis of the relationships, compliance evaluation or any other decision making process that requires the full context of a user, namely all entities related to the user during their interaction with the product. The main goal is to offer an interactive map of our ontology, allowing all our colleagues to untangle complex relations in an intuitive way, tell a story using networks and to discover how we see our customers. It is achieved by providing advanced visualizations [49] of our data and following the architecture described in Fig. 5.

Implementation card

Context Some of the users of our data products do not have insight into our ontology and would like to have a visual representation of the complex relationships the user has with different parts of our products or services. This is a challenge that is tackled by providing advanced network graph visualizations in a web app. it represents our ontology as a network and visualizes the relationships using different colors and icons. The results allow us to rapidly identify the context of a customer (e.g. usage, transactions, interactions, etc.).

Architecture In this architecture, the three main building blocks are in use in the following way:

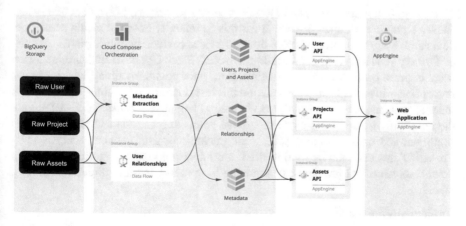

Fig. 5 Generic ontology visualization and business logic compliance

- Data processing block: The pipelines that compute our data ontology are orchestrated by a DAG in Airflow, an extra workflow is used to produce a database of entities (nodes in the network), relationships (edges in the network), and metadata for each entity.
- The three databases are used by three APIs that provide access to the network representation of our ontology (data model).
- A web application that provides an interactive Visualization of the network using D3.js, the app calls the provided APIs and displays the network and metadata. Options for coloring and formatting are given to the user for different use cases, allowing the user to report anomalies or create stories using the data.

Remarks Providing an interactive visualization has had multiple impacts on the way our data products evolve:

- The underlying network and sub-networks can be used as a knowledge graph for downstream machine learning tasks (e.g. recommender system).
- The data visualization allows the team to quickly provide explanations about the data available and the relationship available between different entities.
- Network visualizations are common in fields of compliance, fraud, and money laundering applications; since they provide a direct way for non-data professionals to explore what it is known about our customers (e.g. who have not paid for a service required to contribute to a project in a given company).

4.3   Community Analytics

Today it is commonplace that end users of products or services, software or otherwise, show a significant online presence in various community or social media platforms: dedicated discussion forums hosted by the service provider, or public platforms such as Twitter, YouTube, Instagram, Reddit, Discord, Stack Overflow, and similar. Since users express product opinions—satisfaction, pain points, roadmap wishes, brand perception—in natural language text, and this text is in turn of a volume that prohibits manual analysis, automated analyses via natural language processing (NLP) unlock the possibility to gain insights on the above points at scale, via techniques such as sentiment and opinion mining [50], topic tracking, or user network analysis [51]. These insights can then be used to drive a product roadmap to prioritize user pain points, solve critical problems, or innovate in the most feasible directions, among others.

Implementation card

   Context A user opinion of a company's service or product is of great importance for business success. Since users share their opinions across different platforms and

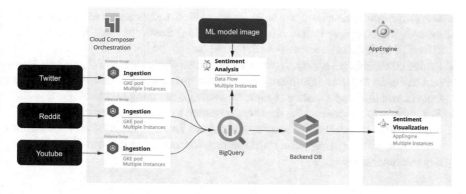

Fig. 6 Architecture for a community analytics application

services, it is a challenging task to have a single source for this insight. Therefore, providing a single interface to explore the company's pulse on social media is vital.

Architecture Providing analytics of user's post in social media that relate to a company is achieved using the following steps, as illustrated in Fig. 6.

- Data ingestion: custom services that ingest the data from common social platforms are orchestrated by Cloud Composer and run periodically (e.g. daily).
- A number of natural language processing (NLP) models are applied on the collected texts at post and thread level: sentiment analysis, topic modelling, semantic similarity, etc. Search indices are built from the dataset and incrementally updated.
- Results of model inference, original posts, and other structural information (threads, subforums, etc.) are transferred to a supporting database in Cloud SQL.
- The data is presented to the user in the form of a web app and a Slack bot, which offer custom search and filters over topics, sentiment, and other structural labels, trend detection, setting custom alerts for particular product features, and similar.
- The front-facing apps above are further used to semi-automatically generate periodicals for reporting the community pulse to internal stakeholders.

Remarks The main impacts we see from this application are:

- The work of community managers in monitoring trends on social media is greatly facilitated, improving their productivity by simplifying their search and analysis of the posts and providing a single searchable source and prioritization by volume and sentiment.
- Product managers are provided with a broad view of various product aspects together with the capability to decompose the overall view all the way down to the level of an individual post to truly understand product feedback both at scale and in intricate detail.

- Marketing managers use our community analytics to track general sentiment trends towards the company brand, aiming at planning, executing, and measuring the effects of targeted engagements with the users.
- Senior executives get a broad, periodical, and data-driven view of the most important topics in the user community for a given time period, together with insight from competitor communities, to inform strategic planning and execution.
- The application exposes APIs that are consumed by various other in-house stakeholder teams which require sentiment analysis-based metrics of product health to combine with existing orthogonal metrics, for managing e.g. quality assurance and customer success.

4.4 Recommendation System

With the ever-growing portfolio of products and content available on marketplaces, from a user perspective, it can be hard to find relevant content without being overloaded with information.

A key challenge for marketplaces is providing personalized content and tailored offerings for their users, easing an end user's quest to find relevant content that a user might find interesting and valuable. This aid can come in the form of a recommender system, which is prevalent for many businesses, like Amazon or Netflix [52], and has become a standard in the e-commerce field to drive revenue and improve user experience.

End users express their level of satisfaction explicitly through, e.g., star ratings, but also implicitly through a user's behavior on the marketplace, including browsing history and search patterns. These implicit ratings allow for collaborative filtering through a latent factor model using matrix factorization (ALS), which ultimately enables the characterization of both items and users. Because each user operates differently and has a variety of special needs, a recommender system allows the marketplace to meet the users' needs through a one-to-one marketing strategy, ultimately enhancing user satisfaction and loyalty and increasing engagement and retention.

Implementation card

Context A marketplace experience is normally accompanied by some tools to enhance the easiness with which the user can find what he or she is interested in. Aside from the search tools, a recommendation system is a great aid to provide the user with relevant content based on their interaction with the marketplace website/app. From many approaches to building a recommendation system, a collaborative filtering approach based on the user's implicit feedback was selected as the baseline.

Architecture The implicit rating is computed as an aggregation of the views, purchase, wishlist, and usage of the item acquired by the user. This is carried out by executing a DAG on Cloud composer (Airflow) that forms the required

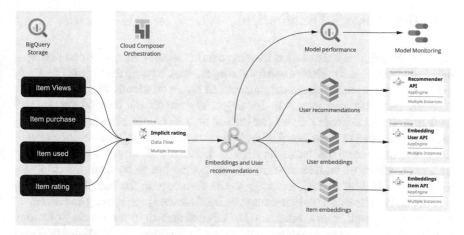

Fig. 7 Generic architecture for serving a recommendation system

aggregations and execute the Spark code that produces recommendations for each user, filters the items (e.g. already bought items) and offers embeddings for other similarity-based recommendations. The overall architecture of the system is shown in Fig. 7. The intermediate results (recommendations and embeddings) are stored in separate instances of Cloud SQL in order to serve the APIs that will be called/exposed by/to the marketplace front-end. The selected architecture allows us to implement this solution using two of our building blocks, namely: Data processing block and APIs serving block.

Remarks The selected architecture permits decoupling of the team's recommender architecture, the model, and the interface with the marketplace. This allow us to make improvements and changes without modifying the interface to the marketplace (provides a user ID and context, and receives recommended items list). Additionally, the API is decoupled from the model and the similarity search—which allows changing the functionality to any of the existing options:

- Database of pre-computed values
- Call for a similarity search engine (e.g. Vertex AI Matching Engine [53])
- Custom code providing the search capability (e.g. custom API running Faiss library [54] to provide similarity search.

A custom API was provided, which allows us to control the costs by running it in Google Cloud Run.

4.5   Data Lineage Visualization

Operating in a distributed and ever changing data landscape comes with complexity—data assets are subject to change and data outages, often beyond the control of a single team. As a result, downstream data assets can break or silently become stale. For a widely used data ontology, this can negatively impact the data products which feed from it and degrade stakeholder data-trust. As the use cases of the ontology grows and the web of dependencies become more cluttered, it becomes increasingly difficult to understand the source data that feeds into the ontology, as well as how the data is consumed downstream. More and more time has to be spent, meticulously assessing the risk of making a change to the ontology, in the fear that something will break downstream. This increasing overhead hinders productivity and drives the contributors to built from scratch, instead of contributing to the existing ontology. Having a full data lineage can mitigate this issue, by giving users an updated overview of how data flows to and from specific data assets. Data lineage can be seen as a directed graph, where each node represent some data asset, e.g., a table, column, dashboard, etc. The edges between the nodes represent a dependency, where the target node has a dependency on the source node. This dependency could, for example, be an aggregated table, reading the rows of an upstream table. Combining all of these dependencies will allow one to get the full overview of all upstream and downstream dependencies. Having this full graph of dependencies can help answer questions such as, what are all the upstream dependencies for a specific data asset, how a single update to a table affects downstream dependencies, how a specific metric was computed, and which downstream dependencies were affected by a specific incident.

> **Context** As a step to ensure data quality, traceability and reproducibility of our pipelines, a data lineage retrieval and visualization tool has been built.
> **Architecture** Building data lineage and structuring it for visualization is achieved using the steps shown in Fig. 8:

- Metadata retrieval block: Relevant workflow metadata is retrieved and staged for later processing. This metadata consists of workflow metadata, for example from our Airflow DAGs and from the BigQuery processing history.
- Dependency retrieval block: Workflow metadata is then parsed to create mappings from data assets to upstream dependencies. Here, a SQL parser, deployed using Cloud Run, is used to retrieve table and column level dependencies. Column transformations are also retrieved during this step. Dependencies are staged on a Google Cloud Storage Bucket. The retrieval of workflow dependencies is executed using the KubernetesPodOperator, where multiple Pods are run on an autoscaling GKE nodepool, for scaling out the dependency building process.
- Dependency graph building block: The dependency mappings are then combined to create entire dependency graphs. This process is executed similarly to the dependency retrieval process. The dependency graphs are pushed to a CloudSQL instance.

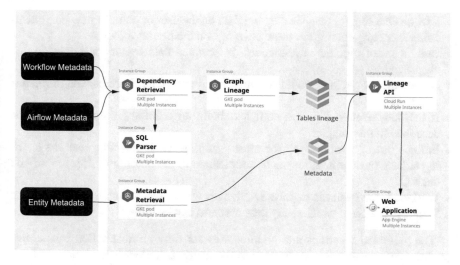

Fig. 8 Architecture for determining lineage and its visualization

- Metadata retrieval block: Metadata is retrieved for all the dependency types. The results are pushed to a CloudSQL instance.
- Data visualization block: A web application is deployed on Google App Engine. The application exposes the dependency graph, shows metadata, and allows users to search through the table and column level dependencies. The graph is exposed to the web application through an API that is deployed on Cloud Run.

Remarks Our data lineage tool provides the following advantages:

- Discovery tool for data professionals (columns and feature dependencies) and fast onboarding of new colleagues
- Overview of the data dependencies and clear view of when a table was updated, which improves the prompt debugging of upstream data quality issues.

4.6 Discussions

The learning from our architecture and solutions seems to allow for a generalization of the development process for a successful data catalogue using serverless offerings. This generalization could be a valuable steps toward a unified methodology on data products development. The following steps were identified:

- Map the relationships between the entities that exist and interact with the user. This mapping allows to create an ontology and an understanding of the user interactions in the context of what a customer experiences.

- Use the Ontology to map the existing data on the user or plan for new instrumentation for data gathering for more coverage on the user as a focus.
- Build a set of pipelines for creating this curated data sets as described in this chapter.
- Build tools for tracing data lineage and discovering data, features, and properties inside the data.
- If the data relationships are complex, build an ontology visualization aid as described in this chapter.
- Enforce use of the ontology for most reporting, analytics, and machine learning tasks. Consider additions and modification to the ontology instead of ad-hoc datasets.
- Work with the business to solve challenges with a reproducible and stable data foundation, atop of which many data products can be built.

The presented solutions and architectures are only an abstraction of the data products described in this work. Due to the nature of enterprise solutions, ad-hoc applications and data integration are expected to accommodate legacy data storage and the existing multitude of PaSS (platform as a Service) and SaSS (Software as a Service) agreements inside the organization. This does not diminish the generality of the proposed building blocks but alerts the reader on existing costs and efforts that are unique to their own organization. One unexpected result was the cost difference between using Google Cloud Run and Google Cloud App Engine, with Google Cloud Run being more cost-efficient overall in our solution.

5   Summary and Concluding Remarks

This work has shown that provided a solid data foundation in the form of an ontology, it is possible to re-organize most data products architectures to re-utilize this ontology and follow the three basic building blocks described in this chapter, namely: data ingestion, storage, and transformation; models and data serving using APIs; and data visualization and user interfaces.

These fundamental building blocks were implemented using existing serverless offerings, including managed Kubernetes, Apache Airflow, Apache Spark, and Apache Beam. The team considers the business context and impact for each use case, showing the multiple ways of re-purposing those building blocks into a diverse data product catalogue.

The data product catalogue had the objective to accelerate the data democratization inside the enterprise, by providing access to a curated set of our data and producing tools that allow any employee to understand the relationships inside the data.

The simplicity of the architectures and building blocks allows a small team to have high velocity in the implementation of new products by focusing their energy on the products and not managing infrastructure.

References

1. Jabir B, Noureddine F, Rahmani K (2022) Big data analytics opportunities and challenges for the smart enterprise. In: Elhoseny M, Yuan X, Krit SD (eds) Distributed sensing and intelligent systems. Studies in distributed intelligence. Springer, Cham. https://doi.org/10.1007/978-3-030-64258-7_70
2. Kozyrkov C (2018) What great data analysts do—and why every organization needs them. Harvard Business Review
3. Anderson J (2020) Data teams: a unified management model for successful data-focused teams
4. Adamson J (2021) Minding the machines: building and leading data science and analytics teams
5. Lakshmanan V (2022) Data science on the Google Cloud platform implementing end-to-end real-time data pipelines from ingest to machine learning. O'Reilly Media, Inc
6. Borek A, Prill N (2020) Driving digital transformation through data and AI. Kogan Page
7. Gilchrist A (2022) Cloud native apps on Google Cloud platform: use serverless, microservices and containers to rapidly build and deploy apps on Google Cloud. BPB Publications
8. Rose R (2020) Hands-on serverless computing with Google Cloud: build, deploy, and containerize apps using Cloud Functions, Cloud Run, and cloud-native technologies. Packt Publishing
9. Venema W (2020) Building serverless applications with Google Cloud Run. O'Reilly Media, Inc
10. Costa R, Baker J (2022) Programming Google Cloud. O'Reilly Media, Inc
11. Sangapu S, Panyam D, Marston J (2021) The definitive guide to modernizing applications on Google Cloud. Packt Publishing
12. Costa R, Hodun D (2021) Google Cloud cookbook. O'Reilly Media, Inc
13. Gruber TR (1993) Toward principles for the design of ontologies used for knowledge sharing. In: Guarino N, Poli R (eds) Formal ontology in conceptual analysis and knowledge representation. Kluwer Academic Publishers
14. Liu L, Özsu M (eds) (2009). Springer, New York, NY
15. Osman MA, Mohd Noah SA, Saad S (2022) Ontology-based knowledge management tools for knowledge sharing in organization—a review. IEEE Access 10:43267–43283. https://doi.org/10.1109/ACCESS.2022.3163758
16. Wijaya A (2022) Data engineering with Google Cloud platform. Packt Publishing
17. Shehabi A, Smith S, Sartor D, Brown R, Herrlin M (2016) United States Data Center energy usage report. Environmental and Energy Impact Division, Lawrence Berkeley National Laboratory
18. Morgan K (2021) The carbon reduction opportunity of moving to the cloud for APAC. S&P Market Intelligence
19. Lakshmanan V, Tigani J (2020) Google BigQuery: the definitive guide data warehousing, analytics, and machine learning at scale. O'Reilly Media, Inc
20. Melnik S et al (2010) Dremel: interactive analysis of web-scale datasets. In: 36th international conference on very large data bases, Singapore, 13–17 Sept 2010
21. Google Cloud Platform (2022) Product documentation for Bigquery. https://cloud.google.com/bigquery. Accessed 14 Oct 2022
22. Google Cloud Platform (2022) Product documentation for Cloud Storage. https://cloud.google.com/storage. Accessed 14 Oct 2022
23. Google Cloud Platform (2022) Product documentation for Cloud SQL. https://cloud.google.com/sql. Accessed 14 Oct 2022
24. Apache Foundation (2022) Product documentation for Apache airflow. https://airflow.apache.org/. Accessed 14 Oct 2022
25. Google Cloud Platform (2022) Product documentation for Cloud Composer. https://cloud.google.com/composer/. Accessed 14 Oct 2022

26. Google Cloud Platform (2022) Product documentation for Google Kubernetes Engine. https://cloud.google.com/kubernetes-engine. Accessed 14 Oct 2022
27. Amazon Web Services (2022) Product documentation for Amazon managed workflows for Apache Airflow. https://aws.amazon.com/managed-workflows-for-apache-airflow/. Accessed 14 Oct 2022
28. Google Cloud Platform (2022) Product documentation for Google Cloud Container Registry. https://cloud.google.com/container-registry. Accessed 14 Oct 2022
29. Apache Foundation (2022) Documentation: KubernetesPodOperator. https://airflow.apache.org/docs/apache-airflow-providers-cncf-kubernetes/stable/operators.html. Accessed 14 Oct 2022
30. Google Cloud Platform (2022) Product documentation for Google Kubernetes Engine node pools. https://cloud.google.com/kubernetes-engine/docs/concepts/node-pools. Accessed 14 Oct 2022
31. Apache Foundation (2022) Product documentation for Apache Spark. https://spark.apache.org/. Accessed 14 Oct 2022
32. Apache Foundation (2022) Product documentation for Apache Beam. https://beam.apache.org/. Accessed 14 Oct 2022
33. Google Cloud Platform (2022) Product documentation for Dataproc Serverless. https://cloud.google.com/dataproc-serverless/docs. Accessed 14 Oct 2022
34. Google Cloud Platform (2022) Product documentation for Dataflow. https://cloud.google.com/dataflow. Accessed 14 Oct 2022
35. Google Cloud Platform (2022) Product documentation for Google Cloud Operations Suite. https://cloud.google.com/products/operations. Accessed 14 Oct 2022
36. Google Cloud Platform (2022) Product documentation for: GKE security best practices. https://cloud.google.com/kubernetes-engine/docs/concepts/security-overview. Accessed 14 Oct 2022
37. FastAPI (2022) Product documentation for FastAPI. https://fastapi.tiangolo.com/. Accessed 14 Oct 2022
38. Google Cloud (2022) Cloud Run. https://cloud.google.com/run. Accessed 14 Oct 2022
39. Google Cloud (2022) Cloud Run: what no one tells you about serverless (and how it's done). https://cloud.google.com/blog/topics/developers-practitioners/cloud-run-story-serverless-containers. Accessed 14 Oct 2022
40. GitHub (2022) GitHub product documentation. https://docs.github.com/en. Accessed 14 Oct 2022
41. Google Cloud Platform (2022) Product documentation for Cloud App Engine. https://cloud.google.com/appengine. Accessed 14 Oct 2022
42. Google Cloud Platform (2022) Google Cloud product documentation: identity and access control—IAM. https://cloud.google.com/iam. Accessed 14 Oct 2022
43. Okta (2022) Product documentation: product integration. https://developer.okta.com/okta-integration-network/. Accessed 14 Oct 2022
44. Google Cloud Platform (2022) Google Cloud product documentation: identity aware proxy—IAP. https://cloud.google.com/iap. Accessed 14 Oct 2022
45. GitHub (2022) GitHub Actions feature overview. https://github.com/features/actions. Accessed 14 Oct 2022
46. The Kubernetes Authors (2022) Kubernetes documentation. https://docs.github.com/en. Accessed 14 Oct 2022
47. Looker (2022) Looker platform/product overview. https://www.looker.com/platform/overview/. Accessed 14 Oct 2022
48. Google Data Studio (2022) Google Data Studio product overview. https://datastudio.withgoogle.com/. Accessed 14 Oct 2022
49. Kirk A (2021) Data visualisation: a handbook for data driven design. SAGE Publications Ltd
50. Liu B (2012) Sentiment analysis and opinion mining. Synthesis lectures on human language technologies. Morgan & Claypool Publishers
51. Albert-László Barabási (2013) Network science. Philos Trans Roy Soc A: Math Phys Eng Sci

52. Bell R, Koren Y, Volinsky C (2009) Matrix factorization techniques for recommender systems. Computer 42(08):30–37
53. Google Cloud Platform (2022) Google Cloud product documentation Vertex matching engine. https://cloud.google.com/vertex-ai/docs/matching-engine/overview. Accessed 14 Oct 2022
54. Johnson J et al (2021) Billion-scale similarity search with GPUs. IEEE Trans Big Data 7(3):535–547. https://doi.org/10.1109/tbdata.2019.2921572. Crossref

QoS Analysis for Serverless Computing Using Machine Learning

Muhammed Golec, Sundas Iftikhar, Pratibha Prabhakaran, Sukhpal Singh Gill, and Steve Uhlig

Abstract Large-scale computing systems are becoming more popular as the need for computing power increases every year. Serverless computing has emerged as a powerful and compelling paradigm for the hosting services and applications because of the rapid shift in business application architectures for containers and microservices. Further, Serverless computing offers economical services and scalability to fulfil the growing demand of computing in a timely manner. Therefore, it is important to analyse the Quality of Service (QoS) of Serverless Computing systems to monitor its performance. In this chapter, we used the latest machine learning models to predict system configurations in Serverless computing environments. Knowing about system configurations in advance helps to maintain the performance of the system by analysing QoS. Further, a no-cost model is proposed to examine and compare different configurations of workstations in serverless computing environments. To achieve this, we deployed Theoretical Moore's, Fitted Moore's, 2-D poly regression and 3-D poly regression machine learning models following Graphics Processing Unit (GPU) requirements and compare the results. The experimental results demonstrated that Fitted Moore was the best model with an R2 score of 0.992.

Keywords Serverless computing · High configuration computer systems · AI predictive models

M. Golec (✉) · S. Iftikhar · P. Prabhakaran · S. S. Gill · S. Uhlig
School of Electronic Engineering and Computer Science, Queen Mary University of London, London E1 4NS, UK
e-mail: m.golec@qmul.ac.uk

S. Iftikhar
e-mail: s.iftikhar@qmul.ac.uk

P. Prabhakaran
e-mail: pratibha.prabhakaran@qmul.ac.uk

S. S. Gill
e-mail: s.s.gill@qmul.ac.uk

S. Uhlig
e-mail: steve.uhlig@qmul.ac.uk

© The Author(s), under exclusive license to Springer Nature Switzerland AG 2023
R. Krishnamurthi et al. (eds.), *Serverless Computing: Principles and Paradigms*,
Lecture Notes on Data Engineering and Communications Technologies 162,
https://doi.org/10.1007/978-3-031-26633-1_7

1   Introduction

Cloud computing is a service that can be defined as Internet-based shared storage services and processing capacity by cloud providers [1]. It entered the world of technology with the launch of Amazon S3, one of its first applications, in 2006 [2]. It is thought by computer scientists that it will replace computer hard disks and shape the future of the Internet. Today it is used by companies, universities, large organizations, and even government agencies globally. Cloud computing consists of four main delivery models. These are IaaS (infrastructure as a service), PaaS (platform as a service), SaaS (software as a service), and Function-as-a-Service (FaaS) [3].

Cloud providers need to be able to successfully meet ever-increasing customer demands in order to provide the best service. For this, a model that can provide dynamic scalability automatically was needed [4]. Also, in traditional computing, it is the customers' responsibility to manage servers, update and perform other operational tasks [5]. This creates additional costs for companies, such as providing qualified experts and necessary equipment [6]. Serverless computing is a new paradigm that has emerged to solve all these problems. Serverless computing is based on the idea that programmers should focus on developing code rather than maintaining the system they are working on [7]. Because cloud providers handle backend operations, organizations are freed from managing, updating, and other operational burdens on servers. Also, serverless-based services have advantages such as dynamic scalability and pay-as-you-go pricing [6]. In this way, customers only pay for what they use and are automatically provided with the necessary processing power if they need it. This article compares serverless computing with traditional computing to better understand, evaluate, and predict deployment requirements and the resulting quality of services (QoS).

The purpose of this research is to examine the graphics processing unit (GPU) setups that can be used to perform large jobs efficiently while minimizing service quality and cost in Serverless Computing. ML models such as regression will be applied to predict configurations. Various algorithms will be extensively analyzed.

1.1   The Trend in CPU and GPU Performance

GPU is the unit responsible for graphics operations in applications [8]. The central processing unit (CPU) is the part of a computer that processes data and is responsible for software commands [8]. Considering the computer architecture, although CPUs can do different tasks quickly, their ability to run several processes at the same time is limited. In contrast to the CPU, GPUs excel at tasks that require a high degree of repeatability and parallelism [9].

A server can have very fast CPUs with 24–48 cores [10]. By adding 4–8 GPUs to the same server, an additional 40,000 cores of processing power can be obtained. Single CPU cores run faster and are more efficient than single GPU cores, but the total

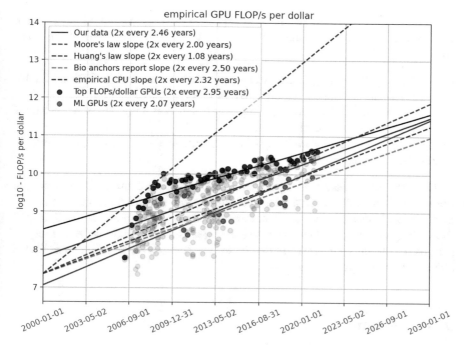

Fig. 1 CPU and GPU over time [11]

amount of GPU cores and the massive amount of parallel processing they provide is enough to close the gap. Hence, in a server, basic algorithms can be fully implemented on the CPU, while GPUs can assist in executing more complex ones.

With the development of graphics-based applications and games, the use of GPUs has also increased rapidly. Figure 1 shows the increase in CPU and GPU usage from 2002 to 2011 [11]. The GFLOPS (GigaFLOPS) shown in Fig. 1 is the unit used to measure the floating point performance of a computer [12]. As can be seen, GPU usage has risen much more steeply.

1.2 Motivation

With the increase in processing power, storage, and other configuration items required by applications, the need for systems with high CPUs and GPUs has also increased. To meet these needs, Serverless computing is invaluable. With its dynamic scalability and payment policies such as pay-you-go, it attracts people from all walks of life, such as software developers, data analysts, architects, visual designers, and AI enthusiasts.

Configuration servers are essential when it comes to deploying applications for small and medium-sized organizations as well as entrepreneurs. Buying a new server is expensive and requires more maintenance. The most cost-effective and efficient way to run a business is to switch to serverless computing, which allows optional scaling of configuration parameter attributes. The main contributions of this work are:

- To predict system configurations in Serverless computing using Artificial Intelligence (AI) and Machine Learning (ML) models.
- To suggest a free model study and compare different configurations of workstations in a serverless environment.
- To provide cost-capacity modeling of GPUs with parallel processing capability in serverless environments.

The rest of the chapter is organized as follows: Sect. 2 presents the literature studies. In Sect. 3, the theoretical background of the proposed work is explained. Section 4 explains the background and mechanism of the proposed study. Section 5 shows the proposed modelling evaluation and results. Section 6 discusses the impact and possible disadvantages of COVID-19. Section 7 concludes the chapter, and Sect. 8 highlights possible future directions.

2   Background

Until recently, applications were installed on physical servers by system administrators. Prerequisites such as memory/disk/processor capacity and required operating system were all the responsibility of system administrators. This is known as the "bare metal setup" [13].

The next form of distribution that emerged was the virtual machine (VM) [14]. Simulated servers are used instead of real ones, giving developers more flexibility. It also gave system admins the flexibility to migrate a VM to new hardware if a hardware component fails. System administrators can use many virtual machines on the same physical server.

Container distribution came to the fore after virtual machines [15]. Some of the emerging containerization technologies are Docker, OpenVZ, FreeBSD regions, and Solaris [16]. Such innovations made it possible for a system administrator to "split off" into an operating system that ran many programs on the same computer. Figure 2 shows the history of software deployment [17].

2.1   Beginning of Serverless

Google App Engine was released by Google in 2008 [18]. This was the first time a designer could deploy software to the cloud without having to think about challenges such as how the server was provisioned or operating system updates. Likewise, a similar service, Amazon Lambda, was announced in 2015 [19].

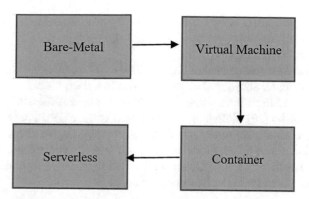

Fig. 2 History of software deployment

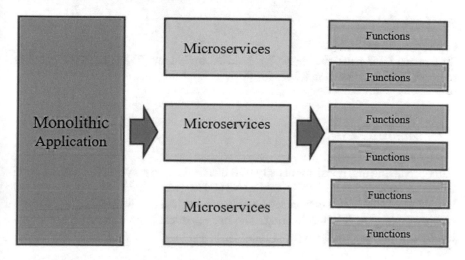

Fig. 3 Monolithic versus microservices versus FaaS

Serverless computing has several advantages over traditional cloud-based or server-centric systems. Serverless computing is a type of cloud computing in which the cloud provider controls the server and provides the resources needed by the customer. Here, the firm is responsible for managing the cloud servers. Figure 3 shows the transition from Monolithic architecture to a serverless architecture.

2.2 CPU and GPU

It is an accepted fact that the CPU is one of the most important units of any computer system. A CPU and GPU (graphics processing unit) work together to improve data throughput and the number of simultaneous computations within a software.

GPUs were originally designed to improve the visual quality of computers and video games [20]. In recent years, GPUs are been using to speed up software that needs large amounts of data.

The most important characteristics that define a CPU are its high clock speeds and increased core count, allowing it to process data in a short time. Because CPUs are primarily designed to perform a single task, they are not as successful in parallel programming as GPUs [20]. With hundreds of tiny cores, the GPU is exceptionally fast for linear algebra and other operations that require a high degree of concurrency, as it uses instruction sets optimized for dimensional matrix arithmetic and floating point calculations. Another reason GPUs are well suited for massively parallel computing is that they are tuned for better bandwidth than latency.

3 Theory

To forecast GPU performance, we've used Moore's Law [21] in machine learning regression models based on R2 scores [22].

3.1 Moore's Law

After predicting that transistor density will quadruple every 2 years in 1965, Gordon E. Moore co-founded Intel (NASDAQ: INTC) [23]. He offered this pronouncement in light of new developments in Intel's chip production. It is also stated by Moore's Law that this growth is accelerating. The mathematical definition of Moore's Law can be expressed in the form [21] (Table 1).

$$mi = m0 \times 2^{(yi-y0)/T} \tag{1}$$

Table 1 Equation explanation (1)

mi	Mean memory
m0	Memory in reference year
y0	Reference year
yi	Year
T	Number of years to double the mean memory

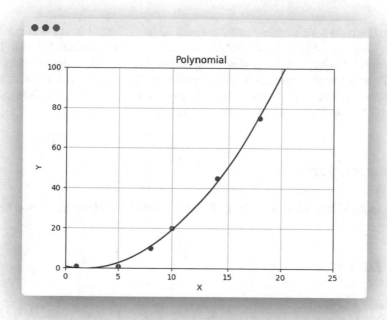

Fig. 4 Two-degree polynomial curve [22]

3.2 Regression Model

For estimating the values of a continuous target quantity, regression is an extremely useful statistical approach. An estimation method known as regression employs independent risk factors to arrive at a target value. The most essential elements in deciding which regression approach to utilize are the number of predictor variables and the type of connection between the independent and dependent variables. Polynomial regression will be used to get the average GPU memory. Polynomial regression uses Linear Regression to turn the original characteristics into polynomial features of a specific degree. Figure 4 depicts the plot of Eq. (2) for the 2-degree polynomial [22].

$$Y = a + bX + cX^2 \tag{2}$$

Figure 4 illustrates that the model remains linear. However, the curve is now quadratic instead of linear, as seen in Fig. 4. Just because of that, the curve gets overfitted whenever we increase the degree to a very high number.

3.3 R2 Score

A regression model's accuracy may be assessed using the R2 score, also known as the coefficient of determination. R2 score can be used to calculate the correlation between two variables. Most models may be said to accurately represent data if the model's anticipated values are within a reasonable margin of error. It's feasible to have a negative score of 1 even if it's the greatest conceivable score. Generally, a higher R2 value denotes a well-fitting model.

4 Methodology

In this section, methods that can be used to forecast future GPU demand are analyzed.

4.1 Dataset

The dataset consists of essential GPU measurements gathered over a 20-year span from four major manufacturers: Nvidia, Intel, ATI, and AMD. Nvidia GPUs from 1999 to 2017, AMD GPUs from 1998 to 2017, ATI GPUs from 2016 to 2017, and Intel GPUs from 2004 to 2017 are all included in this dataset [24]. Figure 5 shows the grouping of GPUs by year. Dataset parameters are shown in Table 2.

Based on a dataset, a bar chart was created to show the GPU production years. It can be easily seen from the graph that the number of new GPUs introduced each year started climbing in 1998, reaching a peak of over 10. With a total of 250 GPUs, the number of GPUs increased dramatically in 2010. In 2012, when the number of GPUs topped 500, there was a massive increase in demand.

Figure 6 shows the comparison of peak memory bandwidth for CPU and GPU by year. Although the difference between GPU and CPU memory bandwidth was not much in early 2008, it seems to have increased almost 8 times towards 2022.

Table 2 ML models' R2 score

Model	R2
Theoretical Moore's	0.56
Fitted Moore's	0.992
2-degree poly regression	0.949
3-degree poly regression	0.985

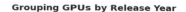

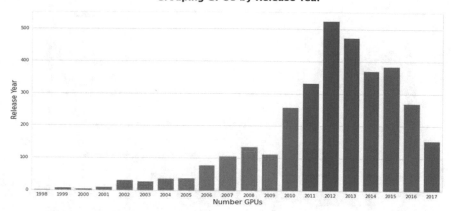

Fig. 5 Grouping GPUs by release years

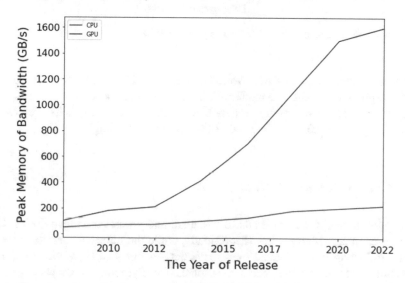

Fig. 6 The comparison of peak memory bandwidth for CPU and GPU

4.2 Preprocessing and Feature Engineering

We should first preprocess the data before executing GPU computations in order to improve the quality of the outputs. Using our dataset's release year as a guide, we compute the dataset's mean and median memory usage. To help us plan for future GPU requirements will be helpful.

GPUs have steadily gotten larger in memory size since their inception in 1998, as illustrated in Fig. 7.

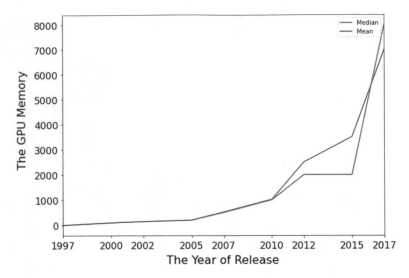

Fig. 7 Mean and median memory GPUs versus year of release

It has maintained at 2000 MB from 2008 to early 2011 and from late 2011 to 2015, despite the fact that the median GPU memory size has climbed to 1000 MB since 2007. Although the average and median memory sizes of GPUs have increased significantly during 2015 to 7000 and 8000 MB, correspondingly.

4.3 Calculation and Fitting of Curve

We applied four different methods to calculate and predict the mean GPUs. One function computes Moore's law theoretically, while the other utilises the curve fit module of Python to fit the Moore's law to predict how many GPUs would be required in the future. The polynomial regression models from Python's sklearn package were also used to fit and forecast the average GPU.

The graph illustrates the mean, median, average, and theoretical Moore's law GPUs throughout the year of release in a logarithmic scale. A comparison of the fitted and hypothetical Moore's law lines shows that both models predict that the GPU would develop at an exponential rate, according to Moore's law. In spite of the fact that the mean and median lines are not straight, they rise with time, showing that the average mean climbed exponentially, but also suffered exponential declines, showing that there was no true average growth or when it remained static.

5 Evaluation and Results

To determine the best-fitting machine learning model based on R2, we'll utilise the predicted mean memory GPUs from the four models we've selected.

5.1 Theoretical Moore's Law

Using the theoretically calculated function of Moore's Law, the year sequence and the minimum year value, the following function is used to arrive at the expected sequence value.

```
# Input:
def calculate_Moores(x, y_trans):
    return memory_median[0] * 2^((x-y_trans)/2)

# Output:
y_pred_moorelaw= calculate_Moores(year_arr,int(year_min))
```

5.2 Fitted Moore's Law

As opposed to the fitted Moore's law function, that has a specified exponential curve function, we receive higher predicted memory values early in the launch year, which slowly decreases in subsequent years.

```
# Input:
def expoCurve(x, a, b, c):
    return a*2^((x-c)*b)


popt, pcov = curve_fit(expoCurve, yeararr, memoryarrmean, p0=(2,
    0.5, 1998))
y_pred_moorelaw = expoCurve(year_arr, *popt)


# Output:
y_pred_moorelaw
```

5.3   Two-Degree Polynomial

Using Python's polynomial regression module, the two-degree polynomial predicted mean memory array value is as follows.

```
# Input:
poly2=PolyFeatures(degree=2,include_bias=False)
X_poly2=polyreg2.fit_transform(year_arr.reshape(-1,1))
linreg2 = LinearRegression()
linreg2.fit(X_poly2, memory_arr_mean)
y_pred_linreg2=linreg2.predict(polyreg2.fit_transform(year_arr.reshape(-1, 1)))

# Output:
y_pred_linreg2
```

Due to its ignorance of output value limitations, the model's output has some negative values. To put it simply, it means that the values are really low.

5.4   Three-Degree Polynomial

The expected memory sequence of a three-degree polynomial is found using the following function. Less GPUs with negative mean memory are present, as opposed to a higher number in the two-degree polynomial.

```
# Input:
poly3=PolynFeatures(degree=3,include_bias=False)
X_poly3=polyreg3.fit_transform(year_arr.reshape(-1,1))
linreg_3 = LinearRegression()
linreg3.fit(X_poly3, memory_arr_mean)
y_pred_linreg3=linreg3.predict(polyreg3.fit_transform(year_arr.reshape(-1, 1)))

# Output:
y_pred_linreg3
```

5.5   Graphical Representation

Figure 8 is created in order to make the data more understandable.

Using the dataset from the year of the release, we fitted four different regression models, as shown in Fig. 8. Memory GPUs began to expand exponentially in 2014, as can be seen from all the models' graphs. Our three-degree model polynomial, two-degree polynomial, and fitted Moore's law model all exhibit the same exponential rise over time. This is because the slopes of the three models are all the same. In contrast, the theoretical Moore's law model has a growth curve that differs significantly from

Fig. 8 Fitting regression
model into dataset

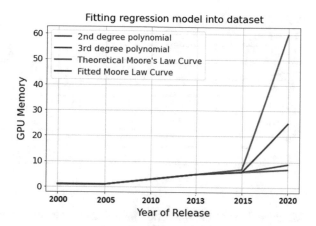

Fitting regression model into dataset

the other models. Since the theoretical Moore's law model predicted the expected
value using a mathematical equation, the dataset has been overfitted.

5.6 Experimental Results

When all four models have been tested, we will select the best model for predicting
real GPU memory. Here, we're using the R2 score of Python's sklearn module, which
takes into account both predicted and calculated memory array values to get an output
value between 0 and 1. Table 2 lists the R2 values for four different models.

An R2 number that is larger than zero is considered to be excellent. Thus, the
Python module's fitted Moore's Law function has an R2 score of 0.992. To anticipate
the future mean memory GPU, we need to select the optimal model based on the
R2 score. After doing the computations and examining the R2 values of our four
models in Table 2, we will select the fitted Moore's law function for prediction. Two-
degree and three-degree polynomial regression models have similar R2 scores, with
a difference of 0.007 and 0.043, respectively, when compared to a fitted Moore's law
function model's R2 score, respectively. R2 values for the theoretically estimated
Moore's law function model were found to be 0.56.

6 Discussion

Using a fitted Moore's law function, it can be predicted that the desire for GPUs
will continue to increase in the coming years. Deep learning, artificial intelligence,
machine learning, and other similar topics will all demand massive volumes of data, or
"big data", which will likely lead to a slew of heavy applications and programs. Using
the customer's year of birth, we can predict the GPU's performance for that year.

Table 3 GPU price rate [25]

GPU (GB)	Cost ($)
30–40	14,547–15,796
40–50	17,046–18,296
50–60	18,751–19,546
60–70	19,671–22,223
70–80	22,199–24,730
80–90	24,780–25,388
90–100	30,668–32,397

Small and medium-sized firms, as well as new ventures, will benefit from this service, which aims to help them determine their technological requirements as they change. It may make sense for a multinational organization to invest in high-configuration servers to achieve better performance. Nevertheless, acquiring servers is not a smart idea for small and medium-sized organizations as well as startups. Server updates, software patching, and software licensing require the appointment of external expert staff. Serverless computing takes on these additional tasks, eliminating the need for specialist staff and thus additional costs. The cost of GPUs for the near future is estimated in Table 3. Aside from their high price, these GPUs will require costly maintenance. 30–40 GB GPU and 60–80 GB GPU are in the 18,000–22,000 range each. And considering the maintenance costs, this amount will be even more for a company.

Serverless computing, on the other hand, does not require a small business to acquire a server. Only the small company needs to examine the compatibility of the machines and their versions. Serverless computing is offered by a growing number of organizations, including Microsoft Azure, IBM, Oracle Cloud, and Google Cloud [26]. The cost of basic cloud installation services is as follows:

- (Pricing—Dedicated Host Virtual Machines | Microsoft Azure, 2021): Microsoft Azure hosting starts at $3.41 per hour [27].
- IBM's virtual server GPU option costs $3.06 per hour to get started [28].
- There is a $2.95/hour minimum charge for using Oracle's cloud computing services [29].
- Based on the parameters of the application, Google Cloud offers services. According to Google Cloud, a simple server with a GPU might cost as little as $1200 per month [30].

In addition to cost savings, serverless computing offers a number of other benefits that vary depending on the cloud service used [31]:

- Additionally, the service provider ensures that applications have the backend infrastructure they want by utilizing its resources horizontally and vertically to satisfy demand, as well as any new resources that could be necessary.
- Real-time resource availability is necessary, even for unforeseen peak demands and disproportionate growth, and this is given by the service provider.

- On-demand serverless computing means that clients simply pay for the resources they use.
- Serverless allows programmers to change a single function at a time rather than the complete program or code as a whole.
- Serverless architectures have a tendency to have a wide range of access points. So because the application's code is not housed on a server, it could be started from any location. Application functionality is executed on servers close to the end user, reducing latency.
- Additionally, serverless computing is less harmful to the environment because it does not require a dedicated server. There will be a considerable increase in the number of physical resources and energy necessary to develop and operate enormous data centres if every business has its own data centre.
- Businesses like online banking and government agencies with changing or escalating bandwidth needs might benefit from serverless computing.

6.1   COVID-19 and It's Impact

A wide spectrum of enterprises have embraced the idea of serverless computing. Many enterprises have used serverless computing as a means of ensuring service availability in the wake of the global COVID-19 epidemic. Although numerous firms are making strong efforts to move to serverless computing following the COVID-19 lockout. Since remote working has become the norm at COVID-19, companies have made the move to the cloud a top priority rather than putting it on the back burner for the long term. Data analysis needs are changing all the time, and serverless computing can keep up. It was especially important in the event of the COVID-19 pandemic, which hit without notice and provided little time for organisations, professionals, or enterprises to create the infrastructure they required. Additionally, serverless's potential to withstand pandemics is enticing businesses to switch to technology. Because of the outbreak, people's freedom of movement was restricted, but server upkeep and data centre needs were significantly impacted as well. Because of this, several companies have experienced losses as a result of customers being unable to access services. This has led many companies to transition to serverless computing, which is more dependable. Deep-learning and AI-based approaches have been in great demand as a result of the COVID-19 epidemic. For spoken translation and automatic hearing, deep learning is used in intelligent devices like smart speakers and smart televisions. High-performance AI processors are employed for bitcoin mining in the rapidly expanding cryptocurrency sector by banks and financial institutions. NLP concepts are being incorporated into AI chatbots by financial institutions, banks, and even government websites to enable self-help customer services. Deep learning requires a fast GPU to speed up the training process for all of these technologies. As a result, the capacity of serverless to adapt grows.

6.2   Drawbacks

Serverless computing provides a number of advantages, but it does have many drawbacks, such as the 'cold-start' of code [32]. Code that hasn't been used in a long
time can be reactivated via this method. The serverless code must be started in this
case. As a result of this delay, termed as "cold-start", the execution's performance
will suffer. A "warm-start" request for a ready-to-use code is referred to when the
service provider keeps the code operational so that it can be utilised regularly [31].
Also, customers' privacy entirely relies on the service provider [33]. The provider
is responsible for all backend services of the application, including security and
privacy. As a result, customers' trust in providers increases. Because each provider
offers unique techniques and services, customers need to find the providers that are
most suitable for them.

7   Conclusions

The main objective of this study to examine the cost of high-configuration serverless computing services. For the ever-increasingly complex and computationally
demanding algorithms, we applied four ML models that include a two-degree polynomial regression model, a three-degree polynomial regression model, a theoretically
calculated Moore's law model, and a model that was fitted to the actual GPU requirements. A model with an R2 score of 0.992 was found to best suit Moore's law, and this
model was chosen as the best prediction model. Due to the rapid development and
increasing interest in the domains of artificial intelligence (AI), deep learning (DL),
and graphics, we anticipate high GPU machines to smoothly perform sophisticated
activities. For this reason, we have gathered cost information for GPUs in servers
with basic configuration. Compared to cloud service providers, these GPUs are far
more affordable and give substantially more benefits. Companies can choose from
a variety of features offered by cloud service providers, depending on their specific
needs. To save money while also benefiting the environment, moving to serverless
computing is a win–win situation for everyone.

8   Future Work

This chapter discusses the GPU configuration parameters of servers. The best model
was identified based on the R2 score parameter of sklearn to forecast these configurations using four different machine learning models: two-degree polynomial regression, three-degree polynomial regression, theoretically computed Moore's law and
fitted Moore's law. It is also necessary to consider storage space, RAM needs, and
performance while looking at other important configurational factors. The service

provider's uptime and robustness should also be taken into consideration in addition to these features. For more precise forecasts, methods other than the machine learning techniques described in this research might be utilised to anticipate future requirements. IoT and serverless computing are two emerging concepts that can work together. COVID-19 pandemic data and files are being maintained online, so that anybody may upload and get data/files from anywhere since remote working is commonplace. The cloud is the ideal place to store all of your data. In lieu of cloud storage, huge data must be stored on a separate storage system, which is more time-consuming and expensive to manage [34]. In order to interpret the data for the study, it's also important to make sure it's dependable and available around the clock, seven days a week. It's much easier to do these things when IoT and serverless are combined. If this method of demand estimation can be expanded, it will be beneficial to a wide range of businesses, including healthcare and IoT [34].

References

1. Jiang L, Pei Y, Zhao J (2020) Overview of serverless architecture research. J Phys Conf Ser 1453:012119
2. Palankar MR, Iamnitchi A, Ripeanu M, Garfinkel S (2008) Amazon S3 for science grids: a viable solution? In: Proceedings of the 2008 international workshop on data-aware distributed computing, pp 55–64
3. Golec M, Ozturac R, Pooranian Z, Gill SS, Buyya R (2021) iFaaSBus: a security and privacy based lightweight framework for serverless computing using IoT and machine learning. IEEE Trans Ind Inform 18(5):3522–3529
4. Cassel GAS, Rodrigues VF, da Rosa Righi R, Bez MR, Nepomuceno AC, da Costa CA (2022) Serverless computing for Internet of things: a systematic literature review. Future Gen Comput Syst 128:299–316
5. Eskandani N, Salvaneschi G (2021) The wonderless dataset for serverless computing. In: 2021 IEEE/ACM 18th international conference on mining software repositories (MSR)
6. Lee H, Satyam K, Fox G (2018) Evaluation of production serverless computing environments. In: 2018 IEEE 11th international conference on cloud computing (CLOUD)
7. Prakash AA, Kumar KS (2022) Cloud serverless security and services: a survey. In: Applications of computational methods in manufacturing and product design, pp 453–462
8. Baek AR, Lee K, Choi H (2013) CPU and GPU parallel processing for mobile augmented reality. In: 2013 6th international congress on image and signal processing (CISP), vol 1. IEEE, pp 133–137
9. Hawick KA, Leist A, Playne DP (2010) Parallel graph component labelling with GPUs and CUDA. Parallel Comput 36(12):655–678
10. Rouholahnejad E, Abbaspour KC, Vejdani M, Srinivasan R, Schulin R, Lehmann A (2012) A parallelization framework for calibration of hydrological models. Environ Modell Softw 31:28–36
11. Trends in GPU price-performance. https://epochai.org/blog/trends-in-gpu-price-performance
12. Lee Y, Waterman A, Avizienis R, Cook H, Sun C, Stojanović V, Asanović K (2014) A 45 nm 1.3 GHz 16.7 double-precision GFLOPS/W RISC-V processor with vector accelerators. In: ESSCIRC 2014-40th European solid state circuits conference (ESSCIRC). IEEE, pp 199–202
13. Kominos CG, Seyvet N, Vandikas K (2017) Bare-metal, virtual machines and containers in OpenStack. In: 2017 20th conference on innovations in clouds, Internet and networks (ICIN)
14. Masdari M, Nabavi SS, Ahmadi V (2016) An overview of virtual machine placement schemes in cloud computing. J Netw Comput Appl 66:106–127

15. Pahl C, Brogi A, Soldani J, Jamshidi P (2017) Cloud container technologies: a state-of-the-art review. IEEE Trans Cloud Comput 7(3):677–692

16. Chen Y (2015) Checkpoint and restore of micro-service in docker containers. In: 2015 3rd international conference on mechatronics and industrial informatics (ICMII 2015). Atlantis Press, pp 915–918

17. History of software deployment. https://dashbird.io/blog/origin-of-serverless/

18. Zahariev A (2009) Google app engine. Helsinki University of Technology, pp 1–5

19. Kiran M, Murphy P, Monga I, Dugan J, Baveja SS (2015) Lambda architecture for cost-effective batch and speed big data processing. In: 2015 IEEE international conference on big data (big data). IEEE, pp 2785–2792

20. Formisano A, Gentilini R, Vella F (2021) Scalable energy games solvers on GPUs. IEEE Trans Parallel Distrib Syst 32(12):2970–2982

21. Moore's law linear approximation and mathematical analysis. https://semiwiki.com/semiconductor-manufacturers/5167-moores-law-linear-approximation-and-mathematical-analysis/

22. Polynomial regression in python. https://pythonbasics.org/polynomial-regression-in-python/

23. Ferain I, Colinge CA, Colinge JP (2011) Multigate transistors as the future of classical metal-oxide-semiconductor field-effect transistors. Nature 479(7373):310–316

24. Intel CPUs EDA. https://www.kaggle.com/trion129/intel-cpus-eda/data?select=All_GPUs.csv

25. GPU price rate. https://www.reddit.com/r/Amd/comments/smlq76/gpu_performance_vs_price_europe/

26. Safonov VO (2016) Trustworthy cloud computing. Wiley

27. Pricing dedicated host virtual machines: Microsoft Azure. https://azure.microsoft.com/en-in/pricing/details/virtual-machines/dedicated-host/

28. IBM compute pricing. https://www.ibm.com/uk-en/cloud/pricing

29. Oracle compute pricing. https://www.oracle.com/in/cloud/compute/pricing.html

30. Pricing overview Google cloud. https://cloud.google.com/pricing

31. Gill SS, Xu M, Ottaviani C, Patros P, Bahsoon R, Shaghaghi A, Uhlig S (2022) AI for next generation computing: emerging trends and future directions. Internet Things 19:100514

32. Golec M, Chowdhury D, Jaglan S, Gill SS, Uhlig S (2022) AIBLOCK: blockchain based lightweight framework for serverless computing using AI. In: 2022 22nd IEEE international symposium on cluster, cloud and Internet computing (CCGrid). IEEE, pp 886–892

33. Golec M, Gill SS, Bahsoon R, Rana O (2020) BioSec: a biometric authentication framework for secure and private communication among edge devices in IoT and industry 4.0. IEEE Consum Electron Mag

34. Gill SS (2021) Quantum and blockchain based serverless edge computing: a vision, model, new trends and future directions. Internet Technol Lett e275

A Blockchain-Enabled Serverless Approach for IoT Healthcare Applications

Mohsen Ghorbian and Mostafa Ghobaei-Arani ⓘ

Abstract This chapter will examine the interaction between serverless computing and blockchain technology for IoT healthcare applications as a novel approach to solving unreliable function problems and resource allocation on serverless computing providers. The existence and execution of functions with malicious code and the lack of confidence and trust in the validity and safety of executive functions can always create concerns for customers during the process of using these functions. In particular, the importance of ensuring the safety and privacy of customers increases when these functions are used in sensitive areas such as health. On the other hand, because there are limited resources available in the Internet of Things devices, serverless service providers must employ a strategy that can provide a basis for addressing requests in the best possible manner by implementing a proper scheduling process. Especially when it comes to health-related programs, this is of utmost importance. We consider the potential and ability of continuous interaction between the two technologies of serverless computing and Hyper Ledger Fabric's private blockchain as a suitable and efficient solution to respond to the needs raised. By enabling the interaction between these two technologies, it is essential to note that this new approach can be suitable for use in the healthcare industry since it can provide the context in which necessary requests can be responded to by performing the appropriate scheduling process. Furthermore, serverless service providers should provide functions that can ensure customers that the service provider will approve the functions they use and that there will be no unreliable origins and functions that use malicious code. Consequently, considering the concerns in sensitive areas such as health, it can be said that this approach tries to respond to these concerns by presenting a novel solution. The proposed approach tries to respond to the sent requests by considering the privacy of its users, along with the appropriate and timely allocation of resources, while providing reliable and trustworthy conditions for customers. It is clear that this provides the basis for good utilization of the resources, which can ultimately increase the satisfaction of the users and the quality of service they receive.

M. Ghorbian · M. Ghobaei-Arani (✉)
Department of Computer Engineering, Qom Branch, Islamic Azad University, Qom, Iran
e-mail: mo.ghobaei@iau.ac.ir

© The Author(s), under exclusive license to Springer Nature Switzerland AG 2023
R. Krishnamurthi et al. (eds.), *Serverless Computing: Principles and Paradigms*,
Lecture Notes on Data Engineering and Communications Technologies 162,
https://doi.org/10.1007/978-3-031-26633-1_8

Keywords Edge-blockchain · Function as a service · Healthcare system · Scheduling · Serverless computing

1 Introduction

IoT technology enables continuous monitoring. Applications of the IoT span various industries, including automotive, telecommunications, and energy. For example, The IoT allows people to control intelligent homes remotely from computers and smartphones. It is also possible to use the IoT application in the healthcare sector to enhance patient monitoring and control [1]. Hence, IoT-based health care is expected to impact population health and improve healthcare performance positively. IoT refers to a wireless, connected, and associated network that can gather, send and store data without requiring direct human-to-human or computer contact. In recent years, it has become increasingly important to understand how the IoT can improve the accessibility of public health systems [2]. One factor that facilitates the success of the IoT is cloud computing services. In cloud computing, clients use services provided over the Internet to accomplish their computing tasks. IoT, in conjunction with cloud computing technologies, has evolved into a form of stimulus. Consequently, IoT and cloud are now inextricably linked [1]. The trouble of processing, accessing, and storing large quantities of data has become more predominant due to technological progress. An important innovation is the common use of IoT and cloud computing. It will be possible to utilize the capability of processing sensorial data streams with new monitoring services [3]. Hence, the data collected by sensors can be uploaded and saved by cloud computing for later use as intelligent monitoring and activation via other devices. A key objective is transforming data into insights, leading to cost-effective and productive action [4]. It is possible to categorize cloud computing into various types based on several factors, including the service type and the service provider's location. Therefore, a cloud computing environment can be classified into three general categories based on location or deployment: Public, Private, and Hybrid cloud computation. On the other side, a cloud computing service, in general, can be classified into four categories based on the services it provides: Infrastructure as a service (IaaS), Platform as a service (PaaS), Function as a service (FaaS), and Software as a service (SaaS). Hence, most other services are built upon these core services [5]. Since these technologies are built on top of one another, they are sometimes called the cloud computing "stack." But, by Understanding their differences and what they are, you will be capable of considering your work objectives more effectively. Throughout this chapter, we will concentrate on the function as a service [6]. Serverless computing allows developers to develop apps without continually managing servers and infrastructure. It means that you, as a developer, will not be responsible for your project's setup, capacity planning, or server management. In serverless architectures, resources are only used during a particular function. This architecture has high scalability and is event-driven. Using Function as a Service is a simple process. Hence, that's an event-based architecture, which means that a given event can trigger its functions, so it's called an event-driven architecture based on events. This kind of architecture is so simple that it is called serverless architecture because of its high level of simplicity [7]. As FaaS is auto-scalable, the developer does not need to worry about things such as deployment, server resources, or scalability as everything is taken attention by FaaS. Therefore, the billing system is based on actual consumption rather than claimed consumption, which translates into a cost reduction for the consumer. Even so, we must realize that one of the most significant drawbacks of FaaS is the security issue, which can significantly impact its usage [8].

A serverless computing system utilizes function-as-a-service technology based on microservice architecture. This architecture uses code fragments called functions rather than a monolithic program. In this architecture, the clients must call and execute functions to perform their processing tasks. In this regard, it is essential to pay attention to the fact that clients want their processes to be executed within a secure environment and with safe functions since they are always concerned with protecting their privacy and preventing the loss of their personal information [9]. On the other hand, when clients request functions from service providers, the system must adopt a method of allocating resources for these requests so it can respond to the requests received. Hence, serverless service providers must automate and optimize the process of allocating resources to requests using a component known as a scheduler. Service providers can use the scheduling component to allocate resources according to the approach they consider. The scheduling process for resource allocation has different approaches and strategies. Therefore, various techniques and approaches have been considered when performing the scheduling process for different use cases. Hence, using the same scheduling strategy for all use cases is impossible. In one type of schedule, the service provider allocates resources based on the amount of energy available, that is, the amount of energy or power stored in batteries in devices such as the Internet of Things. It's important to note, implementing each function requires energy or resources; usually, the amount required is inserted into the function catalogs [7]. Hence, upon receiving a request to execute a function, service providers can compare the amount of energy available on the device requesting the function to the amount of energy required to execute the function by knowing the information related to the available functions. Ultimately, the process is based on considering the provider's service approach will allocate resources to the client. Due to this, in this scheduling approach, it will be essential for the provider to know both the amount of energy available in IoT devices and the resources required to execute each function. Due to its high security, blockchain technology can protect data and provide the basis for meeting needs by storing information related to functions within the blockchain, thereby providing the necessary information. As a result, service providers are responsible for providing this guarantee to service requesters, or, in other words, clients, that the available functions are entirely safe from a provider's perspective and that the providers approve the available functions will be executable [10, 11]. Hence, According to the stated objectives, a mechanism must be proposed to ensure the safety of the executive codes and specify the energy or resources required to execute each function using serverless computing and blockchain technology within the resource allocation process. Hence, the mechanism proposed in this chapter is based on a blockchain-enabled serverless approach in order to meet these objectives and requirements. It is important to note that in this chapter, a scenario in the health field has been proposed to demonstrate how serverless computing interacts with blockchain technology to ensure proper resource allocation and executive function safety.

The main contributions of this work can be summarized as follows:

- We study the interaction between serverless computing and blockchain technology for IoT healthcare applications.
- We propose a blockchain-enabled approach to solve unreliable function problems and resource allocation on serverless computing providers.

This chapter is structured as follows: the first section examines the essential prerequisites, such as Rapid detection and control of the condition of acute patients, the Internet of Things for health, Using the cloud environment and cloud services for data processing, Using serverless computing technology as a cloud service for data processing, Functions available in serverless computing and their use for data processing, Blockchain and its potential for the Internet of Things. The second section of this chapter describes serverless computing, scheduling in serverless computing, and blockchain separately. The third section will examine the architecture of Hyperledger Fabric, Blockchain, serverless computing, and Interacts serverless computing with Hyperledger Fabric blockchain. The fourth section comprehensively examines the combination of blockchain technology and cloud computing. The fifth section of this chapter a hybrid framework for scheduling serverless computing with Hyperledger fabric blockchain is proposed in the context of healthcare. Finally, it is concluded in the sixth section.

2 Preliminaries

In this section, fundamental and essential information and idioms, such as serverless computing and scheduling on serverless computing described. Our objective in this study is to propose a blockchain-enabled serverless approach. Therefore, we will discuss the structure of Blockchain and its application in ensuring the safe execution of executive functions.

2.1 Serverless Computing

An execution model known as serverless computing refers to the process in which a cloud service provider dynamically allocates computing resources and storage resources for executing a specific piece of code and then charges the client for the computing and storage resources necessary to execute the code in question. A developer does not need to know the hardware or operating system on which the code is being executed to work with this model. In other words, serverless providers make code development and deployment easier for clients [12]. It is important to note that, although it is described as serverless, physical servers are still employed in serverless applications by default. Still, developers do not have to be aware that they are present. Actually, serverless means that the developers will not be concerned about servers and can concentrate on their work [13].

It is important to note that serverless computing architecture is primarily based on a Functions as a Service (FaaS) model, which allows cloud platforms to execute code without having to provision infrastructure instances in advance. In other words, serverless computing utilizes functions as a service. As a result, cloud providers offer FaaS service models that are stateless, scalable, and fully managed from the server. Typically, serverless functions are invoked by events, meaning they are only executed when a request triggers them [14]. For this reason, the provider does not charge a flat monthly fee for the maintenance of a physical or virtual server, but only charges for the compute time that execution uses. Depending on the application, these functions can be connected to create a processing channel [15]. They may serve as components within a more significant application, interacting with other code running in containers or conventional systems, or as standalone components. Serverless computing offers two main benefits:

- Instead of worrying about infrastructure, developers can concentrate on writing code that accomplishes business goals.
- Rather than purchasing hardware or renting cloud instances that sit idle most of the time, organizations will pay only for the resources they use.

The benefits of containerized microservices can be accessed using serverless computing without the complexity involved. Serverless functions can coexist alongside containerized microservices within a single application [16]. Despite the numerous advantages that serverless computing offers, some potential disadvantages exist for specific developers and teams. The extent to which these disadvantages can be considered problems depends on what type of product is built. Some of these disadvantages are:

- A function will not retain any stateful information from previous instances when it is started.
- Most serverless providers do not allow your code to run for more than a few minutes.
- Similarly, serverless code can take several seconds to start up, which may not be a problem for most use cases, but may pose a problem for applications that require low latency.
- If a developer grows dissatisfied with the tooling they are using, it can be difficult to switch if they wish.

In the context of serverless computing, Google Cloud Functions, IBM Open-Whisk, Amazon Lambda, and Microsoft Azure Functions are all well-known examples of serverless services on cloud platforms. However, all these platforms are closed-source. On another side, many open-source serverless frameworks have been proposed, including Apache OpenWhisk, Kubeless, Iron Functions, Oracle's FN, Knative, OpenFaaS, and Project Riff, to meet the increasing demands of serverless computing. It should be noted that some of these frameworks are built upon existing cloud services infrastructures to serve applications. In contrast, others are made from the ground up by having the functions deployed directly on the server [17].

2.2   Scheduling in Serverless Computing

Resources scheduling is the methodology by which resources are allocated to tasks. Two methods are used to schedule resources: time-constrained and resource-constrained. For time-constrained scheduling, the time factor is the critical variable. Meanwhile, resource-constrained approaches focus on the limited capacity of resources, emphasizing the resolution of capacity overload issues as part of resource constraints [18]. Generally, Scheduling approaches for tasks can be classified into two categories: centralized scheduling and distributed scheduling, both of which are used to schedule tasks. Centralized schedulers are also called monolithic since they operate on a centralized approach to managing resources and scheduling tasks. But, In contrast to centralized scheduling, distributed scheduling utilizes a scheduler per machine to support a large number of machines in a cluster [19].

Typically, all schedulers employ the same architecture so that task submissions are divided among computing servers by a simple load balancer, and load imbalances are detected and resolved by a scheduling agent per machine. This approach allows selected tasks to be moved from busy servers to lighter ones. Meanwhile, existing scheduling mechanisms are not well suited to serverless computing as they have been designed specifically for cloud computing. It is important to note that the cloud schedulers are typically centralized, so they may not be suitable for serverless environments [20]. Large-scale bursty scheduling requests could result in much higher latency if the platform isn't elastic promptly. Even if burstiness is a serverless feature, it still could cause failure. In addition, it is essential to note that the general task scheduling mechanisms presume that the implementation environment was previously set up among all the servers in the cluster at the time of task scheduling. It is possible to examine this issue from a different perspective: Resource scheduling for cloud workloads can be challenging since the scheduling of resources appropriate will vary according to cloud applications' quality of service (QoS) requirements [21].

Resource scheduling problems arise in cloud computing environments due to heterogeneity, uncertainty, and resource dispersion. Hence, using traditional resource scheduling policies is not an effective method of solving this problem. Despite numerous studies on resource scheduling algorithms, researchers have difficulty identifying an efficient and appropriate algorithm for particular workloads. As a general rule, the scheduling process for serverless functions consists of two phases: runtime configuration and function scheduling, which typically take place on separate platforms. Platforms like OpenWhisk and OpenFaas, which offer serverless services, often generate their runtime configuration by binding the source codes to their running instances. When a serverless platform receives a function call, it first retrieves the source code and then the demanded resource configuration (Memory, CPU, etc.) from a database. In the next step, it allows a balancer to assess the current condition of the platform and then initiate invocations to create containers or send calls directly to pre-warming runtime instances, depending on the QoS requirements [22].

The point that should pay attention is this if the function's lifecycle will proceed to a phase, such as content creation, container orchestration, and container placement, as soon as the current running environment cannot satisfy the developer's requirements. In most cases, requests for container creation are queued in a messaging middleware solution so that they can wait for an invoker, who would either be the host machine or a virtual machine that would complete the request and acknowledge the message [23]. For serverless computing, energy-aware, Resource-aware, Package-aware, Data-aware, Deadline-aware, and Strategies are the most commonly used scheduling strategies. These strategies aim to ensure that resources are allocated to tasks in the best possible condition by using predictions of the system's state, even in a different form. In this chapter, we focused on energy-aware scheduling.

2.3 Blockchain

Blockchain technology is decentralized; the only purpose of blockchain technology is to store and share data. Distributed and public ledgers that allow for recording and storing transactions online. Various blockchain applications can be developed using the technology, including voting systems, exchanges, messengers, games, social networks, storage platforms, online shops, prediction markets, etc. The technology is also the foundation of many cryptocurrencies, such as Bitcoin and Ethereum, which record and transfer information securely. But its applications extend far beyond cryptocurrencies. In blockchain technology, data is stored in blockchain databases, and each block is linked or "chained" to the previous block. This information chain can continue indefinitely so long as the computers that control the database operate, adding successive pieces of information to the previous ones [24]. Blockchain is essentially designed to keep records of all the movements of information. So by its very nature, it's designed to be highly transparent, at least to those with access to the database containing all relevant data. But it is crucial to keep in mind that establishing transparency can only be accomplished by ensuring that the database is secure and resistant to hackers. Since blockchain technology stores information securely, any alterations made to the information need to be recorded on the blockchain database to preserve the record of those changes. Since blockchain collects data over time, they provide a chronological history of the data in the order in which it was recorded irreversibly [25].

Distributed ledger technology utilizes networked computers' redundant power to verify data credit. When a consensus has been reached between the devices involved in storing transactions on a distributed ledger, that record is irreversibly recorded. Therefore, it cannot be removed, altered, or disputed without the knowledge and permission of those who made the record. This process is known as consensus. A consensus mechanism is a fault-tolerant method for achieving a consensus agreement among distributed processes or systems consisting of many agents to achieve a single network circumstance. They can add it to the chain after agreeing on the candidate block [26]. The way of constructing new data blocks is called "mining," a process conducted by computers. The blockchain network runs this mechanism in two ways;

- **Under the Proof-of-Work (POW) system**, the blockchain network nodes compete directly to see who first solves the complex mathematical equation. Using this mechanism, miners earn the reward to mine the next block of the transactions in return for their proof of work. In this way, all participants in the network can have a chance to acquire rewards.
- **Under the Proof-of-Stake system**, a computer algorithm is used to select the nodes of a proof-of-stake system, which employs a specific degree of randomness. This way, it is more likely for nodes that have accumulated an immense amount of the network's currency to be chosen, which rewards extended participation—or their "stake" in the network-over raw computing power per node basis. Hence, those chosen to process a given block are known as validators instead of miners.

When a block is constructed, it is broadcast over the network. All nodes receive the block and accept it only after confirming that all transactions have been completed legally [27].

3 Blockchain for Scheduling in Serverless Computing in the Healthcare Domain

A significant benefit of blockchain technology is its ability to provide clients with excellent security. It can also provide transparency, one of the blockchain network's basic features. On the blockchain network, all transactions are encrypted before they are stored in blocks, using powerful encryption algorithms. Furthermore, it is possible to keep track of all transactions that have been made. This approach is used mainly in the public blockchain network to prevent attacks. Although blockchain technology has the potential to be used for a wide range of purposes, many businesses are unwilling to implement it due to concerns related to privacy [28]. The use of blockchain technology for specific applications requires that it have many distinct characteristics that allow it to be tailored to the organization's objectives. These types of blockchains are commonly referred to as the private blockchain. Hence, an organization or company can establish and operate a private blockchain tailored to its specific requirements. Organizations can benefit from private blockchain in several ways [29].

A serverless computing model provides a wide range of services, and microservice architectures are an approach to developing applications that take advantage of these services. A significant advantage of this approach is that programs used in cloud services are capable of being converted into functions that clients can invoke and use. This step involves creating a function, deploying it, and calling it from the cloud environment [30]. Hence, it is significant for this issue clients to ensure they are safe when using these functions. The Internet of Things bases Serverless computing services can recreate a significant role in health care, providing the benefits of scalability, agility, and efficiency. For instance, doctors can use the Internet of Things technology to investigate illness symptoms to give them valuable information about the state and conditions of ills. Consequently, diagnosing these symptoms quickly and promptly is imperative to provide the patient with the appropriate treatment [31].

In this chapter, the proposed scheme will invoke data processing functions based on the energy available in IoT devices. In such a way that, by having enough energy, the Internet of Things devices can rapidly request desired functions and, after being invoked, execute them locally. However, suppose an Internet of Things device does not have the amount of energy required to execute functions. In that case, it will be unable to process the information received from the sensors, and the patient may be at risk of dying. It is possible to use energy-aware schedulers to solve this problem which can handle this issue automatically. The energy-aware scheduler must quickly determine whether the Internet of Things devices have the capabilities and energy needed to execute their functions by checking the amount of energy available in the devices and utilizing scheduling suitable algorithms. Every function requires a certain amount of resources to be executed, which can be found in the catalog section of every function. By placing the functions' information in the blockchain and allowing the scheduler to query the information about each function using the considered mechanism, the scheduler can make informed decisions regarding each function. In the following, we will describe how serverless computing can interact with the Hyperledger Fabric private blockchain.

3.1 Architecture Hyperledger Fabric a Permission Blockchain

The Linux Foundation sponsors Hyperledger Fabric as one of its Hyperledger projects. It is a new blockchain architecture built on a distributed network to ensure confidentiality, flexibility, and scalability. As one of the first blockchain systems to support the execution of distributed applications written in standard programming languages, Hyperledger Fabric is unique in that it makes it possible to execute distributed applications. The Hyperledger Fabric blockchain system is designed to be a modular and extensible permissioned blockchain system capable of expanding in the future. The Hyperledger Fabric has been used to develop many distributed ledger prototypes, proofs of concept, and production systems. An example, these use cases include healthcare, trade logistics, contract management, food safety, identity management, and settlements through digital currencies.

Each transaction in Hyperledger Fabric is endorsed and executed by a subset of peers, which enables parallel execution of the transaction to achieve high performance. It is important to note that in Hyperledger Fabric, an intelligent contract endorsement policy specifies which peers or how many have to agree to confirm the correct execution of a smart contract [33, 34]. It also supports active replication by documenting the effects of a transaction on the ledger state only after all peers have reached a consensus on how the transaction should be implemented through deterministic validation. Hyperledger Fabric can respect application-specific trust assumptions by relying on the endorsement of transactions. Moreover, the ordering of state updates is delegated to a modular component for consensus. This component is stateless and logically separated from the peers involved in executing transactions and maintaining the ledger. Considering that consensus is modular, it can be used in a manner that is customized to meet the trust assumptions of each deployment. The separation of roles opens up much more flexibility despite the possibility of using peers on the blockchain to implement consensus. It allows for well-established toolkits such as CFT (crash fault-tolerant) and BFT (Byzantine fault tolerance) order to be used in the process. It is now possible for Hyperledger Fabric to be a scalable system for permissioned blockchains that supports flexible trust assumptions due to resolving the abovementioned issues. Blockchain Hyperledger fabrics consist of a network of nodes (see Fig. 1).

There are three nodes in this form of the blockchain network [32, 35]. A Hyperledger Fabric blockchain network comprises several components, as shown in (Fig. 1). Some of these components are created from other components. In this fig illustrates the difference between two-way and one-way interactions using dotted and solid lines. Thus, components with two-way interactions are depicted with dotted lines, whereas components with one-way interactions are depicted with solid lines. This structure has several components, which are examined in the following. These components included:

- **Application**: An application presented suggestions for processing transactions, assisted in orchestrating the execution process, and published transactions for ordering.
- **Peers**: The peer's components are modular and extra components that are able to be created, activated, deactivated, configured, and removed as required. A peer processes a transaction request and validates a transaction.
- **Orderer**: Ordering or Orderer services are included in several nodes. When these nodes are taken into account together, it becomes possible to construct an ordering service. In Hyperledger Fabric, the ordering service determines the general order of all transactions. Every transaction includes the status updates and relationships established during the execution stage and the cryptographic signatures of the

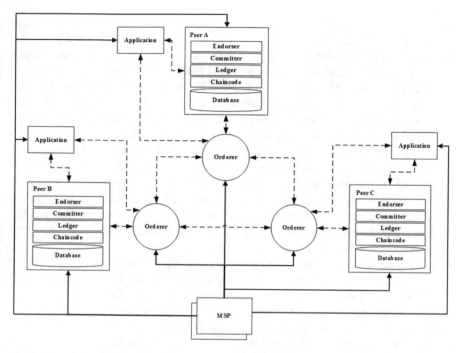

Fig. 1 Hyperledger Fabric network consists of distributed MSPs running three chain codes installed selectively on peers by policy guidelines

peers that endorse it. The orderers are neither responsible for executing nor validating transactions, as they are neither aware of the application status nor involved in its execution.

- **Endorser**: One component of peers is an endorsement. Performing the process of transaction validation is the responsibility of the endorser. The validation phase involves evaluating a policy for compliance with endorsement criteria. Developers of untrusted applications are not permitted to choose or modify endorsement policies. Sometimes, only designated administrators can modify endorsement policies through the system management functions. An endorsement policy is a static library for validating transactions in Hyperledger Fabric. Usually, endorsement policies permit the chain code to specify the endorsers for a transaction in the form of a set of peers necessary to approve the transaction [36].
- **Committer**: This peer component is known as a committing peer. The committing peer is responsible for appending the validated transactions into the ledger unique to the channel in which the transaction has been validated. A peer may only be able to serve as a committer rather than an endorser in more regulated environments. However, in more open environments, a peer may be able to serve both functions [37].

- **Ledger**: During the phase of validation, simulation, and ledger updates, the ledger component at individually peer holds the ledger. It maintains the state of permanent storage to facilitate validation, simulation, and ledger updates. Generally, it comprises a peer transaction manager and a block store. A ledger block store is a collection of files that maintain transaction blocks in the ledger and allows them to be appended only. In an append-only structure, all blocks arrive in the same order and are immutable, so performance is optimized to the maximum extent. It is also possible to randomly access a block or a transaction in a block via the indices that the block store maintains. In peer transaction management, the current state of the transaction is maintained in a versioned key-value store. In addition, any individual entrance key stored by every chain code is stored as a tuple of the form (key, value, version), which contains its most recent value, key, and version. A version comprises two number sequences: the block sequence number and the transaction sequence number. Thus, the version becomes unique and increases monotonically over time [38].

- **Chaincode**: Smart contracts, also known as chain codes, are programs executed in the application's execution phase and are responsible for implementing the logic. It is important to note that chaincode is a core component of a distributed application using Hyperledger Fabric. It can be written by anyone, even by individuals who are not trusted. For the management of the blockchain system, unique chains of codes are required. Collectively, these codes can be considered "system chain codes [39].

- **Database**: The Hyperledger Fabric can be used with two different peer-state databases. A CouchDB database is an alternate state database that permits clients to model the ledger data in JSON form and then execute sophisticated queries based on the data values instead of the keys. Additionally, CouchDB support enables the deployment of indexes in chaincode to enhance query efficiency and make querying large datasets more convenient. In contrast, LevelDB is the state database used by peer nodes by default. In LevelDB, chaincode data is stored as a simple key-value pair and can be queried by key, key range, or composite key [40].

- **MSP**: A membership service provider (MSP) is responsible for maintaining all nodes' identities such as peers, application (client), and OSNs to ensure that they are all authenticated and authorized and for issuing credentials to each to ensure they are both authenticated and authorized. It is important to note that Hyperledger Fabric is a permissioned blockchain. As a result, all communications between nodes must be authenticated, usually employing digital signatures. Each node in the network has a membership service component where transactions get authenticated, transactions verified for integrity, endorsements signed and validated, and other blockchain operations are authenticated. As part of the MSP, tools are provided for managing keys and registering nodes [41].

There is a wonderful synergy between all components of the blockchain structures. Hyperledger Fabric is a permissioned blockchain system that can address many companies' security, privacy, and confidentiality concerns. The important point here to keep in mind is that this type of blockchain does not require clients to pay any fees to carry out their transactions, which will make it economically viable in the long run.

3.2 Architecture Serverless Computing

According to a serverless architecture, the cloud provider runs pieces of code and allocates resources dynamically based on the customer's needs. Also, this model can be used in various scenarios, in which functions can be written that include changes to the configuration of resources necessary to accomplish specific infrastructure management tasks. This means that serverless computing architecture unlocks the complete capabilities of cloud computing since resource allocations and scaling up and down are carried out based on the actual needs of clients in real time, and the client only pays for the resources they use [42].

The system ensures that all resources are automatically scaled down to zero when no client requests are received, and the application is inactive. Essentially, serverless architecture corresponds to a pattern of software design in which third-party cloud vendors manage infrastructure surveillance tasks and provide computing services based on the functions offered by their cloud services. In general, these functions are implemented into ephemeral containers [43]. They can be triggered by various events, such as HTTP requests, cron jobs, monitoring signals, file uploads, database events, and queues. Using this model, clients will not need to worry about servers since the provider's architecture abstracts them from the client's perspective [44]. The serverless computing architecture is shown in Fig. 2. There are several components to the architectural structure, each of which implements a particular concept. As a whole, these concepts comprise the serverless computing architecture, which follows be discussed. The concepts are as follows:

- **The Client Interface**: The client interface is an essential element of serverless functionality. Hence, it is important to note that serverless functionality heavily depends upon the client interface. It is not possible for a client to force-fit a serverless architecture into any given application. An interface should be able to handle short bursts of flexible, stateless interactions and requests to be effective. Furthermore, the interface design must handle data transfers of extremely high or extremely low volume [45].
- **A Security Service**: It is not possible to perform serverless operations without a security service since the application must be able to provide security for thousands of requests at once. Therefore, it is crucial to ensure that there is an authentication process before sending back a response. Its stateless nature makes it impossible to maintain a historical record of previous interactions. It is, therefore, not possible for the application to validate future interactions based on previous interactions. On the other hand, the lack of transparency makes monitoring and tracking serverless models more challenging. Due to the distributed architecture

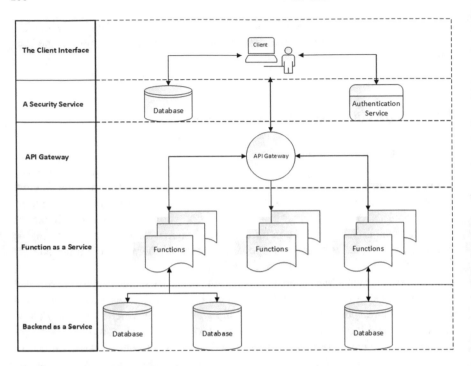

Fig. 2 Serverless architecture's main components

of serverless technologies, several services and vendors may be involved in the process. Hence, securing the entire environment is of the utmost importance. The token service typically provides temporary credentials to clients so that they can invoke functions in a serverless application [46].

- **API Gateway**: The gateway API connects a FaaS service and its client interface. When a client initiates an activity, it is relayed through the API gateway so that it can be used to trigger a function using the FaaS service as specified in the mechanism. The client interface can be connected to several FaaS services through the gateway, increasing the functionality of the client application as a result [47].
- **Function as a Service**: In serverless computing, FaaS is viewed as the most important architecture component because it is responsible for implementing the logic by which resources are assigned in a specific scenario by the requirements. To choose a FaaS service for their cloud environment, clients can select from a variety of open-source and closed-source options, such as Microsoft Azure Functions for Azure, Amazon Web Services Lambda for Amazon Web Services (AWS), IBM Cloud Functions, and Google Cloud Functions for Google Cloud Platform (GCP), OpenFaaS, OpenWhisk, Knative, Kubeless, and Fission for the private or hybrid model. As soon as the client triggers an occurrence, these functions will read the database on the system's back end and obtain and deliver the response [48].

- **Backend as a Service**: The term "BaaS" or Backend as a Service is used to refer to the model of cloud-based computing that manages the backend side of the development process of web or mobile applications. The BaaS component offers a variety of crucial back-end features that enable clients to create an incredibly functional backend application in the most efficient manner possible. BaaS providers include: Azure, Kumulos, Leancloud, and Firebase [49].

Due to this scheme, the application possibly must retrieve and deliver data in several circumstances to function effectively. For instance, when the client creates an HTTP inquiry, the application might have to retrieve and deliver a specific piece of information. Generally, this "if–then" process is referred to as an event in the programming world. A client's action triggers an event once an application is launched and made available to them. The event is triggered in response to the action they initiate once the application has been launched. In this case, the application dispatches event for the cloud service provider. The cloud service is responsible for dynamically allocating the appropriate resources based on the predefined rules for executing the function. After the function has been called, the client acquires the data or any other output specified by the function. Before continuing, it is essential to keep one point in mind. Remember that if the client does not submit a request, no resources will be allocated, and there will be no data storage in an intermediate state. It is, therefore, possible to deliver the most recent and most up-to-date data to the client at any time, on the other hand, enabling an application to be as real-time as possible while reducing the application's cost and storage requirements.

3.3 Interacts Serverless Computing with Hyperledger Fabric Blockchain

The proposed approach seeks to create a dynamic interaction between serverless computing and blockchain technology. We will describe how serverless service providers can communicate with Hyperledger Fabric to take advantage of the particular private blockchain's unique characteristics and interact dynamically.

Hyperledger Fabric is a private blockchain technology developed by the Hyperledger consortium. This private blockchain incorporates enhancements to public blockchain capabilities and additional capabilities that enable it to be utilized in various situations, including interacting with new technologies, such as serverless computing, to make this technology more widely available. Its permission-based functionality is among the features of the Hyperledger Fabric blockchain, making it possible to interact dynamically with serverless computing. Serverless service providers can use this feature of Hyperledger Fabric's private blockchain to design and customize the blockchain according to their needs and desires, as well as guarantee that the information contained within this blockchain will not be accessible to third parties in any way. This feature does not consider available on public blockchains. For example, the storage location of functions in serverless service providers may contain various functions that are potentially unsafe or poorly performed that clients might deploy on the functions registry. As a rule, customized, only those functions will be considered whit serverless service providers that the high-performance and have certified as safe functions and that have been stored on a blockchain. Therefore, applicants will be provided with these functions as authorized and safe in this case. Having submitted a request to the blockchain and received information about the requested functions, the serverless service provider can ensure that the requested functions are among those approved by the provider and then try to respond to the requests. Also, by using clients who have been authenticated and approved by the service provider, it is possible to put functions on the blockchain.

Service providers can benefit from interactive private blockchain by not paying fees for inserting information into the blockchain ledger or recalling information via transactions. This feature makes the service provider, unlike blockchains like Ethereum, not pay any fees for any transaction that includes entering the information of functions or calling their information. As speed is a critical component of serverless computing, using structures like Hyperledger Fabric, which provides a fast and free blockchain for serverless service providers, can be a suitable alternative to other blockchain technologies. For example, in the case of a requester requesting functions from a serverless service provider, if the provider queue up to respond to this request and pay a fee for each request, this can result in service providers being unwilling to utilize blockchain technology. On the other hand, a service provider must establish a dynamic relationship with the blockchain structure so that information about functions in the blockchain can be stored or the information needed can be accessed safely. The Hyperledger Fabric blockchain platform provides these facilities for service providers and is therefore recommended for use by service providers.

4 Combining Blockchain and Cloud Computing Environments

It has become increasingly important to utilize cloud resources in recent years due to their ability to process enormous amounts of information generated by IoT devices. It is estimated that volume will increase along with the expansion of service offerings and the expansion of the network, resulting in a rise in the use of cloud services. It is also important to note that when cloud workloads are heavily loaded, the cloud resource provider will consume a large amount of energy and resources to process the workload [50]. On another side, a wide range of algorithms is being developed as part of the cloud computing environment strategies to optimize the scheduling of resources to reduce energy consumption. An approach such as this utilizes algorithms employed in these systems to optimize scheduling by significantly reducing the time it takes to perform the task, the amount of time spent waiting, and the amount of time spent executing it. The purpose of this chapter is to utilize blockchain technology to assist with resource scheduling and to decrease energy consumption associated with tasks processed in cloud computing environments, especially on serverless computing models. Additionally, because Blockchain is a decentralized system, it can be used to facilitate execution processes in a decentralized manner for about to access functions and information stored on the blockchain ledger [51].

Distributed technology allows the cost associated with energy consumption to be minimized due to the implementation of this distributed approach to schedule resources. It is important to note that serverless computing models attempt to reduce costs associated with their services in several ways, such as building an energy-aware resource scheduling mechanism alongside blockchain-based technology. Due to blockchains' high level of security, some blockchain-based systems may be suitable for use in the cloud computing service especially serverless computing models, to facilitate allocating resources to track the battery charging process. Developing a cloud resource allocation strategy based on energy awareness could increase the efficiency resources rate, thereby improving overall resource and energy consumption. However, it is essential to note that this approach has many disadvantages. Scheduling cloud resources is a challenging task because most of them are NP-hard issues that require various factors to be considered. An example of a scheduling dilemma is often a tradeoff between energy cost and time complexity in scheduling problems [52]. It is important to remember that there are many exhaustive techniques, like liner programming, whose time cost increases exponentially with the parameters used in the program.

Several evolutionary approaches have been suggested to improve scheduling, such as ant colony optimization, particle swarm optimization, and genetic algorithms. Most existing scheduling methods do not perform without latency scheduling because all mechanisms are typically based on a central control hub, which appears to be inflexible in the face of the constantly changing requirements of clients. Due to the security features and decentralized structure of blockchain technology, schedulers can access information regarding executive structures with the assistance of blockchain technology, such as functions, which makes it possible for them to perform resource allocation and scheduling processes more efficiently. Some studies have addressed blockchain's capability to enhance the computation capability of mobile edge computing. Still, no studies have examined blockchain's ability to improve scheduling in serverless computing environments. To reduce the amount of energy consumed by serverless computing models, we propose a scheme that uses blockchain technology to schedule resources.

By adopting an innovative approach, we can examine the relationship between blockchain technology and serverless computing from a different perspective to understand how they work together and, in other words, how to interact together. When these technologies are combined, each has its unique advantages, which makes it possible to increase efficiency. In light of the unique features considered for Hyperledger fabric technology, one can conclude that combining this technology with other technologies, like serverless computing, will allow them to take full advantage of the benefits of blockchain technology and provide a basis for increasing their efficiency.

A ledger subsystem for Hyperledger Fabric comprises two components, the transaction log and the world state, which are the two primary components of the ledger. A world state component indicates the current state of the ledger at a particular point in time, in terms of the ledger state. Every Hyperledger Fabric network where a participant is a member has a copy of the ledger that they can access at any time. Hence, serverless computing can utilize the world state to interact with blockchain technology. Utilizing this component, clients of Hyperledger Fabric can query from the data contained in the world state, such as in relational databases. Using this feature, the serverless service provider can interact with Hyperledger Fabric to receive information regarding the functions stored in the blockchain and use this information to perform processes such as energy-aware scheduling. In this approach, the provider will receive requests for the required functions to execute the process from clients. An initial step in responding to a request regarding the requested functions would be for a provider to submit a query request to obtain information about the requested functions from the world state in the Hyperledger fabric blockchain. Therefore, after getting information from the world state, if the information of the requested function exists in that, a service provider can use this information and be informed of the amount of energy or resource required for each function execution. In this case, the provider can compare the amount of energy available to the clients, which can be considered IoT devices, and the amount of energy that the requested functions require for executing. Thus, the provider determines whether the amount of energy that the client has is sufficient to perform the functions that he has requested. Here, there are two potential scenarios:

- Assuming the client has a suitable energy level, he can execute the function locally, in which case the function is assigned to the client so that he may execute it locally and run it locally.
- Assuming the client does not have enough power and energy to execute the function. Hence, the provider assigns the function to a client with enough energy to execute the requested function.

Additionally, in another aspect of this approach, the service provider can assure requesters that the functions they are using have been verified by themselves. In other words, this will mean that service providers can assure requesters of the accuracy and security of existing functions with the information they receive from operations recorded on the blockchain. Also, it is essential to point out that clients can use these functions to ensure that their privacy is protected according to the terms of the provider's policy. Consequently, malicious code will not be executed due to their use of these functions.

The providers must confirm and authenticate the identity of customers before they can provide them with services. Despite the necessary arrangements, this type of authentication cannot achieve the level of security achieved by blockchain technology. Therefore, combining these two types of technologies makes it possible to ensure very high security for clients' authentication processes in serverless computing. In this way, when clients request registration and authentication in the provider, the provider's system can record the requests in the ledger Hyperledger fabric blockchain rather than storing them in databases specifically designed for this purpose. Therefore, after the completion of this process, when the client requests authentication and execution of functions, the service provider can obtain the information of the requesting client from the blockchain and allow the client to use the service if authentication is performed and send requested functions for them. Using this blockchain feature, the service provider can ensure that unauthenticated clients' cannot send requests and cannot access the services provided. Therefore, it is possible to maximize the benefits of both technologies by establishing interactions between private blockchains such as Hyperledger Fabric and serverless service providers.

It is essential to mention that the scheduler performs the function allocation process. However, we did not intend to examine the timing component. The purpose is only to express the performance and implementation of the proposed approach. Moreover, the referred assumed clients are nodes activated in the network.

5 Scheduling in Serverless Computing with Hyperledger Fabric

In this section, we will try to present our implementation of energy-aware scheduling processes in a framework of scheduling serverless computing on the blockchain using serverless computing technology, blockchain technology, and an Internet of Things in the healthcare field. This scenario will discuss the Corona pandemic within the healthcare sector. It has affected a large number of countries around the world as a result of the pandemic. Consequently, this represents a significant challenge that has disrupted humans' daily lives worldwide. Our scenario is based on the fact that the disease is a pandemic, and many of the people most susceptible to the disease are the elderly, who may have underlying illnesses. The purpose of this scenario is to help take care of the elderly or those who need permanent care at the lowest possible cost through the Internet of Things technology, serverless computing, and blockchain technology. The patient's condition can therefore be continuously monitored, which allows the physician to detect any changes in the patient's condition immediately and to take measures to save the patient's life.

In early December 2019, Coronavirus disease (COVID-19) was first detected in Wuhan, China. It has spread rapidly worldwide, with confirmed cases in nearly every country. COVID-19 is a virus that anyone can contract, regardless of their age. However, it is most common among middle-aged and older individuals [53]. A greater likelihood of developing severe symptoms increases with age. Therefore, those middle-aged and older are at the most significant risk. In the U.S, Nearly 81% of the disease's deaths have been reported in the older population [54]. Additionally, older individuals with other underlying medical conditions are at even greater risk. As a result, these people must be constantly monitored. The length of the treatment period for this disease necessitates a spend quarantine period of 14 days. Hence, the patient and the hospital incur unnecessary costs due to keeping and quarantining these individuals in hospital settings. It is important to note that inessential healthcare costs may result in many undesirable outcomes, which, if possible, can be mitigated by reducing waste [55]. Accordingly, if these people are in stable health conditions, they may be able to continue their treatment at home. Hence, it is essential to monitor the patient's blood oxygen level continuously and heart rate to identify any changes in the patient's condition.

When an individual has a history of underlying diseases, it is essential to maintain continuous control over these conditions. Hence, the Internet of Things technology can help monitor the above factors in the patient. This technology allows us to continuously monitor factors such as the amount of oxygen in the blood and the heart rate of patients by connecting sensors in the form of wearable gadgets to the patients. In the next step of the process, the data collected by the sensors is processed. Lastly, if changes in the condition of the vital factors in the patient, the doctor can find out in the shortest possible period. In this case, it is possible to save the patient's life by taking the necessary measures. An explanation of this scenario is provided by the sequence diagram shown in Fig. 3 to make it as clear as possible. The diagram below shows that the execution process for this scenario consists of numerous steps. Later on, we will provide a more detailed explanation of this scenario. Throughout this scenario, we will have a network of nodes located at the edge level of the network and comprised of several nodes. Hence, the computing layer used in this scenario is edge computing, and the type of service provided is function-as-a-service, which is based on serverless computing. This scenario assumes that wearable devices have been attached to individuals suffering from Coronavirus infection and who also have underlying health conditions. Utilizing these devices makes it possible to continuously monitor and transmit information about the patient, such as heart rate and blood oxygen levels. As a result, It can be argued that these devices are microservices, which require invoke and execute functions to process the workload they generate. Following the creation of workloads by gadgets, the client can now request to receive the functions they will need to process the workloads and the resources they will require to process them.

As mentioned, the processing layer used in this scenario is of the edge type. This means that the system will respond to resource requests from the nodes using resources available at the edge level to provide them with the resources they request. In addition, it is important to keep in mind that the requestors themselves may be able to execute the functions locally if they have the necessary resources. A schedule will determine whether or not applicants will be able to execute self-requested functions according to the scheduler's policy. The schedule intended for this scenario is energy-aware, meaning that if the applicants have sufficient energy, they can execute the functions they have requested locally. In contrast, if they do not have enough resources, the requested functions must be offloaded to other nodes so that they can execute the requested functions. It is important to consider that the amount of energy present on each node will directly relate to the amount of power stored in the battery. Therefore, it is possible to determine the energy required to perform each function based on how much the battery is charged and the number of resources needed. The client or applicant will request by the gateway to receive the functions they need

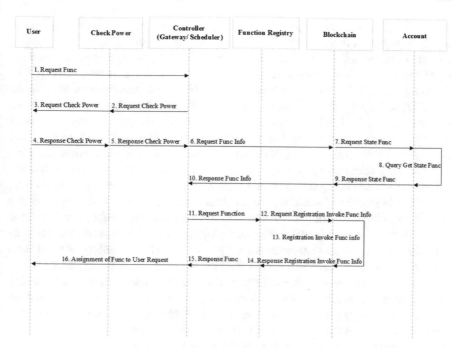

Fig. 3 Displays the sequence diagram of events in the scenario designed using the SSCB framework

to process their self-request. This scenario involves a controller comprised simultaneously of a gateway and a scheduler. Upon receiving a request, the controller gateway transmits it for invocation of the appropriate functions. In other words, the controller receives and processes the client's request through the gateway. It may be necessary to process the quantity of blood oxygen levels and heart rate number in these requests. Generally speaking, the sent requests ask for functions that can be used to determine the patient's condition based on the values sent in the workloads format. When the controller receives a request from the requester, it sends a request to the check power module, which triggers the check power mechanism. Essentially, the controller asks the module to provide information regarding the amount of power or energy in the node's battery that requests these functions. Upon receiving the request, the relevant module sends a request to the node requesting the function. In response to this request, the module will ask the node to specify how much energy or charge is present within the battery. At this scenario stage, the node receives the request and executes a local function to respond to the request and send the result to the module to determine the node's energy level. Check power module sends the controller the response it receives from the node upon receiving the response. The controller will then send a request to the blockchain once it receives the response from the module. A request will be sent to the blockchain by the controller in the next step. It is possible to use the request for two different purposes. Firstly, whether these functions even exist, and if they are, indicates that they are secure and have been authenticated. And secondly, by making this request, the controller attempts to obtain information from the blockchain regarding the requested functions. Among the crucial pieces of information required is the number of resources required for executing the functions. These resources are usually stored in a catalog that is part of the function itself. The scenario is based on employing a permissioned blockchain called Hyperledger Fabric. A unique feature of Hyperledger Fabric is its ability to support two types of ledgers. That means that queries are possible on the part of the Hyperledger Fabric ledger, known as World State, in a similar manner to what is employed with SQL queries. Therefore, by using the desired function selection command like select query in SQL, it is possible to obtain information regarding the functions currently being stored in the blockchain and their current status. Currently, this ledger is stored in CouchDB, the database used for storing it. In response to a request from the controller, the blockchain executes a query to obtain information regarding the requested function. The query by the chaincode is created and sent to the ledger for execution. chaincode is a script specific to the Hyperledger Fabric and is written to execute operations within the framework. In other words, chaincode is what makes up a smart contract in Hyperledger Fabric. After completing the query and receiving the result, the result is provided to the controller with information regarding the energy required to execute the requested functions. This stage involves the controller, who also plays the scheduler role in this scenario, determining whether the node requesting the function has the ability and energy to execute them. As long as the node can execute the functions, which is to say that it has the required energy to execute them, Therefore, the functions will be passed over to the requesting node for execution locally. Sometimes a node cannot execute its functions correctly because it lacks the necessary energy or capability. In these cases, the scheduler will decide to offload those functions to another node. Upon receiving the request, a request is sent to the location where the requested functions are registered and stored. As soon as the function registry receives a request for the desired functions, it will send a request to the blockchain and request that it stores information about the functions requested on its ledger. When the blockchain receives a request for a function, it will store the information about that function once it receives the request. Hence, to notify the function registry that the requested information has been stored in the blockchain, after the blockchain has stored the requested information, it sends a response to the function registry. The next step is for the function registry to present the controller with the function(s) requested by the applicant in response to the controller. The controller, which also functions as the scheduler, is responsible for allocating the functions requested by the node. The allocation of functions can be accomplished in various ways, for example, by executing functions locally or loading them into other nodes. In this scenario, assuming that functions are executed locally. An overview of the structure and architecture of the framework considered in this case can be seen in Fig. 4.

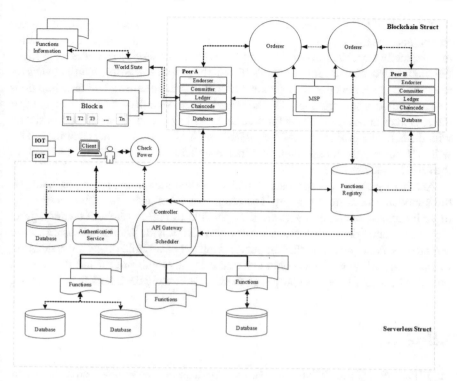

Fig. 4 The proposed framework integrated serverless computing and blockchain technology

6 Conclusion

Currently, serverless computing has been able to be used in a variety of fields, including health, as well as in the form of applications that run on devices that are part of the Internet of Things due to the implementation of microservice architecture. There has always been a concern about using emerging technologies in fields where data and information are critical and where users' privacy is vital. Hence, the use of emerging technologies has always been a very significant challenge in those fields. It is important to note that there is also a concern surrounding using serverless computing in the healthcare industry. The most important concern in this field has always been to avoid using unreliable functions and malicious code to prevent the misuse of customer information. For these reasons, they must ensure that the functions they use are valid and safe. Besides privacy concerns, another concern in this field will be a lack of timely responses to requests. Considering the importance of this area and that any delay in responding to requests can endanger human life, it is, therefore, necessary to respond promptly to the requests they submit. In this chapter, we have tried to provide a new approach that includes the interaction between two serverless computing technologies and Hyperledger Fabric's private blockchain to solve the needs and challenges raised in using serverless computing Service providers in the field of health. The proposed approach, which is the result of combining these two technologies, utilizes the very high level of security provided by private blockchains to completely store the information of all the serverless service provider's approved functions. Also, considering the needs raised, the proposed approach will always try to assure its customers that the executive functions will be safe and reliable functions that the provider has approved. On the other side, service providers will ensure that the resources allocated to the submitted requests have been allocated appropriately with the help of a proper scheduling process, such as energy-aware scheduling. This will help build trust in customers and increase their satisfaction level as a result of the services they receive.

Another aspect of this approach can also be examined in the future, focusing on the importance of customer safety. Therefore, instead of using serverless computing to verify and authenticate its users, it is possible to use blockchain technology to verify and authenticate customers. In turn, this can provide additional assurances to serverless service providers that the services are only available to users whose information is stored on the blockchain and that customers will not misuse the services provided by them for harmful purposes, thereby decreasing the intensity of the attack.

References

1. Biswas AR, Giaffreda R (2014) IoT and cloud convergence: opportunities and challenges. In: 2014 IEEE world forum on Internet of things (WF-IoT). IEEE, pp 375–376
2. Kelly JT, Campbell KL, Gong E, Scuffham P (2020) The Internet of things: impact and implications for health care delivery. J Med Internet Res 22(11):e20135
3. Sadeeq MM, Abdulkareem NM, Zeebaree SR, Ahmed DM, Sami AS, Zebari RR (2021) IoT and Cloud computing issues, challenges and opportunities: a review. Qubahan Acad J 1(2):1–7
4. Patil V, Al-Gaadi K, Biradar D, Rangaswamy M (2012) Internet of things (IoT) and cloud computing for agriculture: an overview. In: Proceedings of agro-informatics and precision agriculture (AIPA 2012), India, pp 292–296
5. Baldini I, Castro P, Chang K, Cheng P, Fink S, Ishakian V, Mitchell N, Muthusamy V, Rabbah R, Slominski A (2017) Serverless computing: current trends and open problems. In: Research advances in cloud computing. Springer, pp 1–20
6. Singh S, Chana I (2016) a survey on resource scheduling in cloud computing: issues and challenges. J Grid Comput 14(2):217–264
7. Castro P, Ishakian V, Muthusamy V, Slominski A (2017) Serverless programming (function as a service). In: IEEE 37th international conference on distributed computing systems (ICDCS). IEEE, pp 2658–2659
8. Lynn T, Rosati P, Lejeune A, Emeakaroha V (2017) A preliminary review of enterprise serverless cloud computing (function-as-a-service) platforms. In: IEEE international conference on cloud computing technology and science (CloudCom). IEEE, pp 162–169
9. Zheng Z, Xie S, Dai H-N, Chen X, Wang H (2018) Blockchain challenges and opportunities: a survey. Int J Web Grid Serv 14(4):352–375
10. Monrat AA, Schelén O, Andersson K (2019) A survey of blockchain from the perspectives of applications, challenges, and opportunities. IEEE Access 7:117134–117151
11. Li X, Jiang P, Chen T, Luo X, Wen Q (2020) A survey on the security of blockchain systems. Future Gen Comput Syst 107:841–853
12. Hassan HB, Barakat SA, Sarhan QI (2021) Survey on serverless computing. J Cloud Comput 10(1):1–29

13. Li Z, Guo L, Cheng J, Chen Q, He B, Guo M (2022) the serverless computing survey: a technical primer for design architecture. ACM Comput Surv (CSUR) 54(10s):1–34
14. Jonas E, Schleier-Smith J, Sreekanti V, Tsai C-C, Khandelwal A, Pu Q, Shankar V, Carreira J, Krauth K, Yadwadkar N (2019) Cloud programming simplified: a Berkeley view on serverless computing. arXiv preprint arXiv:1902.03383
15. Cassel GAS, Rodrigues VF, da Rosa Righi R, Bez MR, Nepomuceno AC, da Costa CA (2022) Serverless computing for Internet of things: a systematic literature review. Future Gen Comput Syst 128:299–316
16. Alqaryouti O, Siyam N (2018) Serverless computing and scheduling tasks on cloud: a review. Am Acad Sci Res J Eng Technol Sci 40(1):235–247
17. McGrath G, Brenner PR (2017) Serverless computing: design, implementation, and performance. In: IEEE 37th international conference on distributed computing systems workshops (ICDCSW). IEEE, pp 405–410
18. Kaffes, K., Yadwadkar, N. J. and Kozyrakis, C. (2019) Centralized core-granular scheduling for serverless functions. In: Proceedings of the ACM symposium on cloud computing, pp 158–164
19. Wu S, Tao Z, Fan H, Huang Z, Zhang X, Jin H, Yu C, Cao C (2022) Container lifecycle-aware scheduling for serverless computing. Softw Pract Exp 52(2):337–352
20. Stein M (2018) The serverless scheduling problem and noah. arXiv preprint arXiv:1809.06100
21. Tariq A, Pahl A, Nimmagadda S, Rozner E, Lanka S (2020) Sequoia: enabling quality-of-service in serverless computing. In: Proceedings of the 11th ACM symposium on cloud computing, pp 311–327
22. Pawlik M, Banach P, Malawski M (2019) Adaptation of workflow application scheduling algorithm to serverless infrastructure. In: European conference on parallel processing. Springer, pp 345–356
23. Zhao L, Yang Y, Li Y, Zhou X, Li K (2021) Understanding, predicting and scheduling serverless workloads under partial interference. In: Proceedings of the international conference for high performance computing, networking, storage and analysis, pp 1–15
24. Feng Q, He D, Zeadally S, Khan MK, Kumar N (2019) A survey on privacy protection in blockchain system. J Netw Comput Appl 126:45–58
25. Belotti M, Božić N, Pujolle G, Secci S (2019) A vademecum on blockchain technologies: when, which, and how. IEEE Commun Surv Tutor 21(4):3796–3838
26. Zhou Q, Huang H, Zheng Z, Bian J (2020) Solutions to scalability of blockchain: a survey. IEEE Access 8:16440–16455
27. Lepore C, Ceria M, Visconti A, Rao UP, Shah KA, Zanolini L (2020) A survey on blockchain consensus with a performance comparison of PoW, PoS and pure PoS. Mathematics 8(10):1782
28. Mingxiao D, Xiaofeng M, Zhe Z, Xiangwei W, Qijun C (2017) A review on consensus algorithm of blockchain. In: IEEE international conference on systems, man, and cybernetics (SMC). IEEE, pp 2567–2572
29. Nandi M, Bhattacharjee RK, Jha A, Barbhuiya FA (2020) A secured land registration framework on Blockchain. In: Third ISEA conference on security and privacy (ISEA-ISAP). IEEE, pp 130–138
30. Fox GC, Ishakian V, Muthusamy V, Slominski A (2017) Status of serverless computing and function-as-a-service (FaaS) in industry and research. arXiv preprint arXiv:1708.08028
31. Lloyd W, Ramesh S, Chinthalapati S, Ly L, Pallickara S (2018) Serverless computing: an investigation of factors influencing microservice performance. In: IEEE international conference on cloud engineering (IC2E). IEEE, pp 159–169
32. Nasir Q, Qasse IA, Abu Talib M, Nassif AB (2018) Performance analysis of hyperledger fabric platforms. Secur Commun Netw
33. Androulaki E, Barger A, Bortnikov V, Cachin C, Christidis K, De Caro A, Enyeart D, Ferris C, Laventman G, Manevich Y (2018) Hyperledger fabric: a distributed operating system for permissioned blockchains. In: Proceedings of the thirteenth EuroSys conference, pp 1–15
34. Thakkar P, Nathan S, Viswanathan B (2018) Performance benchmarking and optimizing hyperledger fabric blockchain platform. In: IEEE 26th international symposium on modeling, analysis, and simulation of computer and telecommunication systems (MASCOTS). IEEE, pp 264–276

35. Brotsis S, Kolokotronis N, Limniotis K, Bendiab G, Shiaeles S (2020) On the security and privacy of hyperledger fabric: challenges and open issues. In: 2020 IEEE world congress on services (SERVICES). IEEE, pp 197–204

36. Cachin C (2016) Architecture of the hyperledger blockchain fabric. Workshop on distributed cryptocurrencies and consensus ledgers, Chicago, IL, pp 1–4

37. Gorenflo C, Lee S, Golab L, Keshav S (2020) FastFabric: scaling hyperledger fabric to 20000 transactions per second. Int J Network Manage 30(5):e2099

38. Li D, Wong WE, Guo J (2020) A survey on blockchain for enterprise using hyperledger fabric and composer. In: 6th international conference on dependable systems and their applications (DSA). IEEE, pp 71–80

39. Elghaish F, Rahimian FP, Hosseini MR, Edwards D, Shelbourn M (2022) financial management of construction projects: hyperledger fabric and chaincode solutions. Autom Constr 137:104185

40. Sharma A, Schuhknecht FM, Agrawal D, Dittrich J (2018) How to databasify a blockchain: the case of hyperledger fabric. arXiv preprint arXiv: 1810.13177

41. Ammi M, Alarabi S, Benkhelifa E (2021) Customized blockchain-based architecture for secure smart home for lightweight IoT. Inf Process Manage 58(3):102482

42. Pérez A, Moltó G, Caballer M, Calatrava A (2018) Serverless computing for container-based architectures. Futur Gener Comput Syst 83:50–59

43. Sewak M, Singh S (2018) Winning in the era of serverless computing and function as a service. In: 2018 3rd international conference for convergence in technology (I2CT). IEEE, pp 1–5

44. Rajan AP (2020) A review on serverless architectures-function as a service (FaaS) in cloud computing. TELKOMNIKA Telecommun Comput Electron Control 18(1):530–537

45. Anderson TE, Dahlin MD, Neefe JM, Patterson DA, Roselli DS, Wang RY (1996) Serverless network file systems. ACM Trans Comput Syst (TOCS) 14(1):41–79

46. Shafiei H, Khonsari A, Mousavi P (2019) Serverless computing: a survey of opportunities, challenges, and applications. In: ACM computing surveys (CSUR)

47. Taibi D, Spillner J, Wawruch K (2020) Serverless computing-where are we now, and where are we heading? IEEE Softw 38(1):25–31

48. Maissen P, Felber P, Kropf P, Schiavoni V (2020) FaaSdom: a benchmark suite for serverless computing. In: Proceedings of the 14th ACM international conference on distributed and event-based systems, pp 73–84

49. Van Eyk E, Iosup A, Seif S, Thömmes M (2017) The SPEC cloud group's research vision on FaaS and serverless architectures. In: Proceedings of the 2nd international workshop on serverless computing, pp 1–4

50. Zhou Z, Yu S, Chen W, Chen X (2020) CE-IoT: cost-effective cloud-edge resource provisioning for heterogeneous IoT applications. IEEE Internet Things J 7(9):8600–8614

51. Gupta A, Siddiqui ST, Alam S, Shuaib M (2019) Cloud computing security using blockchain. J Emerg Technol Innov Res (JETIR) 6(6):791–794

52. Das A, Leaf A, Varela CA, Patterson S (2020) Skedulix: hybrid cloud scheduling for cost-efficient execution of serverless applications. In: 2020 IEEE 13th international conference on cloud computing (CLOUD). IEEE, pp 609–618

53. Islam N, Ebrahimzadeh S, Salameh J-P, Kazi S, Fabiano N, Treanor L, Absi M, Hallgrimson Z, Leeflang MM, Hooft L (2021) Thoracic imaging tests for the diagnosis of COVID-19. Cochrane Database Syst Rev 3

54. Jansen T, Lee CM, Xu S, Silverstein NM, Dugan E (2022) Peer reviewed: the town-level prevalence of chronic lung conditions and death from COVID-19 among older adults in Connecticut and Rhode Island. Prev Chronic Dis 19

55. Shi Y, Wang G, Cai X-P, Deng J-W, Zheng L, Zhu H-H, Zheng M, Yang B, Chen Z (2020) An overview of COVID-19. J Zhejiang Univ Sci B 21(5):343–360

Cost Control and Efficiency Optimization in Maintainability Implementation of Wireless Sensor Networks Based on Serverless Computing

Tinanan Gao and Minxian Xu

Abstract Wireless sensor network (WSN) has been developed for decades and have performed well in the performance, power consumption, and congestion control. However, the following problems have not been addressed, such as inaccurate cost estimation of device's lifecycle, highly-coupled engineering development, and low utilization of hardware and software resources during the life cycle of WSN. Therefore, we first propose the conceptual view of maintainability implementation for WSN based on Serverless Computing. The maintainability implementation refers to the ability to meet the WSN product to consume the minimum resources with a higher probability in configuration, trial production, debugging, batch production, deployment, operation, and maintenance phases. And then, we discuss that Serverless Computing can be realized at the software functional level of WSN to decouple the device operation and functional development, greatly improve the reuse of resources and exclude the hardware interference. From the perspective of maintainability and cost control, the concept of Serverless Computing can be used to build WSN platforms, which can support the functions of data collection and data management into functional development that may benefit from exploration through upfront expenditures, thereby significantly reducing design, manufacturing, and operational costs. Finally, based on existing technologies and smart city scenarios, the idea of a WSN platform for Serverless Computing is given with a case study.

Keywords Wireless sensor network · Maintainability · Serverless computing · Resource optimization

T. Gao (✉) · M. Xu
Shenzhen Institute of Advanced Technology, Chinese Academy of Sciences, Shenzhen, China

University of Chinese Academy of Sciences, Beijing, China
e-mail: tn.gao@siat.ac.cn

M. Xu
e-mail: mx.xu@siat.ac.cn

1 Introduction

Wireless sensor network is a paradigm composed of a large number of stationary or mobile sensors in a self-organizing and multi-hop manner to collaboratively sense, collect, process and transmit information about sensed objects in the geographic area covered by the network and eventually send this information to the owner of the network. Its emergence was based on the industrial need for large-scale, high-efficiency, low-cost control and acquisition and the continued technological advances in radio frequency (RF), processors, nanotechnology and microelectromechanical systems (MEMS) [1]. It has been widely used in industrial monitoring, high-volume intelligent meter reading, parking lot space monitoring and other scenarios, which shows the maturity of WSN paradigm.

Recent research hotspots in this area are centered on media access control (MAC) layer communication protocols, functional boundaries, energy consumption control, security and forensics. The communication protocols of WSN have been well studied. Zimmerling et al. [2] have summarized more than 40 recently used synchronous MAC layer protocols, and the protocols were divided into nine categories according to one, many, all, and two-to-two correspondence (e.g., one-to-many, many-to-many, etc.), and the performance of the protocols in their categories was evaluated. Liu et al. [3] summarized the future 5G environment with multiple input and multiple output (MIMO) antenna conditions, Integrated Sensing and Communication (ISAC) system can be summarized as device-free sensing, device-based sensing, device-free ISAC and device-based ISAC, four kinds of architecture. It is also concluded that the fundamental limitation of the ISAC channel cannot be obtained by employing a simple combination of existing reach performance boundary techniques in sensing and communication systems alone. In terms of energy consumption control, Osamah Ibrahim Khalaf et al. [4] summarized in the paper that some nodes have lower lifetime than other nodes. In the WSN structure of star topology, cluster head nodes, nodes near sink gateway, and multi-hop nodes with high frequency transmission have aggregated low lifetime due to higher operating frequency relative to other nodes, called hotspot issue. And then they summarize historical proposed solutions to mitigate the problem and provide a platform for researchers exploring solutions to these topical problems in the network and proposing novel solutions. In terms of Internet of Things (IoTs) security and forensics, Stoyanova et al. [5] summarize that at this stage, IoT devices such as WSN have evolved to the point where vendors need to provide compliant forensic system designs for their devices, both at the legal level and in terms of technical means. Therefore WSN design solutions that meet the needs of legal forensics may also be an opportunity and challenge for the future.

In engineering applications, WSN has been used in home automation, environmental monitoring, various control systems and other application scenarios. In terms of home automation scenarios, the market is very mature for self-organized communication among various smart appliances, wearable devices and other IoT devices, and mesh networking and smart interconnection among devices have been widely used. In terms of environmental detection, WSN is commonly used as a monitoring execution unit for forest fires, hydrological monitoring, and etc. Taking the production environment in the rolling mill as an example, the application scenario in which its control system is located has the following characteristics: construction workers face high safety risks, high costs of temporary stops during operation, and closely linked production links. Therefore, the sensor information transmission required by its control system should meet the safety production specifications of large-scale collaborative production lines, timely information transmission, and reliable transmission methods. With the low cost of WSN and low coupling with the production environment, stable control and continuous monitoring of the production line can be achieved.

WSNs have following common features. First, the WSN system design is highly targeted, and additional requirements need to be re-established for the next life cycle. Second, the cost of deployment and maintenance is high, therefore, in engineering practices, the developers often use redundant deployment instead of subsequent maintenance to reduce costs. These characteristics guide the implementation strategy often results in WSN material resource waste, channel interference and environmental pollution problems.

In the future direction of WSN, many future directions and challenges [6] have been identified, such as soft computing (edge computing), device security, heterogeneous interoperability, self-organizing protocols, routing schemes for managing IoT networks, data management, etc. These issues are essentially trade-offs between WSN functionality and actual lifetime versus cost and resources. The idea proposed in this paper is to synthesize the above challenges and problems of WSN and to provide an approach to visualize the methods and frameworks into engineering implementation plans based on the concept of *Serverless Computing* in the face of the maintainability assessment methods and decision frameworks that have been constructed in the WSN lifecycle, and then verify whether the expected benefits of maintainability are achieved through engineering practice or not.

In this paper, we illustrate the concept of maintainability in WSN and try to quantify and analyze the links of their lifecycle to achieve controllable expenditures in each stage of their lifecycle in order to improve the expected benefits. And then we use the concepts of Serverless Computing for user participation and control granularity segmentation to build an engineering implementation method to achieve WSN maintainability. The method is, based on the current stage of embedded devices per unit price of performance improvement, cost per unit area of power consumption decline, explore the resulting expected revenue space if it can be achieved with a greater probability, it allows to improve the design phase of expenditure on the use of the system's resources (control, acquisition, computing, WSN reconfigurable architecture, etc.) programmable design to achieve resource-based sale strategy division. The efficient use of WSN resources in the loop can reduce the coupling between the operation and optimize the maintenance phase and the production environment, amortize the cost with pre-taken increased revenue, and achieve true environmental protection through the efficient use of WSN system.

We propose the maintainability implementation of WSN based on the challenges and problems as above and study the strategies for cost resource allocation throughout their lifecycle. In this paper, we first introduce the definition, characteristics, and existing challenges of WSN maintainability implementation in Sect. 2 and the introduction of Serverless Computing in Sect. 3. In Sect. 4, by combining the existing hardware and software technologies, we illustrate the idea of increasing the equipment utilization rate to amortize the research and development (R&D) cost based on Serverless Computing, and propose a feasible model for the whole process of maintainability under the existing technology framework. In addition, the full-flow application is demonstrated in a smart city scenario. Finally, the conclusion and future work are summarized in Sect. 5.

2   WSN Maintainability Implementation

At this stage, the configuration, trial production, debugging, batch production, deployment, operation, and maintenance processes of WSN are still dominated by project requirements, and the coupling in the whole process is so high that the resource utilization in the overall life cycle of WSN is not high and the costs are difficult to accurately predict and control. In order to predict the cost in a certain confidence interval before project initiation, guide the decision making in project practice, reduce the coupling between process links, improve the reusability of resource spending, and provide experience for the next project, we propose the concept of maintainability in WSN. This section covers a practical engineering approach to maintainability called WSN maintainability implementation.

2.1   Definition

WSN maintainability implementation satisfies a prerequisite, which aims to meet a ten-year life expectation of the WSN product throughout WSN maintainability implementation phases conversion. Its phases include configuration, trail production, debugging, batch production, deployment, operation, and maintenance. The phase conversion decision is based on a trade-off between the confidence in the net benefit and the confidence in the successful implementation of the program. The net benefit refers to the expected benefit of the corresponding solution in each phase subtracting the expected cost and resource consumption. Successful implementation means that the implementation of the phase is completed exactly as scheduled and the post-implementation metrics are met, e.g., the WSN product debugging low power current reaches 80% of the chip datasheet within 7 days of the debugging phase. Confidence refers to the probability of achieving the desired goal for both in each phase. We usually expect the cost and resource consumption in each phase to be as low as possible among all possible solutions, and the expected benefits to be as high as possible, and the implementation to be executed successfully, and the confidence level of the above factors to be as high as possible. In practice, however, the tendency of the confidence level of both is often chosen according to the actual needs. The cost refers to the money spent, the resources refer to the manpower, material resources, information channels that can be mobilized in the project process, and there is an intersection between the two. Finally, the resources that can be converted to cost participate in the statistics at the corresponding cost.

2.2 Life Cycle Phase

Based on the tasks and characteristics of each phase, we divided the WSN maintainability implementation life cycle into seven parts.

2.2.1 Configuration

Configuration refers to the analysis of its needs after the project is established, based on previous engineering experience or feedback information resources from other phases of this project, the expected assessment of costs and resources, taking into account the factors of the specific environment, weighing the cost, reliability, upgrade redundancy, technology selection (the cost of technical debt for later upgrades) expenditures, the results of other phases, and repeatedly design the hardware and software options for WSN products. The estimated cost expenditure has a high confidence level. Reliability, upgrade redundancy, and technology selection are the factors that affect the cost. The design process should consider the balance between actual needs and costs, such as high circuit reliability requirements for applications, which should be made in the design phase of circuit redundancy design, up the material cost budget. Upgrade redundancy should consider the hardware needs in ten years can be, should not be too high. Technology selection should also consider the trade-off between reliability and upgrade redundancy, as well as the advantage of increased redundancy due to technology upgrades over a ten-year cycle versus the cost of additional expenditures.

2.2.2 Trial Production

After completing the configuration phase, small-scale trial production in accordance with the design is used as the basis for entering the debug phase after the verification of good products. Trial production is often used to verify the gap between the actual effect of the circuit and the simulation results, and to troubleshoot initial failures, such as Cooja Simulator [7] is the developing and debugging application of Contiki OS, which is an operating system for the Internet of Things. Essentially it's a smaller cost outlay to avoid as much as possible the larger expense of scaling up errors in the mass production phase. For example, small batches are made for each option while the product is tested in the lab for electrical stability.

2.2.3 Debugging

After the electrical performance of the product is verified, the circuit operating parameters are tuned for the datasheet indicators as a benchmark. The essence of the same is to get more guidance information in this project at a smaller cost and resource expenditure. For example, the center frequency offset of the radio frequency (RF) chip, the received signal strength indication (RSSI) can reach the dB indicator of the datasheet. Whether the low-power current of the control part can reach the minimum standard of the datasheet. If not, what is the maximum low-power current that can meet the runtime, and what is the corresponding cost overhead for each reduced current.

2.2.4 Batch Production

After the first three phases are largely completed or the unexpected costs of the later phases are within certain expectations, trade-offs are made between alternative designs, mass production vendors, materials, yields, redundant production, and costs. Based on the trade-offs, batch production is performed. Yield testing and lab environment validation of all products are performed after batch production. In case of large scale problems go back to the configuration phase at a significant cost. For example, radio frequency (RF) modules are to be shielded in places with strong electromagnetic interference. For projects with high maintenance costs in the field, the circuit design should be a highly reliable and slightly more costly solution in exchange for a higher probability of controlling the cost of the maintenance phase at a lower level.

2.2.5 Deployment

After passing mass production verification, package the device to field deployment testing. The process takes full consideration of the actual environment, utilizing the redundant design from the previous phase and prioritizing trying low-cost solutions. For example, we prioritize measuring the communication environment before deployment and switch to the channel with the least interference. For the problem of communication failure in some special locations, priority is given to trying to move the location to exclude the possibility of a multi-path effect.

2.2.6   Operation

After deployment, it works properly, continuously monitors the operational data and analyzes the gap between the actual behavior of WSN packets and the expected behavior. There also needs a corresponding maintenance plan for possible failures. For example, Guo et al. [8] designed a listening method for a WSN protocol based on Dong et al. [9] implemented a packet behavior collection analysis based on the listening method.

2.2.7   Maintenance

Target problem and requirement improvements based on actual packet behavior versus expected gaps until the end of the 10-year life cycle. Add new requirements within hardware cost constraints. Ongoing lessons learned, configuration, debugging, and production phase work can be reused and reduced by subsequent projects, and deployment and operation phases to maximize equipment utilization.

2.3   Resources and Costs

Resources and costs as sufficient conditions for driving links in the WSN life cycle. We give the scope of resources and costs as follows.

2.3.1   Resources

Resources are objectively existing substances that are directly used in the WSN life cycle. Resources include, but are not limited to, information, manpower, and supplier channels. There is a mutual influence relationship between them and costs. For example, information affects the proportion of cost allocated to resources, and cost affects supplier channels. Some of these resources can eventually be counted as cost.

2.3.2   Cost

Cost refers to the time and money spent during the project cycle, and is used as an indicator to evaluate the final maintainability design. Cost control is achieved mainly through the ideas of increasing revenue and cost-cutting. Cost-cutting refers to reducing the budgeted cost of the design under the requirement of stability and reliability, while increasing revenue refers to improving the utilization of resources that have already been paid for. Serverless Computing is an implementation of the increasing revenue idea.

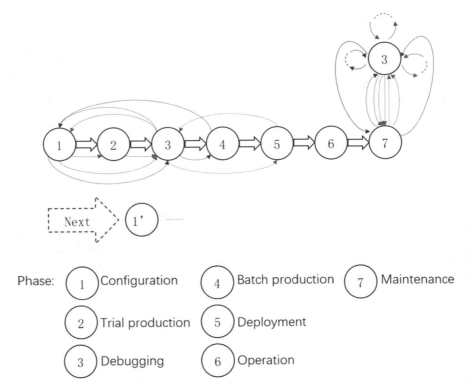

Fig. 1 Life cycle stage flowchart

2.4 Processes and Features

Figure 1 shows a flow chart of the WSN maintainability implementation life cycle stages. The link state body in the life cycle goes as shown by the large arrows. Serial numbers 1–7 represent the seven parts of the maintainability implementation life cycle, each link will incur time costs, small arrows for the failure decision processing feedback links, will generate decisions corresponding to the expected resource consumption, the color along the chromatogram from green to red represents the cost from low to high. The study of maintainability is to study ways to reduce costs in the process of changing the state of this WSN project over a ten-year period. The essence is a resource allocation strategy with a high degree of confidence based on certain a priori information (previous engineering experience, technology trends), combined with the state of the project process.

Take the life cycle of a large-scale meter acquisition WSN system as an example, in the configuration stage, design WSN architecture, MAC layer protocol, hardware selection, inter-chip interface, program architecture, etc. And then along the big arrow of the release into the trial production session, according to the previous design of a small number of purchase materials, Circuit board engraving, SMD, test electrical performance, and RF indicators, estimate the yield rate. If the instability has been ruled out as far as possible (the yield rate is reduced in mass production resulting in a large waste of resources and costs), along the big arrow into the debugging stage, according to the design of the architecture to develop the function, according to low-power indicators to test the low-power current. If a problem is found, such as a capacitor indicator cannot reach the data variance distribution of its data sheet so that it cannot meet the electrical performance of WSN products, at this time, through a certain resource cost (set to the second level corresponding to the yellow path according to the estimated scale, a total of four levels) consumption, return to the design session to choose another program or redesign the material selection, and return to the debugging stage to replace the capacitor of the previous finished product material, and conduct the corresponding index test. If there are key components that can not meet the demand so much as to weigh the later cost and redesign debugging production, the latter is lower, such as a material because of force majeure factors can not meet the demand for mass production, then you can take a third level of resource cost expenditure (orange) and then go to the design debugging trial production link. If the trial production stage targets are completed and possible trial and error overheads that are scaled up in the next stage are excluded as much as possible, enter the mass production session.

We need to be very careful before entering the mass production session, because mass production is an important part of the WSN to amortize costs and maybe a missed session that amplifies errors that bring a lot of invalid expenditures. There-fore, the process expenditure for its failure will be level 4 (red). After successful batch production, enter the deployment link, adjust the performance of WSN to the expected functional indicators (packet loss rate, RSSI, and etc.). In the deployment stage, we install WSN devices on demand in production environments based on pre-existing multi-format, redundant production, so their expenses are often first-tier, so its expenditure is often the first level. And then it is the continuous operation of WSN monitoring and WSN maintenance analysis, troubleshooting, analysis requirements, and upgrade iterations. Switching between maintenance and other stages can be done sequentially according to the failure level and later upgrade requirements and is not limited by the expenditure level. Until the completion of this life cycle after ten years, the lessons learned and reusable resources into the next life cycle.

2.5   Current Situation and Challenges

This subsection summarizes the historical research on maintainability, as well as the work that has been done in the WSN maintainability system, and envisions the challenges that may be faced in future work.

2.5.1 System

Maintainability has been extensively studied in the fields of software engineering [10], mechanical structures [11], and industrial engineering [12]. Maintainability as one of the quality attributes of software engineering, Alsolai et al. [10] surveyed 56 relevant studies from 35 journals and 21 conferences and showed that the most commonly used software metrics (dependent variables) in the selected primary studies were variable maintenance effort and maintainability index, and most of the studies used class-level product metrics as independent variables. Maintainability assessment is the main way to evaluate the maintainability of mechanical product design phase. In order to describe the maintainability of mechanical products, Luo et al. [11] proposed a basic assessment object hierarchy and quantitative assessment methods for maintainability design attributes to improve the accuracy and reliability of assessment results. Thus, the accuracy and reliability of the assessment results are improved thus avoiding the weaknesses of the maintainability design. However, this method does not give improvement measures for maintainability weaknesses. Ahmadi et al. [12] divided the performance of the metro construction material transportation system into reliability, maintainability, and availability (RAM) evaluated separately in terms of time as a unit of measurement. The reliability was analyzed to derive the maximum life time of the system and the availability was analyzed and the time of this attribute was the maximum life time of the system for the part of the reliability above the predefined index. To maintain the system above a certain level of reliability, it is necessary to perform. To maintain the system above a certain level of reliability, maintainability maintenance needs to be performed, and this process also takes time and is called preventive maintenance interval.

The maintainability evaluation system of WSN is still in the exploration stage as a brand-new idea. Among them, Zhang et al. summarized the top-level design and architecture of maintainability [13]. Shen et al. [14] proposed a preliminary evaluation scheme based on subjective expert experience, hierarchical analysis method, and fuzzy comprehensive evaluation. Gao et al. [15] applied an objective sample entropy weight method based on Shen's work to calculate the operational parameters of the project implementation guided by expert opinions in his paper and used it as a basis to update expert experience after the project was completed. Qiu et al. [16] used confidence rule base for evaluation. Wang [17] used expert experience combined with EM algorithm and Bayesian network for evaluation. Future resource cost modeling analysis can be performed for the WSN life cycle conducted in parallel.

2.5.2 Methodology and Engineering Practice

The next step in the development of WSN maintainability adopts an alternating iterative approach of theory and experience. The expert experience summarizes the theory, the theory guides the engineering practice, and the experience is summarized from the engineering practice results and then improves the theory.

WSN maintainability from the time dimension to address two issues:

- How to evaluate the links of this life cycle?
- The data and resources of this life cycle how to guide the next life cycle improvements?

The ultimate goal of solving the problem is to achieve rapid convergence of the corresponding cost to the minimum as the life cycle is rapidly iterated. The way of cost convergence in engineering solutions is still to be explored, and the combination of WSN and Serverless Computing in this paper is an attempt.

The solutions to some problems in maintainability can be based on results from other fields, and in the future can be combined with theories such as optimal control, machine learning, and probability theory to participate in the construction of methodological models. For example, link cost resource consumption and decision process modeling can be attempted in the future using unsupervised learning algorithms to divide attributes and links of factors and explore the contribution of the earlier factors to the later factors at each stage [18], and then transform the machine learning model into an equivalent neural network with layer-wise-relevance propagation (LRP) for the link parameters in which to explain the contribution of the inter-factor effects.

In terms of engineering practice, there is still a long way to go, but it is possible to make bold assumptions by absorbing ideas from other engineering fields. For example, this paper combines the concept of Serverless Computing and tries to treat WSN as the execution unit of Serverless Computing, and sells it as a distributed system resource by dividing its functions into fine-grained and low-latency operations, thus expanding the revenue space to amortize the cost and increase the revenue.

3 Serverless Computing

At this stage, the embedded chip performance has increased significantly and the price has decreased significantly, WSN system design level can be relaxed to consider the performance constraints. In particular, the development of embedded Linux allows the introduction of many excellent design ideas into WSNs, Serverless Computing being one of them. Serverless Computing as, decoupling the development of Serverless Computing application functions from system maintenance abstraction in its WSN domain that is. The following describes the Serverless Computing concepts, WSN maintainability in the use of the features.

3.1 Concept and Application

Serverless Computing [19] is an integration of both BaaS and FaaS as a unified service form to its customers. From the perspective of a functional developer it can be seen as Function as a Service (FaaS), and from the perspective of a Serverless Computing provider it can be seen as Backend as a service (BaaS) [20]. Serverless

Computing allows developers to focus on the development of functional functions without having to focus on the maintenance of the underlying runtime environment. For Serverless Computing vendors, Serverless Computing reduces hardware idle time to improve equipment utilization and achieve greater profitability.

Serverless Computing has been proposed to address the high coupling high cost and inefficiency of the traditional development model lifecycle [21]. Mondal et al. [22] redefined the Serverless Computing paradigm from an administrator's perspective and integrated it with Kubernetes to accelerate the development of software applications. Theoretical knowledge and experimental evaluations suggest that this new approach can enable developers to design software architectures and development more efficiently by minimizing the cost of maintaining hardware facilities for public cloud providers (e.g., AWS, GCP, Azure). However, serverless functions come with a number of issues such as security threats, cold start problems, and inadequate debugging of functions.

In the WSN space, it can be used as an implementation to improve maintainability and better amortize costs by expanding unit hardware utilization.

3.2 Serverless Computing Used in WSN's Maintainability

The process of Serverless Computing consists of two main aspects, function programming and function services. This idea can guide the development model of WSN maintainability design phase, where users only need to call function services to get the required resources, and the development center of WSN developers shifts from implementing a single function to coordinating the front-end and back-end of the system to implement a function programming environment. The implementation of maintainability may borrow many of its features. Therefore, this subsection summarizes some of the features of Serverless Computing that are used in the WSN maintainability development process and explains how to apply them.

- Hostless and elastic: Users do not need to know the specific implementation, operation and maintenance, and troubleshooting of WSN. You can write your own rules to use the acquisition, control, and edge computing resources according to the platform specifications as needed.
- Lightweight: WSN's hardware resource preferences are highly tied to functionality, facilitating fine-grained segmentation and middleware design. For example, edge computing tasks are done by cortex-A performance cores for statistics, and data collection and device control tasks are collected by cortex-M control cores in accordance with the maximum common multiple of multi-user collection frequency capabilities. This feature also allows users to focus on functional logic when writing requirements. However, due to the development of Serverless Computing at this stage, certain restrictions may be imposed on the user writing specification. Such as the number of functions, the number of loop layers, etc.

- Short but variable execution times: In the WSN device is reflected as the response of the ad hoc network instructions to the upper user functions, i.e., as the interpretation language of the functions. At the same time, the instructions should be units that can be billed for metrics, and the embedded device can be billed for the number of acquisitions.
- Burstiness: The task load in Serverless Computing is often difficult to predict and is characterized by burstiness, variable change time, and fluctuating intensity. For the characteristics of WSN distributed, the dynamic design of its topology can be developed in coordination with the front-end serverless service.
- Migration: The automatic scaling of Serverless Computing platforms often comes with the startup cost of a serviceless function, i.e., the system overhead and service latency can be elevated to the point where inter-system migration becomes unattractive. Resource provisioning at the WSN level has a slightly higher energy overhead on the system due to the low-power requirements of energy-constrained devices, but its lightweight firmware makes migration costs low, and ultimately there is still a tradeoff between the expected benefits of migration and the system resource overhead.

4 Implementation Conception

For the needs of multiple parameter acquisition and device control in smart cities, we first introduce the common scheme design patterns and problems, and then elaborate on the idea of increasing the usage amortization cost by abstracting the functions achieved by Serverless Computing in our maintainability design.

4.1 Existing Schemes and Problems

According to some commonly used technology of previous experiments, performance, energy, and cost were limited, and developers had to remove codes to reduce functionality and cost expenditures from all aspects. Each unit will be its own requirements for separate bids, the common development process of the winning bidder is as follows. In the design phase, they develop WSN for controlling a single device or collecting a single parameter. After debugging and production, at the time of deployment, the WSN products usually occupy a common space with the products of other projects. As shown in Fig. 2, it is photographed when we evaluated the deployment environment of our WSN product in a tunnel pipe gallery. There are many parameter acquisition equipments deployed in parallel. The overall vertical division of urban wastes public resources, increases system confusion, and leaves security risks.

Fig. 2 Multiple project
departments are responsible
for the data acquisition
system installation process,
which increases the
unreliable factors in the
industrial environment

4.1.1 Requirement Analysis

In smart city scenarios, there are many control, acquisition, and edge computing needs. For example, the traffic department's timely response to target detection, dynamic control of tidal lane driving direction, and traffic light timing. Environmental protection department for urban environment monitoring, scientific researchers to obtain the city's sound, light, heat, electromagnetic atmospheric pressure, and other physical parameters and other demand scenarios. We integrate these requirements in the same WSN system, and hardware along with software resources can be configured according to fine-grained demands. To establish a unified project approach to fulfill the development of WSN data acquisition and equipment control function service platform, the system can meet the future increase, modification, and deletion needs. Such as the scheme that is not proposed in the tender but may form a profit model in the future with the help of redundant design.

4.1.2 Hierarchical Architecture Design

As shown in Fig. 3, WSN in the resource perspective has the minimum granularity of data collection, device control, edge computing functions. From the perspective of network structure, its minimum granularity for the system's cluster structure can be reorganized on demand. From the hardware selection point of view, its minimum granularity can be achieved by selecting pin-to-pin compatible chips, by selecting resistors and circuit boards to leave redundant material pads. The designed WSN architecture is based on a variant of a star network (consisting of sink gateways,

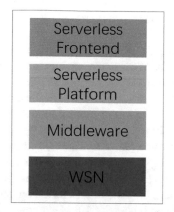

Fig. 3 Layered architecture of smart city WSN system

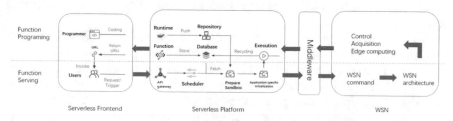

Fig. 4 Hierarchical architecture of smart city system based on serverless computing framework

routers, and nodes), considering the platform requirements for function serviceability and the upgrade redundancy to meet the 10-year life expectancy. The middleware acts as a sandbox for the execution of user-developed functions, responsible for both translating the functions into WSN operations (paid resources disassembled into permutations of WSN minimum granularity operations) and returning the corresponding results to the Serverless platform, which is responsible for user management, resource uploads, account billing, user-developed function property management, etc. Serverless frontend is for users to write and upload resource call functions, view resource call results, and record resource usage billing prices.

4.1.3 Development Process Design

The overall architectural design details are shown in Fig. 4, which is derived from [19]. The specific implementation details of each part in the life cycle will be described from Sects. 4.1.4–4.1.10.

4.1.4 System Configurations

The modular design can meet the needs of trade-offs between different expected benefits, costs, and functions. The WSN hardware unit is shown in Fig. 5. The transceiver refers to the RF chip and its supporting peripheral circuits, which serve as the infrastructure of MAC protocol and are responsible for the communication between self-assembled network devices. The micro-control module is divided into single micro control unit (MCU) and micro process unit (MPU), where the sensing module is responsible for acquisition and control functions, and the power management module designs the corresponding circuit according to whether the energy is constrained or not. In constrained devices (e.g. WSN nodes), it is responsible for low-power control. In unconstrained devices (WSN routers and sink gateways), it is responsible for power outage switching backup power, battery side flush and other functions. Feedback speed of hardware resources, communication between hardware can achieve microsecond level completion speed to meet the needs of Serverless microsecond level tasks.

WSN software design focuses on the hardware abstraction layer, driver layer, real-time operating system layer (RTOS), and application layer. The hardware abstraction layer is a circuit-level abstraction of the hardware unit in the software architecture, defined by Yoo et al. [23], and is intended to solve the error-dead problem when the operating system directly manipulates the chip peripherals, which isolates the system parameter configuration during development from the specific model of peripherals and encapsulates the register behavior of the hardware as a separate function, also known as the board-level support package (BSP). This is to facilitate later development of system functions for specific hardware, i.e., reuse of resources.

The driver layer, as an intermediate link between the operating system layer and the hardware layer, requires efficient and stable code to realize the interactive functions between the upper and lower layers. The operating system layer is often used to make the logic of complex tasks through abstraction and then concise implementation, which also need the efficient execution of program functions. Therefore, the C programming language has become the least used language in the history of embedded development, and the C programming language is a mature technology with rich framework code and error handling. But in the scale of ten years, the project often occurs in human resources changes (engineers level inconsistent, engineering details handover incomplete), tight schedule and other problems. Rust, as a result of recent decades of programming language research, has inherent runtime security. It has a comprehensive package management mechanism, and a secure development process in which code that is not securely specified cannot be checked by the compiler. And it follows a zero abstraction overhead enabling developers to securely control the underlying capabilities. Therefore, we boldly suggest that you can try Rust&C hybrid development in both layers.

The open source, widespread use of Linux systems has enabled the rapid migration and deployment of existing cutting-edge engineering technologies. The increased resource density of microcontroller chips has also enabled complex operating systems like Linux to be applied in embedded devices. Therefore, in our design, arithmetic,

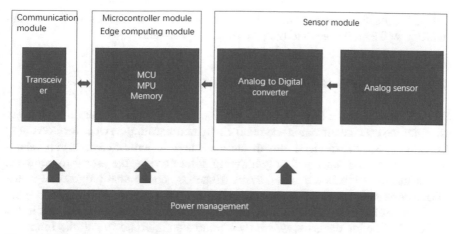

Fig. 5 Hardware architecture for WSN

energy-unconstrained devices are implemented using a mixture of Linux system and RTOS, the former is responsible for arithmetic, business logic, and the latter is responsible for WSN protocol stack, sensor acquisition, and peripheral control. Resource-constrained devices, energy-unconstrained devices use RTOS to implement the sink-gateway routing node function of WSN, and energy-unrestricted devices also use RTOS to implement the node and sensor acquisition function of WSN.

WSN application layer running Serverless Computing middleware backend, i.e., WSN devices with fine-grained control commands, including but not limited to self-organizing network topology adjustment, control, acquisition, and edge computing resource invocation. Users can subscribe on-demand and use on-demand in accordance with the function development.

4.1.5 Middleware Configurations

The middleware runs as a service on the service cluster to realize the conversion of user functions and WSN instructions and the uploading of WSN execution results. Instruction conversion means functions that are involved by multiple users, disassembled into efficient execution of WSN instructions, upload refers to the execution results in accordance with the user function resource calls to split and upload. For example, two user subjects need WSN collection resources to collect air pollutant density, the former according to the frequency of 5 min to 100 nodes of data collection once, the latter according to 10 min to collect 100 nodes, but need to be above the threshold to trigger an alarm and upload data. After aggregating the two functions that need to be executed, the middleware will merge the commands for their execution and send the instructions to WSN for 100 nodes running at a frequency of once every 5 min and set the threshold value to trigger sending, which is the minimum resource consumption for executing all users' commands. The middleware will then upload the results on demand according to its needs, and the Serverless Computing platform will be billed according to the resources invoked.

4.1.6 Serverless Platform Configuration

Serviceless computing is divided into platform and frontend, where the platform side runs the user identification and platform dynamic scaling services for Serverless Computing, and the frontend runs the services used to write functions and visualize results. From a functional programming point of view, the user (programmer) writes functions (cycle execution, event triggering, conditional judgments) in the local environment provided by the frontend (possibly a page in a browser for WSN users) according to specifications we define, and the functions are submitted to the platform's function database, which then returns an interface to call the functions The function is then submitted to the platform's function database, which returns an interface to call the function. From the function service perspective, when a user triggers or invokes a function through the interface, the platform obtains the function from the function database and performs code compliance and security checks, and then sends its instructions to the middleware platform, which aggregates function functions from multiple users and then merges the requirements into one category, breaking them down into WSN instructions in the most efficient way possible.

Considering the design scheme as above, according to the demand analysis of Sect. 4.2, we plan to use 10 energy-unrestricted edge computing WSN hardware responsible for target detection, tidal lane traffic light time dynamic control condition judgment, 40 energy-unrestricted devices (gateway-sink nodes-routing) responsible for, 100 energy-restricted low-power devices to achieve coverage of a small-scale city's major traffic Serverless Computing platform and middleware are deployed as software services on a mature cloud computing platform or on self-built servers.

4.1.7 Trial Production

The pilot production phase follows multiple design options to create separate hardware, purchase servers, and develop a Serverless Computing platform. The purpose is to verify whether the design is feasible, and to stop it in time if it is not. The trial production should fully consider whether the required materials are available and stable in the current market environment, and whether the performance meets its datasheet nominal specifications. It is also necessary to evaluate the tradeoff between the cost and benefit of alternative models, the tradeoff between redundant designs and the benefit of future WSN implementation functions, and the tradeoff between equipment life and maintenance over a ten-year life cycle. Platform development considers the tradeoff between the cost of running the platform and performance redundancy, and the tradeoff between the cost of dynamic platform deflation control and the efficiency of system operation. For example, whether the designed post-paralysis instructions reach the balance interval after the trade-off of resources, cost, and stability. Whether the designed WSN key components meet the data book metrics, i.e., whether they meet the balance between expected life and maintenance costs. In case of unanticipated conditions, the design is adjusted until there are no significant problems. For the needs of smart cities, we produce a prototype of the WSN hardware in the design phase separately by type, and the Serverless Computing platform is developed according to the designed architecture.

4.1.8   Debugging

After completing the trial production, the software and hardware are tested for communication and compatibility respectively. Server-side debugging function development running platform. Independent debugging and joint debugging are selected in turn according to the development progress. In the early stages of development, hardware and software independent debugging, WSN side debugging hardware performance until approximating the datasheet indicators. WSN software based on the QEMU virtual machine simulation of the kernel, peripherals, interrupt development debugging control core, and RTOS functions. The stability of the Serverless Computing platform, security, and the ability to complete the complete isolation of the user from the details of the WSN device. Whether the Serverless Computing server meets the rated amount of concurrent user requests. Whether the middleware correctly subsumes and parses user functions, and whether the results of resource calls are correctly uploaded on demand. WSN hardware performance to meet the data manual indicators, WSN software program operating logic is normal and through the sample test, Serverless Computing platform and middleware of the above indicators are verified and function properly, and all parts of the trouble-free operation time respectively to meet their respective design requirements. After all the respective debugging is completed, the joint debugging of hardware and software can be performed later. With the experience of foregoing practice, WSN devices produced at 30% of the batch production can meet the basic experimental scenario commissioning requirements. The purpose is still to reduce the errors that may be amplified in the mass production phase.

4.1.9   Batch Production

Verify the large-scale communication function to ensure that the communication success rate is above a certain standard, and test the platform function of service-free computing after its verification is successful. Select the program with expected benefits and cost trade-offs to meet the predetermined targets for mass production. And after mass production, all tests should be conducted to determine whether there are failures beyond those expected in the trial production and debugging phases. For example, whether the statistics on RF performance of WSN nodes after mass production are in line with the data book distribution, whether the middleware and WSN protocols can meet the predefined execution efficiency when dynamically expanded, and whether there are problems with the Serverless Computing platform for concurrency of the actual scale.

4.1.10 Deployment, Operation, and Maintenance

The local communication environment in the field deployment phase may have the greatest impact on the system. Therefore, the deployment environment communication channel needs to be measured before deployment, and the WSN switches to the solution with the highest communication success rate, and the server-side Serverless Computing platform comes online to provide services to customers. The operational phase may provide some information to us, which provides a priori information for the selection of the maintenance strategy for the current cycle and experience for the design of the next life cycle iteration. Therefore the fault phenomenon and cause are crucial. So we use a sniffer system here. The maintenance phase also follows the cost reduction and efficiency increase strategy. The sniffer device [8] continuously monitors and analyzes the WSN operational status to improve the WSN protocol stack. Node failure requires maintenance priority to use Sniffer to send a restart command to WSN, if not work then choose the more expensive field troubleshooting to replace the faulty equipment. This is cycled between links until the end of the 10-year life cycle.

4.2 Resource Provisioning Strategy

4.2.1 Initial Time and Time Period

The maximum frequency of equipment acquisition without resource constraint first as the most fine-grained acquisition resource division unit of W. The period is divided into pieces, and the resource-constrained equipment performs as little as possible. For example, assume that the node acquires at most 600 times a minute. Then it can be a time anchor point as a benchmark to 600 users to sell once a minute or 300 users to sell half a minute acquisition task.

4.2.2 Task Preemption Priority and Success Rate

Interruption priority is assigned according to the importance of the task. The high priority is expensive and guarantees a success rate of more than 99%. For example, tidal lane control uses the highest interruption priority highest rate, the rest of the collection is sensitive to the time electric cycle in the second priority second rate (the collection process allows a certain failure rate), and finally the data volume requirements, time is not sensitive to the task of scientific research collection to take the lowest priority interruption lowest rate (in the hardware resource occupation low when the collection, collection time, cycle of the lowest guarantee).

4.2.3   Profit from Prepaid Costs

Functional resource calls create the possibility for continued profitability at a later stage. After we implement the basic function execution framework, this approach allows users to fully explore the most efficient way to achieve their needs and guides us to analyze the needs from the user's perspective. At the same time, the shared ownership of intellectual property between the user and us makes the user willing to put effort into iterating and upgrading the model to enrich the system, and we are able to sell functionality to other customers, thus increasing the resources available in the next lifecycle and expanding our profitability.

5   Conclusions and Future Directions

First, the maintainability of WSN and serviceless computing are introduced. The implementation of serviceless computing in WSN maintainability is introduced for the problem of optimal lifecycle cost and resource consumption of WSN to achieve maximum profitability. Finally, based on the scenario of a smart city gives the whole process solution in the life cycle, firstly, the design phase integrates the requirements of customers from all parties in the same WSN system as far as possible, isolates the system details from the users, and only provides the resource invocation interface that conforms to the specification. Trial production, debugging, mass production phase to improve the hardware and software implementation details. Deployment, operation, maintenance phase in the ten-year life of the premise of the link switch decision and continuous summary of experience, the production of reusable resources.

In the concept of Serverless Computing, hardware systems that meet life expectancy, high reliability, and performance redundancy are designed through the concept of maintainability from the perspective of the infrastructure provider. From the service provider's point of view, the functionality is used in a user-programmable way by reusing hardware resources and paying for them on demand. From the user's point of view, shielding the back-end implementation details only requires the uploading of resource invocation patterns written according to specifications in the form of a web server or sandbox according to their needs. At the same time the user-created and fine-grained WSN operations as well as intellectual property rights and updates. It can reduce the coupling in the production environment, reduce resource waste and improve system security.

As future research direction, although the concept and implementation plan proposed in this paper are innovative, it relies on the development of the existing technology, and the biggest controversy may be in today's industry cost distribution, profit model. However, with the development of embedded hardware and software technology and cost reduction, in the current Moore's Law development, the tendency is also from technical progress to system improvement, that is, to explore higher resource utilization efficiency within the same system, such as pipelining technology of instruction set, virtualization technology of operating system, shadow page table in virtual machine, etc., essentially to improve the resource utilization in each level. In order to meet the needs of various industries and fields, we believe that cost reduction, architecture optimization, and quality improvement will become the future research direction in engineering technology if the cost and technology development is suitable for the concept of this paper.

References

1. Vasanthi G, Prabakaran N (2022) An improved approach for energy consumption minimizing in WSN using Harris hawks optimization. J Intell Fuzzy Syst 43(4):4445–4456. https://doi.org/10.3233/JIFS-213252
2. Zimmerling M, Mottola L, Santini S (2020) Synchronous transmissions in low-power wireless: a survey of communication protocols and network services. ACM Comput Surv (CSUR) 53(6):1–39
3. Liu A et al (2022) A survey on fundamental limits of integrated sensing and communication. IEEE Commun Surv Tutor 24(2):994–1034. https://doi.org/10.1109/COMST.2022.3149272
4. Khalaf OI, Romero CAT, Hassan S, Iqbal MT (2022) Mitigating hotspot issues in heterogeneous wireless sensor networks. J Sens 2022:1–14. https://doi.org/10.1155/2022/7909472
5. Stoyanova M, Nikoloudakis Y, Panagiotakis S, Pallis E, Markakis EK (2020) A survey on the Internet of things (IoT) forensics: challenges, approaches, and open issues. IEEE Commun Surv Tutor 22(2):1191–1221. https://doi.org/10.1109/COMST.2019.2962586
6. Majid M, Habib S, Javed AR et al (2022) Applications of wireless sensor networks and Internet of things frameworks in thISACe industry revolution 4.0: a systematic literature review. Sensors 22(6):2087
7. Aleksandar V, Mileva A (2016) Running and testing applications for Contiki OS using Cooja simulator, pp 279–285
8. Guo X, Gao T, Dong C, Cao K, Nan Y, Yu F (2022) A real-time network monitoring technique for wireless sensor networks. In: 2022 IEEE 12th international conference on electronics information and emergency communication (ICEIEC), pp 32–36. https://doi.org/10.1109/ICEIEC54567.2022.9835059
9. Dong C, Yu F (2015) An efficient network reprogramming protocol for wireless sensor networks. Comput Commun 55:41–50
10. Alsolai H, Roper M (2020) A systematic literature review of machine learning techniques for software maintainability prediction. Inf Softw Technol 119:106214. https://doi.org/10.1016/j.infsof.2019.106214
11. Luo X, Ge Z, Zhang S, Yang Y (2021) A method for the maintainability evaluation at design stage using maintainability design attributes. Reliab Eng Syst Saf 210:107535. https://doi.org/10.1016/j.ress.2021.107535
12. Ahmadi S, Moosazadeh S, Hajihassani M, Moomivand H, Rajaei MM (2019) Reliability, availability and maintainability analysis of the conveyor system in mechanized tunneling. Measurement 145:756–764. https://doi.org/10.1016/j.measurement.2019.06.009
13. Zhang Q, Dong C, Yu F (2022) Maintenance of large scale wireless sensor networks. In: 2022 IEEE 5th international conference on electronics technology (ICET), pp 929–932. https://doi.org/10.1109/ICET55676.2022.9825017

14. Shen J, Qiu C, Yu F (2019) A maintainability evaluation method for wireless sensor networks based on AHP and fuzzy comprehensive evaluation. In: 2019 IEEE 2nd international conference on electronics and communication engineering (ICECE). IEEE, 2019, pp 143–147
15. Gao T, Yu F (2022) A maintainability evaluation method of large scale wireless sensor networks based on sample entropy. In: 2022 IEEE 12th international conference on electronics information and emergency communication (ICEIEC), pp 37–41. https://doi.org/10.1109/ICEIEC54567.2022.9835079
16. Qiu C, Shen J, Yu F (2019) A maintainability estimation method for wireless sensor networks. In: 2019 IEEE 5th international conference on computer and communications (ICCC). IEEE, 2019, pp 604–608
17. Wang C, Guo X, Yu F (2022) Maintenance study based on Bayesian network and expectation-maximum algorithm. In: 2022 IEEE 12th international conference on electronics information and emergency communication (ICEIEC), pp 27–31. https://doi.org/10.1109/ICEIEC54567.2022.9835032
18. Montavon G, Kauffmann J, Samek W et al (2022) Explaining the predictions of unsupervised learning models. In: International workshop on extending explainable AI beyond deep models and classifiers. Springer, Cham, pp 117–138
19. Li Y, Lin Y, Wang Y et al (2022) Serverless computing: state-of-the-art, challenges and opportunities. IEEE Trans Serv Comput
20. Gill S, Xu M, Ottaviani C, Patros P, Bahsoon R, Shaghaghi A, Golec M, Stankovski V, Wu H, Abraham A, Singh M (2022) AI for next generation computing: emerging trends and future directions. Internet Things 19:100514
21. Zhong X, Xu M, Rodriguez A, Xu C, Buyya R (2022) Machine learning-based orchestration of containers: a taxonomy and future directions. ACM Comput Surv 54(10):1–3
22. Mondal SK, Pan R, Kabir HMD, Tian T, Dai H-N (2022) Kubernetes in IT administration and serverless computing: an empirical study and research challenges. J Supercomput 78(2):2937–2987. https://doi.org/10.1007/s11227-021-03982-3
23. Yoo S, Jerraya AA (2003) Introduction to hardware abstraction layers for SoC. In: Embedded software for SoC. Springer, Boston, pp 179–186

Scheduling Mechanisms in Serverless Computing

Mostafa Ghobaei-Arani and **Mohsen Ghorbian**

Abstract Currently, serverless computing is considered a critical event in the information and communication technologies domain. It provides a model with high scalability, pay-as-you-go, and a flexible approach to accessing requests using microservices. Many applications implement a microservice architecture, making them perform better than monolith architecture. Microservices are small pieces of code called functions, each of which can be used to run a series of processes. Consequently, microservices need the resources for execution. Hence, one of the critical issues is the efficient allocation of resources for microservices on the nodes, which are considered by structures called schedulers. Scheduling is the strategy of allocating tasks to resources in time. It increases the serverless domain's performance and efficiency by maximizing resource utilization. This scheduling strategy has to consider restrictions specified by the serverless providers and the careers. Using the scheduler's tasks and maintaining fairness between efficiency and the quality of careers service is complex. Scheduling algorithms are developed considering metrics similar to performance, priority, latency, cost, etc. Therefore, the process of resource allocation is regarded as a critical factor that can be of considerable significance to service providers. In serverless computing, service providers must ensure that their chosen scheduling strategies are satisfactory to service receivers. Although there are various scheduling techniques, it is crucial to point out that no single scheduling technique can accommodate all the requirements of various applicants. An in-depth understanding of the types of schedules and selecting the most effective scheduler for different kinds of applicant requirements is therefore crucial to provide the most efficient allocation of resources. In other words, when selected scheduling is inefficient, several problems can result both for the service provider and the service recipient. Therefore, service providers are forced to increase their costs. As a result, the cost of the applicants is increased. The consequences of this situation are that the requesters are dissatisfied with the poor quality service received and the increased cost they

M. Ghobaei-Arani (✉) · M. Ghorbian
Department of Computer Engineering, Qom Branch, Islamic Azad University, Qom, Iran
e-mail: mo.ghobaei@iau.ac.ir

© The Author(s), under exclusive license to Springer Nature Switzerland AG 2023
R. Krishnamurthi et al. (eds.), *Serverless Computing: Principles and Paradigms*,
Lecture Notes on Data Engineering and Communications Technologies 162,
https://doi.org/10.1007/978-3-031-26633-1_10

have and are less inclined to use a provider's services in the future. Many scheduler strategies are available to providers; therefore, they should become familiar with a variety of them and understand their advantages, characteristics, and disadvantages. In this study, we comprehensively investigate the widely employed schedulers in serverless computing by investigating their advantages, disadvantages, and applications. The purpose of the present study is to present a comprehensive examination of various and effective scheduling techniques that can be a basis for selecting the appropriate scheduling process based on the providers' approach.

Keywords Function as a service (FaaS) · Performance evaluation · Resource management · Serverless computing · Scheduling algorithm

1 Introduction

Resource management is an essential function of any system; hence, managing resources is a crucially important critical factor in evaluating a system's performance, functionality, and cost in any given circumstance. Therefore, resource management can affect performance and cost directly and indirectly by preventing some of the functions from being performed or incurred. A cloud computing system is highly complex with many shared resources that are subject to unpredictable requests and uncontrollable external factors [1]. The establishment of policies and making complex decisions in a way that optimizes cloud resources in a multi-objective manner are crucial elements of optimizing cloud resources. Due to the size of the cloud infrastructure and the unpredictable interactions between the system and such a large number of users, it is challenging to manage resource allocations effectively [2]. Because of the large number of users, it is difficult to predict the type and the intensity of the workload the system will manage. Due to the project's enormous scale, global state information can't be accurate [3]. Consequently, resource management becomes even more challenging when resources are oversubscribed, and users do not cooperate, leading to an even more complex situation. Resource management is not only affected by external factors but also by internal ones, for example, heterogeneous hardware and software architectures, the scale of the system, and the failure rate of the various components [4]. The four basic cloud delivery models are IaaS, PaaS, SaaS, and FaaS. Each of these models provides resources in a manner that differentiates it from the others. In all cases, cloud service providers are required to handle large fluctuations in load, which poses a challenge for cloud elasticity services. In some instances, resources can be provisioned in advance when it is possible to predict a spike in demand, for example, for web services that are susceptible to seasonal spikes in demand [5]. However, the situation becomes a bit more complicated when unexpected demands occur. An auto-scaling system can effectively handle unplanned spikes in workload, providing a pool of resources that can be released or allocated as necessary. When environmental changes are unpredictable and frequent, a centralized management system cannot provide adequate solutions for implementing resource management policies. However, using distributed mechanisms poses challenges since it requires coordination between the entities responsible for controlling the system [6]. Serverless computing is a novel execution model for cloud computing that has gained much traction recently due to its lightweight nature and ease of use. As well as the serverless architecture and serverless technology are also used to describe this model. In this cloud computing execution model, resources are automatically provided to an application at the request of an event to ensure that the application is executed as intended [7]. The serverless architecture allows all operational tasks, such as provisioning, scaling, and scheduling, to be offloaded to the cloud provider, a significant advantage over a traditional cloud architecture based on servers. Unlike serverful computing, developers who use serverless computing do not have to worry about managing infrastructure resources, as the platforms handle such tasks on their behalf [8]. With the lightweight capabilities and automatic management of serverless computing, developers could not only focus on the core logic of applications but also be able to calculate at run time in milliseconds and customize their billing policies. As a result, such an approach can substantially reduce the costs associated with application development [9]. In addition, serverless providers can improve resource utilization by decreasing the number of idle resources to nearly zero. As a result, serverless providers can serve more customers with the same amount of resources, allowing them to provide better customer service [10]. In serverless computing, the functions are written by developers and accessed over the Internet as execution units. The core of serverless computing consists of these functions. FaaS, refers to this service [11]. According to this approach, the modular pieces of the code are executed on the cloud via a serverless approach. Using a service such as FaaS allows developers to update or write code segments on the fly, which are subsequently executed in response to events on a web page, such as clicking an element. Using FaaS can deliver a wide range of benefits to developers, including cost-effectiveness, built-in scalability, and maximum efficiency in the shortest amount of time [12]. Developers consider serverless computing to be FaaS, which means they can create and maintain their applications (or functions) without developing and managing the infrastructure [13]. The serverless computing approach enables developers to create cloud functions in a high-level programming language (e.g., Java, Python) and upload the functions to a serverless platform to perform. And then, it is then possible for them to execute their well-defined computation tasks using the HTTP requests or API return [14]. One of the most pressing issues in serverless computing is scheduling and allocating resources to the requested tasks. Until recently, little research has been conducted on the scheduling process and allocation of resources in serverless computing. The provision of serverless computing services to a large number of consumers can be a challenging task for serverless computing service providers. Scheduling in a serverless computing system is important to minimize execution time, reduce computing costs, and maximize resource utilization [15]. Consequently, it is crucial to notice that the resource allocation process is always one of the critical things that can be of great importance for service providers and requesters. Providers should ensure that the selected scheduling policies ensure the service receivers' satisfaction in server-less computing. Therefore, various scheduling approaches and techniques have been employed. It should be noted that only selecting the scheduler alone cannot satisfy the needs of both the provider and the recipient of the service. In light of this, it is essential to note that no service provider can satisfy all the needs and requirements of a wide range of applicants with a single scheduler technique. The selection of the appropriate scheduler for each type of applicant requirement is essential to provide the highest efficiency in resource allocation. In other words, choosing inappropriate scheduling can pose several problems for both the service provider and the service recipient, increasing the costs for the service provider; as a consequence, requesters are dissatisfied with the service they received, which is of poor quality. Therefore, it is crucial to familiarize oneself with the various scheduler strategies and their characteristics, advantages, and disadvantages. A service provider should become familiar with the different types of schedules to select the appropriate schedule based on the service they provide and their primary goals to improve the quality of their services. As a result, the applicants will be satisfied. There is a lack of comprehensive research on the scheduler strategies in serverless computing, so it has been attempted to examine the widely used schedulers in serverless computing by examining their advantages, disadvantages, and applications comprehensive. This study provides the basis for the appropriate selection of the scheduling process based on the providers' approach.

The main contributions of this review are as follow:

- Providing a summary of the existing challenges according to the resource scheduling mechanisms in serverless computing.
- Presenting a comprehensive review of the resource scheduling mechanisms in serverless computing environment.
- Introducing the open issues and future challenges that the resource scheduling mechanisms can be applied in serverless computing.

This chapter is structured as follows: As a starting point, the first section exam-ines the essential prerequisites, such as using cloud services and cloud computing, implementing serverless technology in the cloud environment, and utilizing server-less functions for processing and scheduling in serverless computing. The second section of this chapter describes the nature and workings of serverless computing and how it can be used. The third section will examine the scheduling process in a serverless environment and analyse the scheduling architecture. The fourth section comprehensively examines the types of schedules used in serverless processing and their advantages. The fifth section of this chapter discusses the challenges associated with scheduling serverless computing systems. The sixth section examines different scheduling algorithms and compares them. Finally, the seventh section concludes with a conclusion. Moreover, Table 1 shows the commonly used abbreviation in this work.

Table 1 Abbreviations

Meaning	Abbreviation
Functions as a Service	FaaS
Platform as a Service	PaaS
Software as a Service	SaaS
Virtual Machine's	VMs
Google Cloud Functions	GCF
Command Line Interface	CLI
Amazon Web Service	AWS
Custom Resource Definitions	CRDs
Service Level Agreement	SLA
Application Programming Interface	API
Hypertext Transfer Protocol	HTTP
Quality of Service	QoS
Google Cloud Functions	CGF
High-performance computing	HPC
Command-Line Interface	CLI

2 Preliminaries

Serverless computing is an extended model of cloud computing that provides computing resources on request to run applications. This is a type of software design pattern where third-party services host applications, eliminating the need for developers to manage server hardware and software. Serverless architecture (also known as serverless computing or FaaS) is a concept in which third-party services host applications. With Serverless computing, developers can develop and execute code remotely from anywhere in the world without worrying about the server environment on which they are running their code [16]. Using a microservices architecture is the basis for serverless computing. Consequently, monolithic architectures traditionally implemented in servers cannot be applied to serverless architectures. On the other hand, the microservice architecture approach divides programs into smaller functions instead of the large programs typically used in traditional programming approaches. As a result, each of these functions can be invoked and executed independently [17]. When hosting a software application on the internet, server infrastructure is typically required to support the application. Usually, this means it must configure and manage a virtual or physical server, an operating system, and other web server hosting processes required for the application to run correctly. Using a virtual server provided by a cloud provider such as Google or Amazon eliminates the need for physical hardware maintenance [18]. However, there is still the issue of implementing the operating system and managing the software processes on the web server. The developer can focus exclusively on the individual functions within its code using a serverless architecture. The following will describe some of the essential advantages of the serverless architecture approach, which will be described as follows:

- **Cost-effectiveness**: Serverless computing architecture simplifies the development process and ignores the need to pay charges for idle computing time in the development phase while simultaneously improving productivity. Thus, Serverless architectures are easier to use and more cost-effective for the client since they allow them to obtain services at the lowest possible price [19].
- **Scalability**: Scalability is one of the most wonderful aspects of serverless computing architecture. The system's high reliability eliminates the need for developers to worry about heavy use and high traffic contingencies. Because, compared to previous architectures, this architecture can handle all scalability concerns more effectively [20].
- **Increase Developer's speed**: Through serverless computing architecture, developers can devote more time to writing code for the websites and applications that they are developing, thus increasing their speed in the best possible way, and they can in more impressive efficiency in their program development process. By doing this, developers will spend a lot less time deploying and can easily get a faster development turnaround, increasing their productivity [21].

It is important to note that the serverless computing architecture has its disadvantages, just like any other emerging technology. As a result, serverless computing is far from perfect. The following will provide examples of its most important disadvantages to illustrate this point. These disadvantages include:

- **Complex testing process**: Developing a serverless architecture and integrating its code into a local testing environment can be challenging since it's not always straightforward to integrate all of the code elements, leading to a complex application testing process [22].
- **Cold start**: When apps have not been used for a long period, the initial startup and initial processing of requests may take longer than usual. One clear thing about this approach is that it cannot be an effective and efficient starting point since it could limit the possibilities in specific scenarios. Evidently, this is not an option that most clients would choose [23].
- **Reduced controllability**: Third-party services can be an effective way to provide a lesser level of managerial control over the system. Due to this, clients will not be able to understand what is happening in the system entirely. This is because they will not be able to see everything going on [24].

Even though serverless computing has some significant disadvantages, such as a complex testing process, a cold start, and reduced controllability, which have been discussed in detail above, it has managed to attract a lot of attention. Therefore, scientists and researchers are constantly trying to provide a basis for more effective and efficient use of this technology by fixing the existing defects.

Cloud computing has led to the development of a new cloud computing model FaaS. Cloud service providers that advertise the possibility of deploying FaaS in an organization should be able to help ecosystems that wish to use FaaS as a platform for executing cloud services. FaaS service delivery models eliminate the need for developers to maintain application servers as part of their development process. This model is established on serverless computing architectures and technologies, allowing developers to deploy applications to the cloud without worrying about managing servers. FaaS can run small, modular fragments in the form of functions via a serverless architecture, using a function-based approach. These services enable software developers to execute functions that clients need [25]. It is important to note that when a function is invoked, a server starts, the function is executed, and then the server is shut down. As opposed to other models in which software developers run their applications on dedicated servers, serverless architecture only activates when an application is being run. Hence, once the function execution has successfully been completed, it can quickly be terminated to devote the same amount of computing resources to other tasks. Software developers can utilize FaaS to access a platform capable of executing application logic on a demand-based basis and where all application resources are coordinated and managed by a service provider to ensure that they are secure throughout the execution process. Therefore, whenever cloud service providers manage servers for the deployment of applications, software developers can benefit from simplified processes, cost savings, and scalability, allowing them to focus more on developing application code and less on managing servers [26]. A FaaS model is generally appropriate for simple, repetitive tasks, such as processing queued messages, scheduling routine tasks, or processing web requests. An application's functions are a collection of tasks that can be defined separately as pieces of code and executed independently. Therefore, the most efficient use of resources is to scale a single function since it is not necessary to deploy an entire application or even an entire microservice when scaling a single function [27].

The cloud computing model consists of many resources federated into a single machine, and it shares some characteristics with parallel computing, including cluster and grid computing. Still, its main characteristic is the use of virtualization for managing resources in cloud computing. This approach permits computing resources to be scheduled so that they are provided as a service to the clients. Among the most significant developments in the world of computing has been the availability of enormously elastic computing nodes, which can be employed anywhere and at any time and only paid for when utilized [28]. It would not have been possible to achieve this goal with traditional computing or data centers. It is important to note that idle, underutilized, and inactive resources contribute significantly to energy waste. Scheduling resources optimally in the cloud is a challenging task. The number of resources to allocate to cloud workloads will depend on several factors, including the quality of service requirements for the applications hosted in the cloud and the energy consumption of the computing resources. It is not appropriate for existing solutions for resource allocation in a cloud environment to address the problems caused by uncertainty, dispersion, and heterogeneity of resources. As a result, increasing energy efficiency and resource utilization is very difficult for heterogeneous cloud workloads [29]. A cloud-based platform that automatically schedules computational resources by evaluating energy consumption as quality of service criterion is needed to address this issue. In autonomous systems that are self-optimizing, autonomous resource scheduling can enhance resource utilization and user satisfaction. Self-optimization refers to the system's ability to allocate resources in the most efficient way possible. As a result, Allocating cloud workloads to appropriate resources is necessary to improve QoS parameters such as energy consumption and resource utilization [30]. Different scheduling algorithms are employed in cloud computing to schedule resources according to user needs. A scheduling algorithm has been designed to maximize the throughput as much as possible. Hence, Specific performance metrics are used to evaluate the cloud scheduling algorithm's performance [31].

3 Scheduling in Serverless Computing

Essentially, a serverless execution model involves shifting responsibility for managing application resources from developers to cloud service providers. Since the serverless model uses real-time resource allocations, providers must manage resource allocations autonomously during the execution of the application. In contrast, if the provider uses a serverless model, such as infrastructure as a service, the user would be responsible for configuring the environment and assigning resources before the application's execution. It is difficult for serverless platforms to make informed decisions concerning the allocation of resources during the initial allocation of resources as they lack sufficient information regarding the resource needs of various functions [32]. For example, when executing a function via AWS Lambda and GCF platforms, the user can specify how much memory will be allocated to the function, and CPU power will be allocated linearly based on the quantity of memory. It has been observed in a review of serverless environments that CPU usage is often a source of contention, mainly in the case of computationally intensive applications, which ultimately results in high application latency [33]. Consequently, any arbitrarily determined resource allocation policy may result in subsequent resource contention for applications during their runtime, resulting in the user's SLA being violated. Therefore, it is necessary to develop techniques for dynamically managing resources. According to the serverless deployment model, clients will only be required to pay for the resources utilized during the application's execution. Despite this, the cloud service provider will provide ongoing support for the underlying VMs throughout their lifecycle [34]. It is important to note that the resource time allocated by a virtual machine during its lifetime includes the setup time and the entire active time during which one or more functions use resources on the virtual machine. In light of the popularity of the serverless model and its increasing use for long-running tasks, including parallel tasks and the like, it has become increasingly common to use partial resources. Therefore, the cloud service provider needs to maximize the utilization of its virtual machines at any given moment to reduce the maintenance costs associated with too many unused machines [35]. In contrast, when an application has a deadline to complete its execution (part of the service level agreement). In that case, the decision to place a function instance on a VM should be determined by optimizing the use of resources within the VM without compromising the deadline. Generally, existing serverless platforms utilize various strategies for managing their underlying infrastructure. An example is Docker Swarm, which uses a spread placement algorithm to distribute workload evenly between nodes in the cluster [36].

3.1 Scheduling Architecture in Serverless Computing

Utilizations of serverless computing require that developers be able to create cloud functions in a high-level language and then make the triggers that will pull those functions into action whenever they are called. In this regard, the serverless provider of the infrastructure is responsible for ensuring that everything is configured to execute the user's function appropriately. According to Fig. 1, a typical serverless environment can be seen by examining a typical scenario for a serverless platform. When clients request a new function, they are sent via HTTP protocol to the scheduler so that it may be invoked in the server for execution.

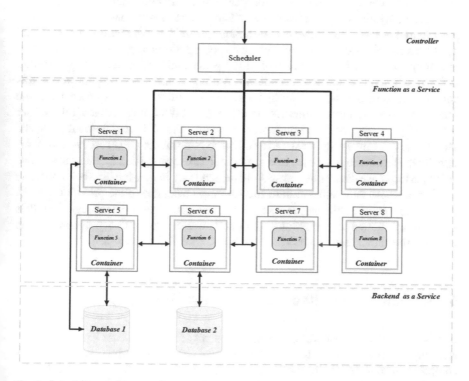

Fig. 1 Scheduling architecture for serverless computing

In the following, the scheduler will choose the host, also known as the server, on which the function will be executed; the server, also known as the host, will typically be a physical server or a virtual machine. In response to the client's request, eight functions were invoked and executed based on the client's requests, which can be seen in Fig. 1, which illustrates the process. So, in this case, the server executed these eight functions as soon as the client called them. In other words, these functions are executed using containers on servers to facilitate their implementation by executing their containerization. An application could potentially utilize supplementary back-end services required to execute its functions, such as an object store, database, etc. Generally, serverless platforms can host and execute their functions on a server in a variety of ways, but they typically use containers or other sandboxing mechanisms to do so; in the context of Serverless computing, a server is either a virtual machine or a physical machine that hosts the containers that contain functions. In a serverless environment, it is necessary to maintain an acceptable level of latency for function requests made as part of the serverless process. Due to this, one of the critical concerns in the serverless environment is maintaining an acceptable response time for function requests. It is important to note that a cold start can significantly impact an application's latency. Therefore, it is essential that providers carefully distribute user requests among servers, so that cold starts are minimized and guarantee satisfactory performance. When serverless workloads scale, providers must maintain resource efficiency. As a result, providers competing in the market must increase resource efficiency. The provisioning resource process in a serverless environment differs from that in VM cloud provisioning resource in that users do not have to provide detailed information regarding their resource requirements; this complicates the provisioning process for the service provider. For example, when the user specifies a memory requirement for GCF and AWS Lambda that is all the user can specify. In contrast, it is impossible to specify the CPU requirements, and the service provider internally makes use of that to allocate the CPU resources proportionally [37]. In some situations, users may need to overprovision memory to achieve better performance. Since client workload demands are increasing, this behavior may result in a severe underutilization of the resources available to the provider. Hence, to maintain resource efficiency and keep costs low, providers must manage the CPUs and memory allotted to applications. Since serverless workload demands can be quite variable, the above requirements become even more challenging [38].

4   Scheduling Algorithms in Serverless Computing

In a cloud computing system, computing resources are presented as virtual machines, reducing the costs of resource use. Essentially, virtualization involves but collect computing resources, whether software or hardware, into virtual machines and delivering them as services from a provider to a consumer. Scheduling algorithms have an important role to play in cloud computing by reducing response times and optimizing resource utilization [39]. Therefore, by dynamically allocating computing resources based on the consumer's requirements and preferences, computing resources can be tailored to the consumer's requirements. Several different approaches and strategies have been considered to achieve the best possible results regarding scheduling resources. Depending on the goals set, various approaches or strategies are used when considering how to schedule resources. Managing resources efficiently requires a range of techniques, which is as complex as scheduling resources [40]. For this reason, it is important to understand the mechanisms behind these schedules to select a scheduling strategy that will achieve the desired results for the system. An essential part of a research or implementation project is choosing the best scheduling strategy based on enough knowledge, guaranteeing a better and more effective outcome. This section examines some of today's most widespread and vital scheduling strategies. Despite serverless computing being regarded as an emerging technology, it is essential to realize that compared with other technologies, it has little depth in employing scheduling strategies. This means that there have been very few research studies conducted in this field compared to other fields, and therefore the number of researches is relatively tiny [41].

4.1 Energy-Aware Scheduling in Serverless Computing

This part describes related features to the energy-aware method that is considered a scheduling strategy in serverless computing. Energy-Aware Scheduling allows the scheduler to foresee its decision's influence on the power consumed by processors in the systems. On another side, the scheduler is trying to choose the task that is anticipated to efficiently the most helpful energy consumption without damaging the system's throughput. But in a cloud environment, the principal concept of energy-aware Scheduling is to posit passive Structures called containers or the runtime environment in a cold-state mode to decrease energy consumption [42]. When the container state converts from cold-state to warm-state mode cause latency in the execution of invocation functions which may range outside the deadlines the client considers. This can cause problems and also reduce the level of satisfaction of consumers. Therefore, It will be possible for an energy-aware scheduler to profit from postponing Tasks that are not highly ranked in terms of sensitivity Or, in other words, more trivial tasks to decrease overall energy consumption. In other words, the scheduler can prioritize sensitive and vital tasks and allocate resources [43]. Using energy-aware scheduling in serverless computing can provide providers with several benefits. Some of these benefits are as follows:

- minimizing the overall response latency
- improves container utilization
- reduced the total energy consumption
- reduced the execution time
- improve throughput
- response time acceptable

- increase the performance of the energy model.

Due to serverless computing technology's emerging nature, employing energy-aware scheduling in this technology includes limited cases. Therefore, in the following, we will try to mention some case studies for this scheduling.

Aslanpour et al. [44] tried to get ideas from scenarios implemented in the cloud; based on this idea, to eschew a cold start of functions, a function is invoked by transmitting forgery requests, considered a warm function. Then, they proposed a schedule-driven technique, warm scheduling, that pre-schedules a function towards a lower position node on peers with a further step. The proposed algorithms demonstrate that an energy-aware scheduler can ease up and run throughput and usability over the criterion algorithms that are close to optimal while preserving QoS, response time, and acceptability. It can be said that the proposed algorithm's goal in this research is to reach the highest possible level of the up-and-running approachability of edge nodes while to the lowest possible level, the change thereof sans compromising the throughput and QoS [44].

Gunasekaranet et al. [45] propose a solution to microservice-agnostic scheduling and colossal container over-provisioning, resulting in poor resource utilization, especially during workload fluctuations. Hence, they proposed Fifer. The main idea is through function-aware container scaling and intelligent request batching, bin-packing jobs are efficiently distributed into fewer containers and utilized more efficiently. Fifer has been described as a framework that efficiently deploy function chains on serverless platforms using an adaptive resource helm mechanism. Also, Fifer avoids cold starts by proactively spawning containers, maintaining SLO compliance while minimizing overall latency. With these benefits combined, Fifer improves container utilization and cluster-wide energy consumption without sacrificing SLOs, compared to most serverless platforms' state-of-the-art schedulers. However, it is essential to note that in energy-aware strategies, the implementation is designed to minimize the number of transports between containers [45].

4.2 Data-Aware Scheduling in Serverless Computing

In this part, an overview of the data-aware scheduling strategy in serverless computing is provided. The pliable and varied nature of the cloud environment presents challenges for effective resource management, including resource provisioning and allocation, especially when data motion is crucial among multiple areas. It is essential to consider these points to enable effective resource management in cloud environments [46]. When a substantial workload is involved, insufficient resources can result in poor performance, and excessive resources can result in inessential financial expenditures. Therefore, whit using data-aware scheduling (data affinity), we can application tasks can be intelligently scheduled. In addition, this feature can contribute to meeting the challenges associated with addressing the latency requirements in real-time applications. Also, using data-aware scheduling for workflow applications can significantly improve performance and guide efficient resource provisioning and allocation based on variable workload characteristics. Also, performance can be improved by considering data location when dispatching them [47]. Usage data-aware scheduling in serverless computing can provide providers with several benefits. The following are some of these benefits:

- reduces execution cost
- reduces overall application latency
- reduce response time
- fewer latencies of function interactions
- data transmission rate between resources is improved
- execution speeds are faster
- reduces bandwidth
- reduce time spent on data movement.

The use of data-aware scheduling in serverless computing technology is limited due to its emerging nature. Therefore, we will present some examples of case studies of this scheduling in the following.

Rausch et al. [46] propose an efficient scheduling system for containers that enables platforms to take advantage of edge infrastructures to the fullest extent possible. To achieve this, they developed Skippy, an open-source container scheduling system that lets available container orchestrators, such as Kubernetes, support serverless edge functions based on available containers. Skippy achieves this by presenting scheduling constraints based on the nodes' potential, the application's data flow, and the network topology. And in ending, they have demonstrated that the trade-off between data and computation movement is the most factor consideration when placing functions in data-intensive serverless edge computing [46].

Wu et al. [48] propose a computing model that eliminates the convolutions associated with cloud programming for remote sensing data analysis while taking advantage of efficient multiplexing techniques. The proposed computational strategy is based on the idea of on-demand execution, which allows instant response and multitenancy. On the other hand, they demonstrated that the model of on-the-fly computing for analyzing remote sensing data attainable to be incorporated into a serverless architecture with high efficiency [48].

4.3 Deadline-Aware Scheduling in Serverless Computing

One of the advantages of serverless computing models is that they can implement deadline-aware scheduling to maximize the efficiency of their operations in terms of the number of tasks they can complete before the deadline. As a result of maximizing efficiency in this schedule, the selection strategy may end up biasing towards task types that, on average, have shorter expected execution times. Since these tasks are more likely to be successful, their expected execution times are generally shorter, increasing their likelihood of success [49]. Meeting latency deadlines is one of the priorities of this scheduling. A scheduling policy should be implemented that minimizes the likelihood of many deadlines being missed in the future. Many instances have occurred in which management has been unfair, certain task types have been neglected, and deadlines have exceeded expectations consistently. The serverless computing provider will use deadline-aware scheduling to maximize the number of service requests that meet the deadline constraints within the resources they have at their disposal [50]. This will allow them to maximize the number of service requests that meet those deadline constraints within the resources they have at their disposal. Deadline-aware scheduling in serverless computing can provide providers with several benefits. The following are some of these benefits:

- low latency request execution in a serverless setting
- improves energy efficiency
- reduce the incurred cost
- maintaining fair share allocation guarantees
- reduce the serverless function's latency
- significant decrease in response latency.

Using deadline-aware scheduling in serverless computing technology includes limited cases because of the technology's emerging nature. We will therefore attempt to provide some examples of this scheduling in the following.

Singhvi et al. [41] propose a serverless platform capable of executing low-latency requests. Every application in this platform is viewed as a DAG of functions, each with its latency deadline. In the Archipelago platform, for achieving per-DAG latency goals, a cluster is partitioned into smaller pools of workers, and each pool is assigned a distributed scheduler (SGS). Also, a layer of load balancing routes requests from various DAGs to the correct SGS utilizing a latency-aware scheduler and preventative sandbox allocation to minimize overhead. It also automatically scales the numeral of SGSs within the DAG as required. Finally, with this platform, they were able to achieve test results that showed that it met the latency requirements for more than 99% of practical workloads for applications and significantly reduced overall latency figures to 36× as compared to the current serverless platforms [41].

Mampage et al. [36] present a policy for function placement and dynamic resource management for serverless computing applications from the perspective of the provider and the user. Considered policies are designed to minimize the costs associated with service provider resource consumption while meeting the needs of users, e.g., deadlines. The suggested approaches in this research are sensitive to deadlines while increasing resource efficiency for the provider while dynamically allocating resources in order to improve the response time to functions. They put into practice and evaluated the proposed approach through simulation using the Container CloudSim tool. Compared with conventional scheduling approaches, the suggested function assignment policy can decrease resource utilization by as much as three times. Another side, they demonstrated when compared with a static resource assignment policy and a dynamic CPU-share policy, the dynamic resource assignment policy results in a 25% improvement in meeting deadlines for the required functions [36].

4.4 Resource-Aware Scheduling in Serverless Computing

To schedule inbound user functions in a serverless environment, the provider needs to choose the appropriate invoker, in other words, the invoker that is to be selected to handle incoming user functions. Several serverless applications are available, and their resource consumption methods are quite varied. Since different applications consume vastly different resources based on resource consumption patterns, so different scheduling algorithms must be tailored to each application's needs and diverse applications. For example, intermittent activities, embarrassing parallel services, and short-lived applications require different scheduling algorithms [51]. It is important to note that computing functions can be classified in various forms. For example, scientific computing is typically CPU-demanding, whereas analytical applications are usually memory-demanding. The co-location of many CPU-demanding functions on one physical node may cause resource contention, which can delay the running of those functions. In addition to CPUs, numerous other computing resources, including servers, networks, and disks, adhere to the same basic principles. A resource-aware scheduler assigns functions to the appropriate computation nodes to provide the required resources in the proper time. By doing so, Management ensures that several different applications, or functions within an application, can consume resources. In other words, ensuring workloads and functions that are needed are prioritized on time. It is, however, common practice to miss opportunities to implement a more awareness scheduling policy in challenging situations, such as those where current resources cannot meet incoming demands [52]. Further, limits and challenges are a very useful tool for developers to manage costs and better understand expected workloads' characteristics. A resource-aware scheduling approach can provide several benefits to serverless computing providers, including the following:

- reduced runtimes
- adapts to varying load conditions
- Automated resource allocation for real-time stream processing
- reduce the cost of various application services
- improves the resource efficiency
- reduce application latency
- reduce the average response time.

As serverless computing is an emerging technology, resource-aware scheduling in this technology includes limited cases. In the following, we will discuss some examples of this scheduling method.

Cheng and Zhou [51] present an ARS (FaaS) framework that automatically schedules and allocates resources to streaming applications. Their primary platform is an HPC Cloud platform, on which FaaS is explored for stream processing. Their proposed ARS (FaaS) framework provides high-efficiency HPC Cloud computing resources and accelerates real-time and stream analytics. With this work, they tried have improved the flexibility and efficiency of scheduling, making resources available autonomously on demand and responding in real-time to unpredictable conditions and fluctuations in streaming data [51].

Suresh and Gandhi [52] present a framework for function-level scheduling that minimizes the costs associated with providing resources while meeting customers' performance needs. FnSched scheduling operates by carefully controlling the resource utilization of the concurrent functions within the invoker and by auto-scaling capacity based on distributing the load on fewer invokers in order to respond to varying traffic levels. As part of the project, they implemented a working model of FnSched. They demonstrated that it significantly increased resource efficiency by an estimated 36–55% while maintaining an acceptable level of application latency compared with existing baselines [52].

4.5 Package-Aware Scheduling in Serverless Computing

The advantages of cloud computing include launching functions quickly because they are running in pre-allocated containers and VMs. It is important to note that when these functions are dependent upon large packages to function, they will start slowly. This reduces the elasticity of the application, as it takes a long time for the application to respond to sharp bursts of load during the day. Additionally, time-consuming function launches directly impact the quality of the results of serverless applications being offered as a service, and finally, they directly impact the quality of the results. Therefore, the program cannot be entirely executed due to the delay [53]. The computation node is typically responsible for fetching application libraries and related packages, when it receives an invocation request for them, among the serverless technologies currently being used. For instance, OpenLambda has the computation node fetching and installing application libraries and their related packages. The worker nodes with preloaded packages can reuse the execution environments. As a result, the task launch time and, therefore, the turnaround time for the task is shortened, thus making them more efficient. As well as further investigating package-aware research scheduling in the future, there is also the potential to use this scheduling strategy for machine learning approaches to predict future libraries that will be required based on the currently installed libraries [54]. This problem is similar to the basket recommendation problem, as it predicts future libraries based on the available libraries. Several benefits can be gained by serverless computing providers from a package-aware scheduling approach, including the following:

- reduce the delay of pod startup
- ensuring the balance of node resources
- Minimize data transfers

- reduce average response time
- un usage over-use of the CPU resources
- avoid worker overload.

Due to serverless computing technology's relatively new nature, hence employing package-aware scheduling in this technology includes limited cases. Therefore, in the following, we will try to discuss some case studies for this scheduling.

Kim [53] sought to develop an approach based on scheduling algorithms that would accomplish the scheduling process without the drawbacks associated with pod-by-pod scheduling. In this proposal approach, they tried to take advantage of the concurrent pod scheduling mechanism implemented in the Serverless cloud paradigm to improve the effectiveness of pod scheduling. According to the experimental results, the new scheduling algorithm performs better than pod-by-pod scheduling, and the start-up latency for pods has significantly been reduced compared to pod-by-pod scheduling. Consequently, they increased the effectiveness of pod scheduling by optimizing it further [53].

Bai et al. [55] propose a new Attribute-aware Neural Network Approach (ANAM) to deal with problems where a user's appetite for items is not explicitly tracked. The existing approaches do not consider essential attributes of an item, such as its category. Hence, the ANAM approach employs an awareness mechanism to specifically model users' evolving appetites for items and a hierarchical structure to incorporate attribute information into the model, which helps to model the evolution of the user's appetites. This approach attempted to use a persistent neural network to approach the sequential behavior of the user over time; in particular, there are two levels of hierarchy in this method. By using consideration weights distributed across baskets at both levels, the system relays the user's preference for their attributes and items to the next basket based on the user's preference for their attributes and items. According to their findings, the purpose ANAM model effectively recommends the next basket base [55].

4.6   Cost-Aware Scheduling in Serverless Computing

Several factors have contributed to the increasing heterogeneity, availability, and geographical dispersion of computing resources. Hardware specialized computing can be efficient in terms of enhancing the performance and productivity of tasks that must be processed but will frequently require the requirement to dispatch computations to remote machines to be executed, negatively affecting performance and efficiency. Having to allocate such a large amount of resources could be extremely expensive. Despite the difficulties associated with remote resource coordination and the increasing availability of cloud computing and various remote computing services, selecting the most appropriate resource for the job has become increasingly complex, and coordinating this resource is becoming increasingly important. Serverless computing can increase interoperability among these resources and transparency, including FaaS frameworks that allow a generalized expression of computing tasks and seamless remote executions. Due to the growing utility of the serverless computing model, it is necessary to support increasingly heterogeneous computing environments and to provide a method of selecting resources from a pool of available resources [56]. As a result of the differences in the available resources in a heterogeneous environment, it is extremely challenging to select the appropriate execution location within the project scope. Additionally, selecting an execution environment suitable for a given project is nearly impossible because these resources are limited in terms of availability and cost. Therefore, it is necessary to utilize a scheduling strategy that allows users to allocate functions across heterogeneous and distributed FaaS nodes while accounting for such allocations' computational and cost implications. The scheduling strategies aim to optimize resource allocation to achieve a single goal in the shortest time. Through a cost-aware scheduling method, schedules can offer multidimensional estimations regarding costs and budgets, so specific metrics can be specified by the scheduler so that efficient resource allocation decisions can be made, taking into account the respective trade-offs [57]. The primary objective of cost-aware scheduling is to reduce costs by altering the cost metric. Providers can reap several benefits from using cost-aware scheduling in serverless computing. Here is a list of some of these benefits:

- reduce cloud monetary cost
- reduces bandwidth
- minimize the resource consumption cost
- Improve the performance of serverless workloads with negligible costs
- reduce operational costs
- cost savings.

Due to the relatively new nature of serverless computing technology, the use of cost-aware scheduling in this technology includes limited cases. Therefore, we will discuss some case studies related to this scheduling in the following sections.

Bisht and Vampugani [56] propose a new version of the min-min algorithm for workflow scheduling in a heterogeneous environment that considers cost, energy, makespan, and load balancing to optimize load balancing and task scheduling in Cloud-Fog-Edge collaborations between servers. Their goal is to reduce costs by addressing traditional problems associated with resource constraints, such as load balancing and scheduling. Several offloading scenarios were implemented and evaluated using their proposed algorithm, including cloud-only offloading, fog-only offloading, cloud-fog-edge offloading, and cloud-fog-edge edge offloading. Finally, they showed that the proposed method performed better, resulting in a shorter duration, reduced energy consumption in conjunction with load balancing, and marginally lower costs compared to ELBMM and min-min techniques [56].

4.7 History-Aware Scheduling in Serverless Computing

There are several different forms of caching, but content caching is the most basic. In content caching, the provider builds a cache that regularly stores content frequently accessed in a location near where the users are making their requests. According to the total response time and the inter-component latency of serverless architectures, there are significant latency issues due to network delays between the serverless components. Despite the cold start delay being ignored, there is a significant difference in the latency of serverful as opposed to serverless systems. A serverless environment is characterized by the fact that each program is composed of multiple components. It is common for an application to have no unique computing process that makes it unique, and even if it did have such a process, it would generate the same results regardless of the input. Consequently, finding the optimal caching policy for the network can significantly impact performance. For caching, the serverless architecture uses a concept known as function caching, for which the results of each function are saved to speed up the system's performance. Therefore, in serverless architectures, it is possible to utilize a caching-based policy to maintain the state of the functions throughout their execution. Doing so can minimize the time it takes to execute machine tasks and acquire data from external storage devices. It's important to mention that a machine can flush its memory [58]. Hence, a repeating task will be considered inappropriate to be sent to that machine. The costs of a cache miss can be equivalent to the costs associated with a cold start-up. So it would make sense to identify the popular functions and cache the execution requirements for them on a machine to execute them more efficiently. A different type of caching, known as history caching, begins with selecting the machine that will handle an incoming request. In all cases, the scheduler responsible for the assignment will determine the appropriate machine to use based on the availability of resources at the time that the task is executed. It is important to have this advanced level of proficiency to improve the quality of decisions made and the average response time for each decision [59]. However, despite this, the decision-making process's complexity in scheduling might increase profits from the division of tasks. Using history-aware scheduling in serverless computing has several benefits for providers. The following are a few of these benefits:

- increase the average response time speed
- reduce the time it takes for a node to start up
- Reduces the duration of overall execution
- Reduce costs associated with operations.

Since serverless computing is relatively new, the use of history-aware scheduling in this technology includes limited cases. Thus, we will examine some case studies related to this scheduling in the following.

Silab et al. [58] propose a multi-step process that prudently allocates tasks to machines, resulting in shortened average response times for users. For this work, they used the controller equipped with a small cache, which could be used for various purposes, including recording the history of task deployments and maintaining a hash of task outcomes. Finally, the evaluation results indicate that the presented cache-aided serverless approach can accelerate the decision-making procedure by as much as 21%, on average, by using and processing simultaneously [58].

Banaei and Sharifi [59] propose a predictive scheduling approach that can reduce response times for invocations and increase workers' productivity and resource utilization. They called this platform ETAS. In the ETAS approach, functions are scheduled based on their processing times, estimated by the history of their previous executions, arrival times, and the status of their containers. Finally, they implemented ETAS in Apache OpenWhisk. Also, they demonstrated that it outperforms other scheduling schemes by reducing average waiting times by 30% and improving throughput by 40% compared to OpenWhisk [59].

5 Open Issues and Future Directions

The serverless computing model is an emerging technology that allows on-demand provisioning of event processing services. On another side, the serverless platform provider is responsible for managing resources in serverless environments, so users' objectives in serverless analytics are distinct from those in server-centric environments. Using serverless computing, users do not have to rely on providers to dispatch and execute events on their behalf as they would with traditional virtual machine rental solutions. Instead, users can handle dispatching and executing events themselves via their functions. The charge for serverless computing is based solely on how many events are dispatched and the resources consumed in executing those events. Therefore, Deployment models have been transformed from the traditional resource allocation model to the on-demand execution model. It is important to note that data center efficiency has previously been criticized as contributing to comatose and idle servers. Therefore, by shifting from a rental model to a pay-per-use model, providers and customers can reduce the cost of deploying services closer to the level of demand they can serve realistically. Although minimizing job completion time remains the most critical objective, metrics such as resource isolation and utilization are now largely the cloud provider's responsibility. By shifting goals, serverless analytics users face a new schedule issue at the task level [60]. Hence, users must determine the optimal method of executing each job based on the cumulative task duration presented in this shift in goals. When it comes to the change of the goals, it is important to recognize that this flexibility and change come with a cost. In addition, it is essential to note that Serverless utilizes a stateless worker model, which means that each function is responsible for externalizing all the information that needs to be maintained across two consecutive calls. However, there is no guarantee that the context will remain active after the event has concluded. During the execution of a function, the amount of time devoted to data access and code initialization is another aspect of the billed execution time [61].

This amount is smaller than the time spent on virtual machines but is still significant compared to the entire time spent on the function. In the past, orchestrating cloud-native applications would demand the design of application-independent engineering and load balancers to determine a satisfactory trade-off between data locality, proactive resource allocation, and response time guarantees. Hence, it is the provider's responsibility to resolve these scheduling challenges under their service model. Serverless systems are expected to simplify operations automation and orchestration challenges since they can decompose the application into a series of stateless non-blocking executions that are structurally similar. Therefore, orchestration and automation will be easier to implement. Typically these are similar structured jobs that can be scaled separately and are also capable of decomposing into a functional set of stateless executions (e.g., events or data-driven). The provider challenge in serverless computing is to develop a comprehensive schedule that fits and is compatible with all applications' performance specifications. For stateless, short-lived executions, it is necessary to adjust the scheduling so that the reduced resource access latency and time can be accommodated to maximize resource allocation efficiency, resulting in reduced energy consumption, response times, deadline, and cost. The practice of preference scheduling is common in high-throughput computing. It is not uncommon for providers to strive for high utilization of resources while customers strive for a minimum response time. However, this can be challenging due to local conditions. It is not uncommon for providers to strive for high utilization of resources while customers strive for a minimum response time [62]. However, this can be challenging due to local conditions. In many cases, a function will require a different amount of computing power or memory according to its requirements. For example, serverless providers may charge a fee based on the consumption of the resources for short times or impose a strict time limit on the execution of the function. When scheduling for created workloads' processing, employ functions, and allocated resources, users should also consider the overheads, parallelism constraints, and the mechanism by which the data is accessed. However, this illustrates that the scheduling of cloud functions remains a significant challenge [63]. To determine which approaches are most likely to be effective with serverless models, it is necessary to rethink and investigate the scheduling problem. There are some of these critical challenges and solutions that will be discussed in the following in this section. These challenges and solutions include:

- The scheduling of heterogeneous resources is an important problem in managing resource utilization, and it depends on various factors, such as optimization criteria, constraints, time, and cost. In the best scenario, all functions scheduled for execution begin to execute after they are submitted, and there is no delay between the submission of each function and its execution. In this scenario, an idealized infrastructure may be assumed. Therefore, it is possible to distinguish between dynamic and static scheduling approaches. Consequently, schedules can be created dynamically and statically, and the type of scheduling used depends on the application, the goal, and the environment. The static approach involves allocating tasks to computing resources before executing them, for example, based on information obtained from prior operations. As far as utilizing resources is concerned, this approach is inarguably inefficient on the serverless. According to dynamic approaches, the schedule is monitored during execution so that it can be adjusted as necessary at runtime. For instance, assigning tasks to faster functions may be possible if a deadline is at risk. In comparison to static approaches, dynamic approaches are more complex. The scheduler in this approach needs to be set based on awareness of various environmental events; otherwise, implementation costs will be increased.

- As a result of the specific constraints imposed by the providers of FaaS on the time required for the execution of cloud functions, inevitable delays may occur during the execution of these functions. A situation may arise in which specific processes require more resources than the virtual limit that has been set, which can result in difficulties in allocating resources and, consequently, in their execution. Due to this issue, other processes cannot complete their tasks due to a reduction in resources. Therefore, it is imperative to employ a mixed-model execution approach when executing tasks whose duration may exceed the limit, which entails using standard virtual machines to execute the tasks. It remains, however, to address the problem of virtual machine provisioning, which involves determining the initial size of the virtual machines and the timing of their start-up and shutdown.

- This challenge arises from the fact that the current infrastructure and the ideal model differ significantly. Because the cloud's performance is variable, execution may deviate from the schedule that has been employed, by an awareness-based scheduling algorithm, as a consequence of the variability of cloud performance. Therefore, many of the defined constraints may result from this execution. Consequently, this event would probably be even more detrimental once the maximum number of concurrent users has been reached. Having to perform tasks competing with each other may result in a significant delay in the workflow, simultaneously. As it turns out, it is possible to address this type of challenge by evaluating the performance, structure, and mechanisms of providers of FaaS services that offer this type of service in a variety of manners. Several methods can facilitate this process, including systematically monitoring the functions that are called and used.

- Keeping in mind that cloud functions are typically stateless, it is necessary to use an external storage device to store intermediate data that the cloud function has not yet processed. The infrastructure model provides a solution for this problem. Therefore as a solution, the cloud can be used for object storage. Unlike the traditional scheduling of heterogeneous sources, in which data is transmitted directly between resources, data, in this case, isn't transmitted directly between resources.

Utilizing serverless frameworks and implementing them can break free from the complex trade-off about cost and completion time that eager and lazy solutions present [64]. It is possible to eliminate the complex trade-off demonstrated by eager and lazy solutions by designing fine-grained task-level schedulers that invoke functions on-demand at fine-grained timings to eliminate the complex trade-off. It is possible to accomplish this type of scheduling by using fine-grained scheduling, which launches each task at precisely the right time, optimizing both cost and time. These gains may also be even more significant for multi-stage tasks with a complex set of dependencies within each stage of the job regarding production workloads, response times, and cost savings. Thus, utilizing the optimal schedule strategies, there are expected to be significant improvements in the time required to complete the work and the cost of completing the tasks.

6 Discussion and Comparison

The investigation has shown that there have been very few studies conducted on the use of scheduling to improve the performance of a serverless computing system. This section reviews and analyzes a selection of research articles and papers and presents the results in Table 2. As part of this research, we investigate the strengths and weaknesses of each scheduling strategy in the serverless computing model and the parameters used to evaluate them. Following is a detailed description of the evaluation parameters used in this section.

6.1 Programming Language

There has been considerable interest among developers in serverless computing, a widely accepted technology because of its unique features, which have not yet been made available in other forms of computing. Hence, It is imperative that serverless computing be supported by a wide range of famous, powerful, and popular programming languages. Some of the most popular and well-known providers of serverless computing models offer support for the majority of these languages. For example, the Azure Functions service provides a more extensive range of programming languages, including TypeScript, F#, Batch, C#, Java, PHP, JavaScript (Node.js), PowerShell, Bash, and Python. Using the AWS Lambda service provider, you can develop functions in JavaScript (Node.js), Java, C#, Python, Go, Ruby, and PowerShell. The GCF service provider supports Python, Go, and JavaScript (Node.js). Cloudflare Worker's executing functions are developed in JavaScript or a language capable of being compiled into WebAssembly. Also, the IBM Cloud Functions service provider supports Swift and JavaScript (Node.js). The value of this parameter indicates which programming language has been employed to develop the functions [65].

6.2 Cloud Execution Layers

In recent years, there has been a great deal of innovation in cloud computing, fog computing, and edge computing. These technologies are being employed to

Table 2 Review and comparison of types of serverless computing scheduling

Used method	Advantage	Disadvantage	Performance metrics	Programming language	Cloud execution layers	Frameworks
Data-aware [68]	The data transmission rate between resources has been improved, and the execution speed has been increased	The lack of a proper data storage mechanism can impede the running of big data applications quickly without a server	The execution time and CPU utilization	Python	Cloud	AWS
Deadline-aware [69]	Reduce the cost incurred, and reduce response times by a significant amount	The research approach lacks robust evaluation mechanisms and does not support the parallel execution of multiple tasks	Capacity and average response time	Golang	Cloud	OpenWhisk
Resource-aware [70]	Reduction in the cost of application services and provision of a secure distributed environment	Lack of application of appropriate approaches such as dynamic learning and ineffectiveness of the approach presented as a result of diverse demands	The cost of execution and the total execution time	Python Java	Fog	AWS
Package-aware [71]	Using CPU resources efficiently and avoiding data overloading	An insufficient evaluation of the proposed schedule against workloads with a high volume and variety	The cost and throughput of execution	Golang	Cloud	AWS
Cost-aware [45]	Reduces bandwidth consumption and costs associated with resource consumption	Lacked deployment in public cloud systems and evaluated for real-time workloads from a variety of applications	The average utilization of VMs over a given period and CPU utilization	Java	Cloud	AWS
History-aware [58]	Improves the average response time speed and reduces the overall execution duration	The lack of use of high-efficiency techniques, such as machine learning, to improve performance	Implementation of tasks and delivery of outcomes	Python Golang	Cloud	Open-FaaS

increase the efficiency and speed of the collection and processing of information and bring cognitive capabilities closer to sensors and actuators. Cloud computing, fog computing, and edge computing differ in how they are used, where, and when the information can be collected, processed, and stored. In contrast to other types of layers, cloud storage and processing environments are located further away from endpoints. It has the highest bandwidth, latency, and network requirements. However, the cloud is an effective global solution capable of handling large amounts of processing data and tasks because a more significant number of resources and servers are engaged. Hence, this product is ideal for analyzing big, historical data and storing large amounts of data for long periods. Fog is an intermediate layer between cloud computing and edge computing, providing advantages from both. In addition to relying on and working closely with the cloud, fog is located closer to the network's edge. It utilizes storage resources and local computing to provide quick responses and real-time analytics to events when necessary. In a similar manner to the edge, fog is decentralized, containing a large number of nodes. However, as opposed to the edge, fog is structured as a network. As fog nodes are interconnected, they can redistribute storage resources and computing to carry out tasks more effectively. Edge computing has the fastest response and the lowest latency to data and is closest to end devices. In this layer, there is limited capacity to process small volumes of data directly on applications, devices, and edge gateways. Edge nodes usually have a sparsely connected structure, interacting independently with data. This feature distinguishes an edge from a network-based fog [66].

6.3 Frameworks

Currently, Several open-source and closed-source serverless frameworks are available to accommodate the ever-increasing demand for serverless computing. Some frameworks are developed based on existing cloud infrastructures in follow:

- Kubernetes: Kubernetes, for serving applications. On the other hand, some frameworks are developed from scratch and deployed directly on servers instead of embedded within existing cloud infrastructures. The implementation methods, there will be a significant impact on the schedule performance of such frameworks. Kubeless is an Open-source serverless framework that builds on core Kubernetes features such as services, ConfigMap, and deployment. Kubeless uses CRDs to create functions as custom Kubernetes resources and employs a controller to manage and execute those resources. In Kubeless, most of the existing scheduling logic codes within the Kubernetes kernel are simplified, so there is no need for developers to rewrite them. Implementing this architecture balances the platform's efficiency and the convenience of designing a serverless platform.
- OpenWhisk: The OpenWhisk is a robust and open-source framework for serverless computing sanctioned and promoted by the Apache Software Foundation, which offers a variety of options for deployment. Many essential components

in Apache OpenWhisk contribute to providing sufficient scalability and quality. Comparatively to other open-source projects, OpenWhisk has many participants and high-quality features. Despite this, its mega tools challenge developers and users who must learn these tools. In addition, the complexity of the components adds a level of performance overhead, causing performance issues, especially in scenarios with high concurrency.

- OpenFaaS: Unlike OpenWhisk, OpenFaaS is an open source, cloud-free application (native) server framework. It includes API gateways for invocation tools, auto-scaling components, and asynchronous calls. Considering its lightweight feature, OpenFaas is suitable for deployment in private and public cloud environments. As well as being able to be deployed on bare metal servers, it can be deployed in various cloud management systems.

- AWS: AWS is one of the market's most established and mature service providers of serverless frameworks. It is also one of the most popular and well-established providers. The number of AWS services that Lambda integrates with is growing every day, and, indeed, this list will only continue to grow in the future as Lambda expands. IoT and mobile developers prefer Lambda for its flexibility and power, which makes it an ideal platform for developers. Known as one of the most popular platforms for creating applications that can be activated by voice for Amazon Echo devices, it is of particular interest to developers because of its strong compatibility with Alexa Skills Kit. The command line tools and an interactive console are available on Amazon for uploading and managing snippets.

- GCF: GCF is a service that provides serverless computing on the Google Cloud platform. The GCF platform provides not only a serverless execution environment but also an SDK for developing and managing functions throughout their entire lifecycle, as with other serverless platforms. GCF have become increasingly popular for several reasons; they have simplified the complex development process for developers. Also, they are highly scalable and have been shown to reduce the complexity involved with infrastructure deployments; finally, their deployment requires significantly fewer steps than Lambda. Only one step is involved in the deployment process using this serverless provider, so users can integrate with GCF to manage their deployment infrastructure needs. The CLI tool makes it relatively easy to deploy the serverless application when it is ready for deployment. This parameter determines the type of framework that will be used to implement the scheduler [67].

6.4 Performance Metrics

Researchers find research metrics essential in measuring the quality and effectiveness of their research about their defined research goals. They provide evidence of the researcher's performance in a particular area of inquiry and can also be used to demonstrate the researcher's work value to the research process. It is essential to define this parameter according to the goals considered during the implementation process of the scheduler.

Serverless computing has attracted great interest in recent years due to its unique features that make it attractive to researchers. Unfortunately, this technology has not yet reached a sufficient level of maturity. Consequently, any approach implemented for this technology, such as a scheduling process, will likely bring many challenges and difficulties. It is impossible to decide which scheduler is better than the other because this involves taking into account many factors and variables simultaneously. In this regard, determining and implementing the appropriate scheduler will largely depend on the purpose for which it is considered. Moreover, it is crucial to consider that different environments and frameworks can result in different results when implementing a scheduler. Selecting a scheduler for any project requires consideration of various factors, so it is crucial to choose one that is both appropriate for the purpose and suitable for its implementation.

7 Conclusion

Despite being a relatively new technology, serverless computing has already attracted the attention of many researchers within a short period. A key aspect of this technology is that it considers its application and goals to accomplish two of the most important issues simultaneously. As a first issue, there is the issue of users. Users are always looking for the best and highest quality service providers to get the best services at the lowest possible cost and respond to their requests quickly. This is why users always look for the best and highest quality service provider. Moreover, there is a second issue involving all service providers wishing to provide the same quality services to their customers at a lower cost so that they may continue to maintain their customer base and remain profitable. Due to this, it becomes increasingly important to manage the allocated resources properly. A scheduling structure is one of the most effective elements for solving defined issues in serverless computing. When it comes to maximizing the utilization of resources, it is imperative to select a scheduler capable of allocating resources in a way that maximizes their utilization and, in turn, benefits both the user and the provider of the service. Therefore, selecting a scheduler capable of accomplishing this task with the most significant degree of efficiency and effectiveness is critical. Thus, selecting and implementing a scheduler can be fraught with challenges and problems, which require more extensive research to discover solutions to the existing problems and challenges.

In the future, it is possible to move the scheduling process in a direction that can be more effective in dealing with various variables, including volume, variety, and number, by conducting more thorough research and utilizing other technologies, such as artificial intelligence. Therefore, it is crucial to design a flexible scheduling system that can accommodate strategies appropriate to different conditions, depending on the type of conditions and environment, and implement the proper strategies.

References

1. Mustafa S, Nazir B, Hayat A, Madani SA (2015) Resource management in cloud computing: taxonomy, prospects, and challenges. Comput Electr Eng 47:186–203
2. Younge AJ, Von Laszewski G, Wang L, Lopez-Alarcon S, Carithers W (2010) Efficient resource management for cloud computing environments. In: International conference on green computing. IEEE, pp 357–364
3. Bi J, Li S, Yuan H, Zhou M (2021) Integrated deep learning method for workload and resource prediction in cloud systems. Neurocomputing 424:35–48
4. Haber MJ, Chappell B, Hills C (2022) Cloud computing. In: Cloud attack vectors. Springer, Berlin, pp 9–25
5. Khan Y, Varma S (2020) An efficient cloud forensic approach for IaaS, SaaS and PaaS model. In: 2nd international conference on data, engineering and applications (IDEA). IEEE, pp 1–6
6. Lorido-Botran T, Miguel-Alonso J, Lozano JA (2014) A review of auto-scaling techniques for elastic applications in cloud environments. J Grid Comput 12(4):559–592
7. McGrath G, Brenner PR (2017) Serverless computing: design, implementation, and performance. In: IEEE 37th international conference on distributed computing systems workshops (ICDCSW). IEEE, pp 405–410
8. Pérez A, Moltó G, Caballer M, Calatrava A (2018) Serverless computing for container-based architectures. Futur Gener Comput Syst 83:50–59
9. Mahmoudi N, Khazaei H (2020) Temporal performance modelling of serverless computing platforms. In: Proceedings of the sixth international workshop on serverless computing, pp 1–6
10. Suresh A, Gandhi A (2021) Server more: opportunistic execution of serverless functions in the cloud. In: Proceedings of the ACM symposium on cloud computing, pp 570–584
11. Shahrad M, Balkind J, Wentzlaff D (2019) Architectural implications of function-as-a-service computing. In: Proceedings of the 52nd annual IEEE/ACM international symposium on microarchitecture, pp 1063–1075
12. Sánchez-Artigas M, Sarroca PG (2021) Experience paper: towards enhancing cost efficiency in serverless machine learning training. In: Proceedings of the 22nd international middleware conference, pp 210–222
13. Raza A, Matta I, Akhtar N, Kalavri V, Isahagian V (2021) SoK: function-as-a-service: from an application developer's perspective. J Syst Res 1(1)
14. Scheuner J, Leitner P (2020) Function-as-a-service performance evaluation: a multifocal literature review. J Syst Softw 170:110708
15. Kaffes K, Yadwadkar NJ, Kozyrakis C (2019) Centralized core-granular scheduling for serverless functions. In: Proceedings of the ACM symposium on cloud computing, pp 158–164
16. Werner S, Girke R, Kuhlenkamp J (2020) An evaluation of serverless data processing frameworks. In: Proceedings of the sixth international workshop on serverless computing, pp 19–24
17. Hellerstein JM, Faleiro J, Gonzalez JE, Schleier-Smith J, Sreekanti V, Tumanov A, Wu C (2018) Serverless computing: one step forward, two steps back. arXiv preprint arXiv:1812.03651
18. Choi S, Shahbaz M, Prabhakar B, Rosenblum M (2019) λ-nic: interactive serverless compute on smartnics. In: Proceedings of the ACM SIGCOMM conference posters and demos, pp 151–152
19. Carver B, Zhang J, Wang A, Anwar A, Wu P, Cheng Y (2020) Wukong: a scalable and locality-enhanced framework for serverless parallel computing. In: Proceedings of the 11th ACM symposium on cloud computing, pp 1–15
20. Pu Q, Venkataraman S, Stoica I (2019) Shuffling, fast and slow: scalable analytics on serverless infrastructure. In: 16th USENIX symposium on networked systems design and implementation (NSDI 19), pp 193–206

21. Castro P, Ishakian V, Muthusamy V, Slominski A (2019) The server is dead, long live the server: rise of serverless computing, overview of current state and future trends in research and industry. arXiv preprint arXiv:1906.02888

22. Larrucea X, Santamaria I, Colomo-Palacios R, Ebert C (2018) Microservices. IEEE Softw 35(3):96–100

23. Solaiman K, Adnan MA (2020) Wlec: a not so cold architecture to mitigate cold start problem in serverless computing. In: IEEE international conference on cloud engineering (IC2E). IEEE, pp 144–153

24. Al-Ali Z, Goodarzy S, Hunter E, Ha S, Han R, Keller E, Rozner E (2018) Making serverless computing more serverless. In: IEEE 11th international conference on cloud computing (CLOUD). IEEE, pp 456–459

25. Sewak M, Singh S (2018) Winning in the era of serverless computing and function as a service. In: 3rd international conference for convergence in technology (I2CT). IEEE, pp 1–5

26. Oakes E, Yang L, Zhou D, Houck K, Harter T, Arpaci-Dusseau A, Arpaci-Dusseau R (2018) {SOCK}: rapid task provisioning with {serverless-optimized} containers. In: 2018 USENIX annual technical conference (USENIX ATC 18), pp 57–70

27. Aytekin A, Johansson M (2019) Exploiting serverless runtimes for large-scale optimization. In: IEEE 12th international conference on cloud computing (CLOUD). IEEE, pp 499–501

28. Basu S, Kannayaram G, Ramasubbareddy S, Venkatasubbaiah C (2019) Improved genetic algorithm for monitoring of virtual machines in cloud environment. In: Smart intelligent computing and applications. Springer, Berlin, pp 319–326

29. Gouda K, Radhika T, Akshatha M (2013) Priority based resource allocation model for cloud computing. Int J Sci Eng Technol Res (IJSETR) 2(1):215–219

30. Singh S, Chana I, Singh M (2017) The journey of QoS-aware autonomic cloud computing. IT Professional 19(2):42–49

31. Xu B, Zhao C, Hu E, Hu B (2011) Job scheduling algorithm based on Berger model in cloud environment. Adv Eng Softw 42(7):419–425

32. Alqaryouti O, Siyam N (2018) Serverless computing and scheduling tasks on cloud: a review. Am Acad Sci Res J Eng Technol Sci 40(1):235–247

33. Niu X, Kumanov D, Hung L-H, Lloyd W, Yeung KY (2019) Leveraging serverless computing to improve performance for sequence comparison. In: Proceedings of the 10th ACM international conference on bioinformatics, computational biology and health informatics, pp 683–687

34. Zhao L, Yang Y, Li Y, Zhou X, Li K (2021) Understanding, predicting and scheduling serverless workloads under partial interference. In: Proceedings of the International conference for high performance computing, networking, storage and analysis, pp 1–15

35. Yuvaraj N, Karthikeyan T, Praghash K (2021) An improved task allocation scheme in serverless computing using gray wolf optimization (GWO) based reinforcement learning (RIL) approach. Wireless Pers Commun 117(3):2403–2421

36. Mampage A, Karunasekera S, Buyya R (2021) Deadline-aware dynamic resource management in serverless computing environments. In: IEEE/ACM 21st international symposium on cluster, cloud and internet computing (CCGrid). IEEE, pp 483–492

37. Lloyd W, Vu M, Zhang B, David O, Leavesley G (2018) Improving application migration to serverless computing platforms: latency mitigation with keep-alive workloads. In: IEEE/ACM international conference on utility and cloud computing companion (UCC Companion). IEEE, pp 195–200

38. Gramaglia M, Serrano P, Banchs A, Garcia-Aviles G, Garcia-Saavedra A, Perez R (2020) The case for serverless mobile networking. In: IFIP networking conference (Networking). IEEE, pp 779–784

39. Pawlik M, Banach P, Malawski M (2019) Adaptation of workflow application scheduling algorithm to serverless infrastructure. In: European conference on parallel processing. Springer, Berlin, pp 345–356

40. García-López P, Sánchez-Artigas M, Shillaker S, Pietzuch P, Breitgand D, Vernik G, Sutra P, Tarrant T, Ferrer AJ (2019) Servermix: tradeoffs and challenges of serverless data analytics. arXiv preprint arXiv:1907.11465

41. Singhvi A, Houck K, Balasubramanian A, Shaikh MD, Venkataraman S, Akella A (2019) Archipelago: a scalable low-latency serverless platform. arXiv preprint arXiv:1911.09849
42. Yao C, Liu W, Tang W, Hu S (2022) EAIS: energy-aware adaptive scheduling for CNN inference on high-performance GPUs. Futur Gener Comput Syst 130:253–268
43. Kallam S, Patan R, Ramana TV, Gandomi AH (2021) Linear weighted regression and energy-aware greedy scheduling for heterogeneous big data. Electronics 10(5):554
44. Aslanpour MS, Toosi AN, Cheema MA, Gaire R (2022) Energy-aware resource scheduling for serverless edge computing. In: 22nd IEEE international symposium on cluster, cloud and internet computing (CCGrid). IEEE, pp 190–199
45. Gunasekaran JR, Thinakaran P, Kandemir MT, Urgaonkar B, Kesidis G, Das C (2019) Spock: exploiting serverless functions for slo and cost aware resource procurement in public cloud. In: IEEE 12th international conference on cloud computing (CLOUD). IEEE, pp 199–208
46. Rausch T, Rashed A, Dustdar S (2021) Optimized container scheduling for data-intensive serverless edge computing. Futur Gener Comput Syst 114:259–271
47. HoseinyFarahabady MR, Taheri J, Zomaya AY, Tari Z (2021) Data-intensive workload consolidation in serverless (Lambda/FaaS) platforms. In: IEEE 20th international symposium on network computing and applications (NCA). IEEE, pp 1–8
48. Wu J, Wu M, Li H, Li L, Li L (2022) A serverless-based, on-the-fly computing framework for remote sensing image collection. Remote Sens 14(7):1728
49. Krishna SR, Majji S, Kishore SK, Jaiswal S, Kostka JAL, Chouhan AS (2021) Optimization of time-driven scheduling technique for serverless cloud computing. Turkish J Comput Math Educ 12(10):1–8
50. Wang B, Ali-Eldin A, Shenoy P (2021) Lass: running latency sensitive serverless computations at the edge. In: Proceedings of the 30th international symposium on high-performance parallel and distributed computing, pp 239–251
51. Cheng Y, Zhou Z (2018) Autonomous resource scheduling for real-time and stream processing. In: IEEE smart world, ubiquitous intelligence and computing, advanced and trusted computing, scalable computing and communications, cloud and big data computing, internet of people and smart city innovation (Smart World/SCALCOM/UIC/ATC/CBDCom/IOP/SCI). IEEE, pp 1181–1184
52. Suresh A, Gandhi A (2019) Fnsched: an efficient scheduler for serverless functions. In: Proceedings of the 5th international workshop on serverless computing, pp 19–24
53. Kim YK, HoseinyFarahabady MR, Lee YC, Zomaya AY (2020) Automated fine-grained cup cap control in serverless computing platform. IEEE Trans Parallel Distrib Syst 31(10):2289–2301
54. De Palma G, Giallorenzo S, Mauro J, Trentin M, Zavattaro G (2022) A declarative approach to topology-aware serverless function-execution scheduling. In: 2022 IEEE international conference on web services (ICWS). IEEE, pp 337–342
55. Bai T, Nie J-Y, Zhao WX, Zhu Y, Du P, Wen J-R (2018) An attribute-aware neural attentive model for next basket recommendation. In: The 41st international ACM SIGIR conference on research and development in information retrieval, pp 1201–1204
56. Bisht J, Vampugani VS (2022) Load and cost-aware min-min workflow scheduling algorithm for heterogeneous resources in fog, cloud, and edge scenarios. Int J Cloud Appl Comput (IJCAC) 12(1):1–20
57. Shafiei H, Khonsari A, Mousavi P (2022) Serverless computing: a survey of opportunities, challenges, and applications. ACM Comput Surv 54(11s):1–32. Article No: 239. https://doi.org/10.1145/3510611
58. Silab MV, Hassanpour SB, Khonsari A, Dadlani A (2022) On skipping redundant computation via smart task deployment for faster serverless. In: ICC-IEEE international conference on communications. IEEE, pp 5475–5480
59. Banaei A, Sharifi M (2022) ETAS: predictive scheduling of functions on worker nodes of Apache OpenWhisk platform. J Supercomput 78(4):5358–5393
60. Van Eyk E, Iosup A, Seif S, Thömmes M (2017) The SPEC cloud group's research vision on FaaS and serverless architectures. In: Proceedings of the 2nd international workshop on serverless computing, pp 1–4

61. Kijak J, Martyna P, Pawlik M, Balis B, Malawski M (2018) Challenges for scheduling scientific workflows on cloud functions. In: IEEE 11th international conference on cloud computing (CLOUD). IEEE, pp 460–467
62. Li Y, Lin Y, Wang Y, Ye K, Xu C-Z (2022) Serverless computing: state-of-the-art, challenges and opportunities. IEEE Trans Serv Comput. https://doi.org/10.1109/TSC.2022.3166553
63. Gadepalli PK, Peach G, Cherkasova L, Aitken R, Parmer G (2019) Challenges and opportunities for efficient serverless computing at the edge. In: 38th symposium on reliable distributed systems (SRDS). IEEE, pp 261–2615
64. Aslanpour MS, Toosi AN, Cicconetti C, Javadi B, Sbarski P, Taibi D, Assuncao M, Gill SS, Gaire R, Dustdar S (2021) Serverless edge computing: vision and challenges. In: Australasian computer science week multiconference, pp 1–10
65. Kritikos K, Skrzypek P (2018) A review of serverless frameworks. In: IEEE/ACM international conference on utility and cloud computing companion (UCC Companion). IEEE, pp 161–168
66. Wang H, Liu T, Kim B, Lin C-W, Shiraishi S, Xie J, Han Z (2020) Architectural design alternatives based on cloud/edge/fog computing for connected vehicles. IEEE Commun Surv Tutorials 22(4):2349–2377
67. Martins HJM (2019) Plataformas de computação serverless: Estudo e benckmark. Universidade de Coimbra
68. Das S (2021) Ant colony optimization for mapreduce application to optimize task scheduling in serverless platform. National College of Ireland, Dublin
69. Zuk P, Rzadca K (2022) Reducing response latency of composite functions-as-a-service through scheduling. J Parallel Distrib Comput 167:18–30
70. Lakhan A, Mohammed MA, Rashid AN, Kadry S, Panityakul T, Abdulkareem KH, Thinnukool O (2021) Smart-contract aware ethereum and client-fog-cloud healthcare system. Sensors 21(12):4093
71. Totoy G, Boza EF, Abad CL (2018) An extensible scheduler for the OpenLambda FaaS platform. In: Min-Move'18

Serverless Cloud Computing: State of the Art and Challenges

Vincent Lannurien, Laurent D'Orazio, Olivier Barais, and Jalil Boukhobza

Abstract The serverless model represents a paradigm shift in the cloud: as opposed to traditional cloud computing service models, serverless customers do not reserve hardware resources. The execution of their code is event-driven (HTTP requests, cron jobs, etc.) and billing is based on actual resource usage. In return, the responsibility of resource allocation and task placement lies on the provider. While serverless in the wild is mainly advertised as a public cloud offering, solutions are actively developed and backed by solid actors in the industry to allow the development of private cloud serverless platforms. The first generation of serverless offers, "Function as a Service" (FaaS), has severe shortcomings that can offset the potential benefits for both customers and providers—in terms of spendings and reliability on the customer side, and in terms of resources multiplexing on the provider side. Circumventing these flaws would allow considerable savings in money and energy for both providers and tenants. This chapter aims at establishing a comprehensive tour of these limitations, and presenting state-of-the-art studies to mitigate weaknesses that are currently holding serverless back from becoming the de facto cloud computing model. The main challenges related to the deployment of such a cloud platform are discussed and some perspectives for future directions in research are given.

Keywords Cloud computing · Resources allocation · Task placement · Hardware heterogeneity · Scaling · Caching · Isolation · QoS · SLA

V. Lannurien (✉) · J. Boukhobza
Lab-STICC, UMR 6285, CNRS, b<>com Institute of Research and Technology, ENSTA Bretagne, Brest, France
e-mail: vincent.lannurien@ensta-bretagne.org

J. Boukhobza
e-mail: jalil.boukhobza@ensta-bretagne.fr

L. D'Orazio · O. Barais
CNRS, Inria, IRISA, b<>com Institute of Research and Technology, University of Rennes, Rennes, France
e-mail: laurent.dorazio@irisa.fr

O. Barais
e-mail: olivier.barais@irisa.fr

© The Author(s), under exclusive license to Springer Nature Switzerland AG 2023
R. Krishnamurthi et al. (eds.), *Serverless Computing: Principles and Paradigms*,
Lecture Notes on Data Engineering and Communications Technologies 162,
https://doi.org/10.1007/978-3-031-26633-1_11

1 Introduction

In 1961, John McCarthy imagined that the time-sharing of computers could make it possible to sell their execution power as a service, just like water or electricity [50]. Due to the democratization of high-speed Internet access in the mid-2000s, McCarthy's idea was implemented in what is known as *cloud computing*: companies and individuals can now drastically reduce the costs associated with the purchase and maintenance of the hardware needed to run their applications by delegating responsibility for the infrastructure to service providers. This model is called "Infrastructure as a Service" (IaaS) [78].

Over the years, new trends appeared with the aim of reducing the customer's responsibilities. For example, in the "Platform as a Service" (PaaS) model, customers do not have direct access to the machines that support their applications and perform most of the management tasks via specialized interfaces. In these models, the customer pays for resources that are sometimes dormant. This is because reserved resources must often be over-provisioned in order to be able to absorb the surge in activity and handle hardware failures [52].

The serverless model represents a paradigm shift in the cloud: as opposed to traditional models, serverless customers do not reserve hardware resources. The execution of their code is event-driven (HTTP requests, cron jobs, etc.) and billing is based on actual resource usage. In return, the responsibility of resource allocation and task placement lies on the provider [106].

While serverless in the wild is mainly advertised as a public cloud offering (Table 3), solutions are actively developed and backed by solid actors in the industry to allow the development of private cloud serverless platforms (Table 4).

The first generation of serverless offers, "Function as a Service" (FaaS), has severe shortcomings that can offset the potential benefits for both customers and providers—in terms of spendings and reliability on the customer side, and in terms of resources multiplexing on the provider side. Circumventing these flaws would allow considerable savings in money and energy for both providers and customers. This chapter aims at establishing a comprehensive tour of these limitations and presenting state-of-the-art studies to mitigate weaknesses that are currently holding serverless back from becoming the de facto cloud computing model.

This chapter is organized as follows: after an introduction, some background and motivations related to serverless computing are given. Those include an introduction to cloud computing and virtualization technologies. Then, the serverless paradigm is introduced: we discuss its characteristics, benefits and offerings for both public and private cloud. A state of the art related to the challenges on resources management for serverless computing is then drawn. It is built around six issues: cold start, cost of inter-function communication, local state persistence, hardware heterogeneity, isolation and security, and programming model and associated risks of vendor lock-in. Finally, we introduce perspectives and future directions for research in the field of serverless computing.

2 Background and Motivations

In this first section, we introduce characteristics of cloud computing; its service models and associated technologies. We also take a glance at recent transformations in the ways developers program for the cloud, and the consequences in terms of applications deployment.

2.1 The Promises of Cloud Computing

The NIST definition of cloud computing [82] gives five essential characteristics for the cloud computing model, as opposed to on-premises and/or bare metal deployments:

- **On-demand self-service**—Customers can book hardware resources through e.g. a web interface rather than by interacting with operators. In return, they usually have limited control over the geographic location for these resources;
- **Broad network access**—These resources are immediately made available over public broadband connections by the provider;
- **Resource pooling**—Compute time, storage capacity, network bandwidth are all shared between customers. Virtualization techniques are used to ensure isolation between workloads;
- **Rapid elasticity**—Applications can benefit from increased or decreased computing power through scaling, as resources are dynamically provisioned and released to absorb variations in demand;
- **Measured service**—Cloud providers instrument their infrastructures so as to precisely monitor resource usage. Customers can then be charged in a fine way according to their needs.

These characteristics are found in three cloud service models, as shown in Fig. 1:

- **Software as a Service** (SaaS)—Targets the end user by offering access to fully managed software (the application);
- **Platform as a Service** (PaaS)—Targets developers and DevOps teams who want to deploy their applications quickly without managing servers, generally through the use of containers. Customers deploy their applications on top of a runtime environment that is managed by the provider;
- **Infrastructure as a Service** (IaaS)—Targets architects and system administrators who want fine-grained control over their infrastructures. IaaS customers are in charge of their own servers, usually virtual.

It is debatable how much traditional IaaS and PaaS cloud offerings hold up to their promise regarding rapid elasticity and measured service. SaaS is out of the scope of this study, as it targets end users rather than application developers.

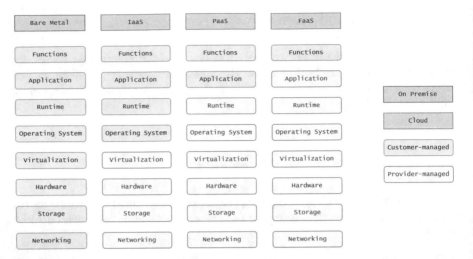

Fig. 1 Comparison of the different cloud service models in terms of customer responsibility [inspired from Red Hat's documentation (https://www.redhat.com/en/topics/cloud-computing/iaas-vs-paas-vs-saas)]

Elasticity, in the sense of "automatic scaling", intrinsically cannot be offered by IaaS or PaaS deployments. Developers are responsible for planning ahead and specifying their needs, that is, booking an adequate quantity of resources [4]. These are usually called "instances" by cloud providers. Cloud instances are typically distinguished according to their specifications in resource type and capacity: for example, there can be instances with many CPU cores, whereas others provide access to a GPU or some other hardware accelerator.

The choice of instance type(s) and quantity for an application depends on (a) the nature of the computations it runs, and (b) the acceptable latency and the desired throughput [134]. However, it is the customer's responsibility to not over-provision beyond their actual need.

This offering design has further consequences. First, it means that billing is done coarse-grained: per instances booked rather than per resources actually used. Besides, idle resources are always paid for, and scaling to zero cannot be achieved in this setting, because an application will always require at least one instance running to handle an incoming request.

Cloud computing platforms have to accommodate for an important number of customer jobs, leading to massive multitenancy which requires adequate isolation and virtualization techniques. Those are introduced in the next section.

2.2 *Virtualization Technologies*

Multitenancy is a defining characteristic of cloud computing. It is the ability for a cloud provider to share resources between multiple customers to secure cost savings [130]. As resources are pooled and different applications make use of them, it creates direct channels between processes in the user space: multitenancy comes with the responsibility for the provider to guarantee privacy and security across the different workloads of the customers [124].

To meet these guarantees, providers must resort to protection measures providing airtightness between processes that are not supposed to be aware of each others. This mechanism of transparently presenting separate execution environment with distinct address space, filesystem and permissions is called isolation [40]. For this purpose, providers may rely on virtualization technologies.

Virtualization is an isolation technique that allows one to run an application within the boundaries of a secure execution environment, called a *sandbox*, by introducing a layer of indirection between the host platform and the application itself [113].

Virtualization of the host resources can be done with virtual machines (VMs) or containers. These sandboxes give the underlying processes the impression of having an entire machine at hand, but while VMs virtualize the physical resources of the host, relying on the CPU's architecture to achieve isolation, containers depend on the host operating system's system call API to isolate workloads [77].

These techniques are beneficial both on the provider and the developer side. The former leverages virtualization to achieve isolation of customers' workloads as well as flexibility to manage scaling of sandboxes given a finite amount of hardware resources. The latter organizes their development pipelines so as to replicate a production-like environment during development stages, and to deliver and deploy their products.

Virtualization became such a cornerstone in cloud computing that Kubernetes [30], an orchestration system that leverages containers[1] to manage applications from deployment to scaling of services, is increasingly referred to as "the operating system for the cloud" [103].

When choosing the isolation model they want to rely on to achieve multitenancy, cloud providers have to trade off between performance and security. Containers are frequent targets of privilege escalation attacks [16, 137] but they perform multiple orders of magnitude better than virtual machines: containers startup time is in the hundreds of milliseconds, while VMs boot in seconds [77]. Designing lightweight virtual machines that offer performances comparable to containers is a critical research topic [1, 10].

---

[1] Note that Kubevirt [31] aims at making Kubernetes suitable for VM workloads.

2.2.1 Virtual Machines

Virtual machines virtualize the physical resources of the host: hardware-assisted virtualization allows multiple full-fledged *guest* operating systems to run independently on shared physical resources, regardless of the nature of the *host* operating system [64].

From a VM point of view, the execution sandbox is seen as a complete platform, while it actually is a subset of the host computer resources, determined by the hypervisor (or VMM for Virtual Machine Manager), a low-level software that can run on bare metal or as a process of the host operating system.

The hypervisor has the responsibility to manage VM lifecycle: creation, execution, destruction, and sometimes migration of virtual machines are handled by the hypervisor.

Hypervisors exist in two different abstractions, as shown in Fig. 2:

- Type-1 (bare-metal) hypervisors run directly on the host machine's hardware. They rely on the host's CPU support for virtualization. Given that they do not depend on an underlying operating system, they are considered more secured and efficient than their hosted counterpart. Common examples of type-1 hypervisors include VMware ESXi [127], KVM [71], Xen [70] and Hyper-V [85];

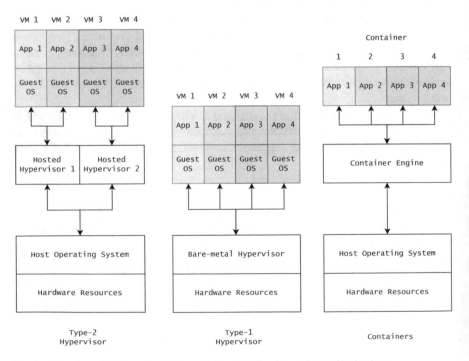

Fig. 2 Overview of different isolation models: virtualization and containerization

- Type-2 (hosted) hypervisors run on top of an operating system. These hypervisors are consumer-grade products that provide a convenient way for end users to run systems or programs that would otherwise not be supported by their hardware or OS. Examples of type-2 hypervisors include QEMU [102] and Oracle VirtualBox [94].

2.2.2   Containers

Containerization is an OS-level virtualization technique. The host operating system's kernel is responsible for the allocation of resources. Containers virtualize the OS: they give the containerized process the impression of having the entire machine at hand, while actually being constrained and limited regarding resource utilization by the host kernel [17].

From the running application's point of view, the execution platform behaves as if it were bare metal. However, its allocated resources are actually a virtualized subset of the host's hardware resources.

Containers constitute a lightweight isolation mechanism that relies on the kernel isolation capabilities of the host system, as shown in Fig. 2. Namely, under Linux:

- `chroot` changes the apparent root directory for a given process tree. It allows a container to operate on a virtual/directory that could be located anywhere on the host's filesystem;
- `cgroups` create hierarchical groups of processes and allocates, limits, and monitors hardware resources for these groups: I/O to and from block devices, accesses to the CPU, memory and network interfaces;
- `namespaces` are an abstraction layer around global system resources, such as networking or IPC. Processes within a namespace have their own isolated instances of these resources.

The idea behind containers is to sandbox an application's execution in a process isolated from the rest of the system. That process is bootstrapped from an image which contains all dependencies needed to either build and/or execute the application.

Among the container ecosystem, Docker [36] in particular has seen important traction since its inception in 2013. Docker was instrumental in specifying industry standards for container formats and runtimes through the Open Container Initiative (OCI) [69].

The OCI specification is a Linux Foundation [120] initiative to design an open standard for containers. It defines container image specifications—guidelines on how to create an OCI image with its manifest, filesystem layers, and configuration— and runtime specifications regarding how to execute application bundles as they are unpacked on the host OS.

This is how that specification translates in the case of Docker:

- `dockerd` is the daemon that provides both the Docker Engine application programming interface (API) and console line interface (CLI), capable of building distributable images representing the initial state of future containers. It is the high-level interface through which the user can either programmatically or interactively manage networking, storage, images and containers lifecycle;

- `containerd`, a Cloud Native Computing Foundation (CNCF) [41] initiative, handles containers lifecycle (hypervision, execution) and manages images (push and pull to and from image registries), storage and networking by establishing links between containers namespaces;
- `runc` implements the Open Container Initiative specification and contains the actual code allowing a container's execution. It creates and starts containers, and stops their execution.

2.3 From Monoliths to Microservices

Cloud computing saw the birth of new development techniques. "Cloud-native development" means building applications for the cloud from the ground up, with scaling capacities in mind [37, 43].

When an application grows, there are two ways of making room for new requests by scaling:

- **Vertically**: attaching more hardware resources to the servers that support the application. It may mean moving data to new, more powerful servers, thus impacting application availability;
- **Horizontally**: increasing the server count for running the application. It may mean introducing a load balancing mechanism to route requests and responses between users and the multiple instances of an application, thus impacting complexity.

A monolithic application is built as a single unit. There is no decoupling between the services it exposes, as they are all part of the same codebase [126]. When scaling for a monolith, adding more resources (scaling vertically) does not solve the problem of competing priorities inside the application: when the popularity of a monolithic application increases, some parts of the codebase will be solicited more than others, but the strain will not be distributed across the application. On the other hand, spinning up more instances of the monolith (scaling horizontally) can prove cost ineffective, as not all parts of an application suffer from load spikes at the same time: in Fig. 3, an increase in authentication requests means scaling the infrastructure for the whole application.

The Twelve-Factor App methodology, a set of guidelines for building cloud-native applications, recommends "[executing] the app as one or more stateless process" [131]. This is called a microservices architecture—arranging an application as a collection of loosely-coupled services. Each of these services runs in its own process, communicates with the others through message-passing and can be deployed to scale

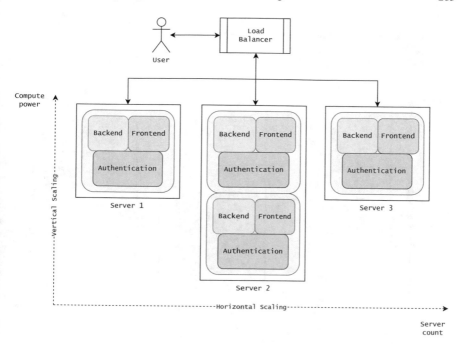

Fig. 3 Scaling out a monolith-architectured application requires replicating the monolith on multiple servers

independently on heterogeneous servers to meet service level objectives: in Fig. 4, an increase in authentication requests can be absorbed by scaling the infrastructure for the authentication microservice alone.

However, a microservices infrastructure implies a complex layer of centralized management, which either drives costs in operations or in DevOps teams. It relies on long-running backend services (databases, message buses, etc.) that also have to be monitored and managed. In IaaS and PaaS settings, developer experience in particular is not satisfying: cloud deployment comes with a burden of systems administration [62]. Microservices alone do not solve the deployment problem: writing container images recipes does not add value to the product, as it relates to operational logic rather than business logic.

Continuous Integration (CI) and Continuous Delivery or Deployment (CD) are foundational building blocks of the DevOps culture—the idea that, at any given stage of an application's lifecycle, its codebase is in a working state [68]. This can be achieved through the automation or running unit and integration test suites (CI), and automated deployment of the main code trunk to a staging (or pre-production) environment with heavy use of monitoring and reporting [109]. It allows developers to ensure no regression is introduced by the addition of new features.

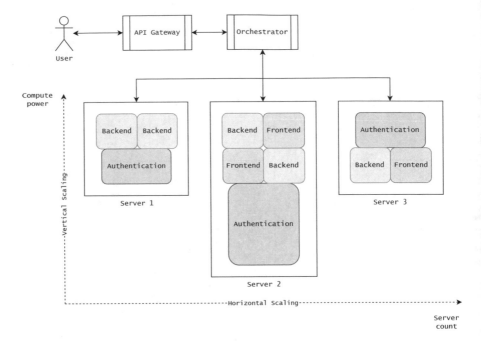

Fig. 4 Scaling out a microservices-architectured application allows distributing and replicating each microservice independently

Continuous practices align neatly with the microservices architecture, where incremental modifications to the application as a whole can be deployed as microservices updates. Popularity of both DevOps practices and the microservice architecture led to transformations in the cloud landscape. To some extent, the microservice architecture can be mapped to a cloud programming and service model, *Function as a Service* [60].

3 Serverless, A New Paradigm

In this section, we introduce the serverless model of programming for, and deploying applications to the cloud. We will go through essential characteristics of serverless platforms and highlight the tradeoffs that both service providers and application developers have to consider when targeting serverless infrastructures. This section also proposes a description of serverless offerings in public cloud solutions, and open source platforms for private cloud architects.

3.1 Characteristics of Serverless Platforms

Serverless refers at the same time to a programming and a service model for cloud computing. In a serverless architecture, developers design their applications as a composition of stateless functions. Stateless (or "pure", side-effect free) means that the outcome of the computation depends exclusively on the inputs [20]. These functions take a payload and an invocation context as input, and produce a result that is stored in a persistent storage tier. Their execution is triggered by an event that can be described as the notification of an incoming message, be it an HTTP request, a cron job, a file upload, etc. As such, serverless is an event-, or demand-driven model [106].

The aforesaid design is illustrated by an example serverless application in Fig. 5: when the user's request hits the serverless platform's API gateway, it triggers the execution of different functions according to the requested HTTP endpoint—these functions are not *daemons* listening for events (e.g. on an opened socket), they are executed on demand.

In commercial offerings, serverless is usually referred to as *Function as a Service* (FaaS). It has been believed that functions as an abstraction over cloud computing are part of a first generation of serverless offerings, and might change later on [54].

Serverless does not mean that servers are no longer used to host and run applications—much like PaaS, from a customer's perspective, serverless refers to an abstraction over computing resources that allows engineers to leave out thinking of the actual servers supporting their applications. Thanks to auto-scaling mechanisms, they do not have to consider the optimal number of instances needed to run their workloads in a capacity planning fashion. Serverless platforms are designed to handle scaling requirements and address fluctuations in demand, therefore freeing customers from the burden of having to define explicit scaling strategies. In a layered vision of cloud deployments, the SPEC Research Group [115] presents a FaaS reference architecture that shows that serverless development allows developers to be as close as possible to business logic [123].

Developing for serverless platforms requires re-thinking an application's architecture. Indeed, long-running backend servers are relegated to *serverful* solutions that provide always-on servers [80], such as IaaS offerings.

As Shafiei et al. [108] pointed out in their 2022 survey on serverless computing, there is no formal definition for the concept of serverless computing, although we can identify various essential differences between the serverful and serverless models (summed up in Table 1).

A major difference between PaaS and FaaS is that FaaS achieves scaling to zero: providers only bill customers when their application actually uses hardware resources, i.e. when functions are executed on the platform. That is made possible because, in the FaaS paradigm, applications are designed as a collection of short-running microservices.

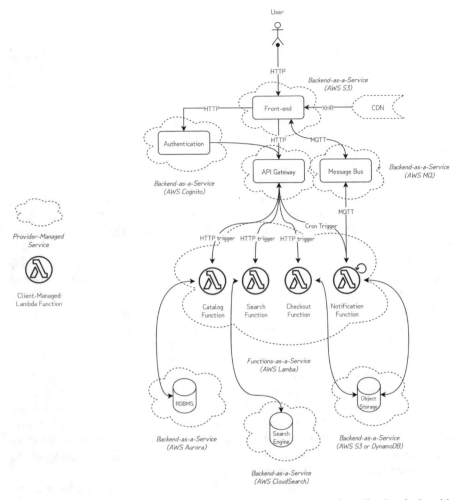

Fig. 5 Fictional reference architecture for a serverless e-commerce web application deployed in the Amazon web services ecosystem

Backend as a Service (BaaS) solutions are commercial, managed offerings of backend services, made available to application developers through a unified API [104]. Backend software usually consist in stateful, long-running services that cannot be scaled down to zero. In order to maintain a consistent pricing model, providers must offer these services in the same pay-as-you-go manner as they do serverless functions. These third party services constitute the backbone infrastructure of serverless applications by handling the state of the functions deployed by developers, through e.g. key-value stores or file storage; providing authentication to application endpoints; allowing communications between functions using message buses; etc. Figure 5 shows possible dependencies between serverless functions and BaaS soft-

Table 1 Comparison of key characteristics in serverless and serverful service models

Characteristic	Serverful (IaaS, PaaS)	Serverless (FaaS)
Provisioning	Customer responsibility	Fully **managed** (i.e. by the provider)
Billing	Pay for **provisioned** resources	Pay for **consumed** resources
Scaling	Customer responsibility	**Auto-scaling** built in
Availability	Depends on **provisioned** resources	Code runs in multiple high availability zones
Fault tolerance	Depends on deployment strategy	Backend services are fully **managed** and retries are guaranteed
Concurrency	Depends on **provisioned** resources	Virtually **infinite**

ware: the example application relies on provider-managed authentication, message bus, relational database, search engine and object storage, and is accessed by the user through the provider's API gateway. This situation introduces a high degree of coupling between the application and vendor-specific services, potentially tying developers to their initial choice of service provider.

Serverless allows reduced developer overhead by abstracting away server management, while enabling providers to share physical resources at a very fine granularity, thus achieving better efficiency. The fine level of granularity presented in the FaaS model enables the provider to offer perfect elasticity: scaling out and scaling in are event-driven, in a typical *pay-as-you-go* pricing model.

This abstraction makes it desirable for providers to deploy code in multiple geographic zones. This fail-over mechanism guarantees availability in case of outage in one deployment zone and decreases the risk of function failure cascading through the application [117].

Furthermore, as function instances are spun up on-demand by the provider, the concurrency model offered by FaaS platforms means that an application's performance can scale linearly with the number of requests [81].

Various authors [54, 63, 108, 123] already consider serverless to be the future of cloud deployment. However, FaaS adoption seems to be stalling among cloud developers [76], and the Cloud Native Computing Foundation (CNCF) even reports decreasing figures [29].

In the next section, we provide insights regarding workloads for which serverless is particularly desirable. In Sect. 4, we provide an overview of technical challenges that are still holding serverless back from becoming the go-to cloud subscription model.

3.2   Suitable Workloads

In their 2018 white paper [95], the Cloud Native Computing Foundation (CNCF)—a Linux Foundation initiative supported by more than 800 industrial members involved in cloud services—identifies characteristics for serverless use cases, including:

- "Embarrassingly parallel" workloads: asynchronous and concurrent, with little to no communication and no synchronization between processes;
- Infrequent with unpredictable variance in scaling requirements, i.e. event-driven or interactive jobs rather than batch jobs;
- Stateless and ephemeral processes, without a major need for instantaneous start time.

We can argue that these conditions are too restrictive for general computing: for example, it implies that long-lived jobs cannot be mapped to FaaS deployments. Providers have introduced mechanisms such as Step Functions or Durable Functions [7, 48, 86] to implement serverless workflows: an orchestration function maintains state across the application's stateless functions to create stateful workflows [20].

Developers are already deploying part of their applications' logic to serverless functions. According to a survey [96] led in 2018 by the Serverless framework publisher among a panel of their users, examples of such logic include data transformation pipelines, high-availability alerting platforms, ETL (*Extract, Transform, Load*, batch data manipulation) tools, media transcoding, etc. These applications are a subset of computer programs that produce output that only depends on the program's input: they apply purely functional transformations to data.

Problems that are conveniently split up in batches of sub-tasks would also benefit from the virtually infinite level of concurrency offered by the serverless model [46]. In [1], the authors identify use cases for serverless in massive scale video encoding, linear algebra and parallel compilation.

As there are fundamental similarities between a microservices-architectured application and an application devised for FaaS deployment [60], full-fledged applications can be designed with FaaS in mind. Figure 5 gives an example architecture for an e-commerce web application. The application's business logic comprises three serverless functions (catalog, search and checkout) that are triggered by user navigation, and one function (notification) that is scheduled to run periodically. As these functions are spun up and down according to the application's load, state has to be stored in persistent storage, i.e. a relational database and an object store that are both managed by the provider. The application further relies on provider-managed services: its search engine is powered by a BaaS solution, as is the notification function and the authentication mechanism.

Using AWS Lambda in 2022, this kind of application could scale from zero resources used to 150 TB of RAM and 90 000 vCPUs in less than two seconds [58], allowing for timely reaction to load variations.

3.3   Tradeoffs in Serverless Deployments

When we consider serverless as a programming model, the immediate benefit is a reduced development cost for teams leaning on BaaS offerings: instead of implementing in-house backend services such as authentication or notifications, developers merely introduce boilerplate in their codebase so as to connect frontend applications to their cloud provider's BaaS APIs. However, that degree of coupling means developers can find themselves locked-in in a vendor-specific environment, ultimately losing control over deployment costs [13].

From a customer's point of view, FaaS coupled with managed BaaS achieve perfect scaling. Customers are charged at a fine grain, only when resources are actually used, and for the exact duration of execution. From a provider's point of view, increased tenancy creates an opportunity to achieve better resource multiplexing, allowing for increased efficiency and thus greater profits. This auto-scaling mechanism has a cost in terms of latency: spinning up new sandboxes for incoming requests can create situations of *cold starts* where initialization times dominate execution times of functions [61]. Serverless auto-scaling has further implications in terms of throughput: since function state has to be persisted in disaggregated storage, applications that display patterns of extensive communications between functions can suffer from the shipping time of data to compute nodes [89].

A side effect of the FaaS service model is the increase in job count per customer. While the rise in concurrency, density and resources usage is a selling point for both FaaS providers and customers, these short-lived jobs have to be isolated from each other to prevent the leaking of secrets across customers, or at the scale of a single application comprising multiple individual functions [124].

To build on the model's strengths, customers and providers have to consider the tradeoffs that are associated with serverless deployments. Table 2 summarises key takeaways when using serverless to deploy applications to the cloud.

Table 2   Considerations regarding the FaaS service model

Pros	Cons
Reduced **development costs** for teams leaning on BaaS offerings	Can we map **any application** to the FaaS architecture?
Reduced **costs in operations** thanks to fully managed infrastructure	Risks of **vendor lock-in** due to high degree of coupling with BaaS offerings
Perfect scaling allows billing granularity close to actual use of resources	Increase in **latency** due to cold starts, and decreased **throughput** from communications through slow storage to handle statefulness
Providers can achieve **better efficiency** in resources multiplexing leading to increased profits	Massive multitenancy might involve **security** threats

3.4   Description of Current FaaS Offerings

In this section, we provide an overview of current FaaS offerings from public cloud providers and open source solutions for private cloud.

Table 3 presents a summary of major FaaS offerings regarding their pricing models and properties; including Alibaba Function Compute [3], Amazon Web Services Lambda [6], Microsoft Azure Functions [83], Google Cloud Functions [47], IBM Cloud Functions [57] and Oracle Cloud Functions [93].

Table 4 presents a summary of self-hostable FaaS platforms regarding their project status and adoption, and corporate backers; including Apache OpenWhisk [11], Fission [100], Fn [92], Knative [42] and OpenFaaS [128].

3.4.1   Commercial Solutions

To position each offering among the commercial offerings, we chose to compare them in terms of:

- Pricing model: the manner in which a customer can expect to be billed for product usage;
- Properties: the limits imposed by the provider regarding resource usage.

3.4.2   In the Open Source Community

To measure a project's status and adoption, we chose two indicators publicly available at GitHub [84]:

- GitHub "stars" indicate how many GitHub users chose to keep track of a project;
- Contributors are people who pushed 10 or more *git commits* (modifications to the codebase) to the repository.

In this section, we introduced FaaS as a deployment and a programming model, available for both public and private cloud. Serverless sparked interest in academia and industry, as a solution for customers to reduce their development and operation costs, and for providers to maximize resource usage. However, despite attractive pricing with extensive free plans in commercial offerings, and a various panel of open source solutions targeted at the major cloud orchestrators, FaaS has not yet become the go-to cloud subscription model: some challenges are still open and have to be addressed before serverless can become ubiquitous. Those are described in the next section.

Table 3 Cloud customers are faced with a diversity of FaaS offerings

Service	Pricing model		Properties			
	Free quota per month	Pay-as-you-go cost	Code size	Memory	Execution time	Payload size
	# of invocations/compute resources [GB s]	1M requests/1 GB s compute [USD]				
Alibaba Function Compute	1,000,000/400,000	20[a]/0.000016384	500 MB	3 GB	24 h	128 kB (request), 6 MB (response)
AWS Lambda	1,000,000/400,000	0.2/0.0000166667	10 GB	10,240 MB	15 min	6 MB (synchronous), 256 kB (asynchronous) for requests and responses
Azure Functions	1,000,000/400,000	0.2/0.000016	N/A	1.5 GB	10 min	100 MB (request)
Google Cloud Functions	2,000,000/400,000	0.4/0.0000025	500 MB	8 GB	9 min	10 MB for requests and responses
IBM Cloud Functions	5,000,000/400,000	N/A/0.000017	48 MB	2048 MB	60 s	5 MB for requests and responses
Oracle Cloud Functions	2000,000/400,000	0.2/0.00001417	N/A	2048 MB	5 min	6 MB

[a] Billed per 10,000 requests (for USD 0.02)

Table 4 Open source solutions allow cloud providers to devise their own FaaS offering

	Project status and adoption	Corporate backer
Apache OpenWhisk	5.8k GitHub stars, 34 contributors ($\geq$ 10 commits)	IBM (Apache Foundation)
Fission	7.3k GitHub stars, 10 contributors ($\geq$ 10 commits)	Platform9
Fn	5.3k GitHub stars, 21 contributors ($\geq$ 10 commits)	Oracle
Knative	4.5k GitHub stars, 55 contributors ($\geq$ 10 commits)	Google
OpenFaaS	22.2k GitHub stars, 13 contributors ($\geq$ 10 commits)	VMware

4 Problems Addressed in the Literature

In order to achieve flexibility and performance comparable to PaaS or IaaS solutions, FaaS providers need to tackle major problems that hinders the progress of serverless in becoming the norm in cloud computing. Serverless is a lively topic in cloud computing and many authors are contributing toward mitigating these setbacks: the number of published papers around serverless almost doubled between 2019 and 2020 [53]. The following sections will describe each problem and provide a set of state-of-the-art solutions that have been proposed in the literature. A real challenge in addressing these shortcomings is to avoid "serverful" solutions to the problem of dynamic allocation of resources, i.e. allocating additional stable resources that purposefully do not scale to zero [54].

4.1 Cold Start Delays and Frequency

As serverless containers must spend a minimum amount of time in an idle state, they are spun up and down very frequently as compared to PaaS containers or IaaS VMs. Each time a function is called and has to be scaled from zero, the container or virtual machine hosting the function's code has to go through its initialization phase: this is called a "cold start" [73]. Cold starts can incur latency penalties, aggravated by delays snowballing during composition of functions in the context of complex applications [88].

Functions are typically invoked in bursts—the AWS Lambda execution model can maximize concurrency by instantiating a function in hundreds to thousands of sandboxes across different geographical locations [8]. Minutes after handling a request, a function's sandbox is freed from the execution node; moreover, future new instances are not guaranteed to be created on the same node. This leads to situations in which a function's environment is not cached on the node. Code and associated libraries have to be fetched and copied to the filesystem again, resulting in cold start latency.

A "naive" approach would consist in pre-allocating hardware resources in order to keep a pool of function containers ready for new requests. This is not acceptable [75] as it strides away from the possibility of scaling to zero.

Vahidinia et al. [122] propose a comprehensive study of the position and strategies of various commercial FaaS offerings regarding cold start. While serverless computing suggests spinning up disposable instances of functions to handle each incoming request, the authors note that commercial actors such as AWS, Google and Microsoft all re-use execution sandboxes to some extent, keeping them running during a timeout period in order to circumvent latency costs incurred by cold starts.

4.1.1 Reducing Initialization Times

Different approaches can be implemented by cloud providers to shrink the initialization time of function sandboxes. This is a crucial work as function invocations follow mostly unpredictable patterns [110]. This section explores contributions from the literature that focus on bridging the gap in latency between serverless and serverful models.

Sandbox Optimization Approach

In [89], the authors propose Lambada to address the cold start problem in the context of distributed data analytics by batching the invocation of workers in parallel. They identify a bottleneck in the invocation process of new workers: in their evaluation, they show that invoking 1000 AWS Lambda workers takes between 3.4 and 4.4 s. In their contribution, each worker is responsible for invoking a second generation of sandboxes, which will in turn invoke a next generation of workers until the scaling process is complete. This technique allows to spawn several thousands of workers in under 4 s.

In [77], the authors propose LightVM to put VM boot time in the same ballpark as containers. The authors show that instantiation times grow linearly with image size: creating a sandboxed environment for an application to run is an I/O-bound operation. By redesigning Xen control plane and using lightweight VMs that include a minimal environment needed to run the sandboxed application, they achieve boot times comparable to the performances of the `fork`/`exec` implementation in Linux.

In [2], the authors propose that full-blown isolation mechanisms such as containers are needed to isolate workloads among customers, while at the granularity of a single application, processes are enough to isolate functions. In SAND, they implement an isolation mechanism on top of Docker that enables resources-efficient, elastic, low-latency interactions between functions.

In [1], the authors present Firecracker, which grew to become the de facto virtualization technology for serverless, being used at AWS Lambda. They tackle the trade-off in isolation versus performances by introducing lightweight VMs (or MicroVMs) in lieu of containers as a sandboxing mechanism for serverless workloads. Firecracker achieves boot times under 125 ms by replacing QEMU with a custom implementation of a virtual machine monitor that runs on top of KVM and allows to create up to 150 MicroVMs per second and per host with a 3% memory overhead.

Snapshotting Approach

In [38], the authors argue that startup overhead in virtualization-based sandboxes is caused by their application-agnostic nature. Indeed, they show that the application initialization latency dominates the total startup latency. In Catalyzer, the authors show that sandbox instances of one same function possess very similar initialization states, and thus present a snapshotting solution that allows restoring a function instance from a checkpoint image, effectively skipping the application's initialization phase when scaling from zero. They build a solution based on Google's gVisor [49] that consistently outperforms state-of-the-art technologies such as Firecracker [1], HyperContainer and Docker by one order of magnitude.

In [121], the authors present vHive, a benchmarking framework for serverless experimentation that allows them to show that high latency can be attributed to frequent page faults during sandboxes initialization, with very similar patterns among executions of a same function—97% of the memory pages being identical across invocations of studied functions. They propose REAP to create images of a sandbox memory layout that enable prefetching pages from disk to memory, eagerly populating the guest memory before function execution and thus avoiding the majority of page faults at initialization time. This technique allows a cold start delay speedup of 3.7 times on average.

In [112], the authors propose Faaslets, a new isolation mechanism based on software-fault isolation provided by WebAssembly. Faaslets allow restoring a function's state from already initialized snapshots. These snapshots are pre-initialized ahead of time and can be restored in hundreds of milliseconds, even across hosts. Faaslets take advantage of the WebAssembly memory model: linear bytes arrays can be copied without a lengthy (de)serialization phase.

Caching Approach

In [91], the authors argue that while serverless allows costs savings through increased elasticity and developer velocity, lengthy container initialization times hurt latency performances of deployed applications. They identify bottlenecks in Linux primitives involved in containers initialization, with package dependencies being the major culprit in I/O operations during sandboxing. They propose SOCK as a container system optimized for serverless tasks that builds on OpenLambda and rely on a package-aware caching system, and show that their solution offers speedups up to 21 times over Docker.

In [44], the author show conceptual similarities between object caching and function keep-alive, allowing them to devise policies that reduce cold start delays. Building on that analogy, they propose a keep-alive policy that is essentially a function termination (or eviction) policy. By keeping functions warm as long as possible (i.e. as long as server resources allow it), FaasCache manages to double the number of serviceable requests in their OpenWhisk-based implementation.

The ability for serverless platforms to scale a function to zero replicas in order to avoid billing customers for idle resources is a key difference with regards to traditional cloud service models. Seeking techniques that minimize the impact of a cold start on function latency is a critical research topic, as prohibitive initialization times hinder the potential for FaaS to compete with PaaS platforms.

4.2 Data Communications Overhead

In FaaS offerings, functions are non-addressable: composition is done through storing results in a stateful slow storage tier that is usually not collocated with the computing tier.

As functions of the same application cannot share memory or file descriptors to achieve IPC, they have to establish communication through message-passing interfaces, introducing overhead when data need to flow through the application.

This problem is particularly concerning when data-hungry applications have to work with cold data, i.e. data that are sparsely accessed and therefore not cached, and stored on lower performing storage such as hard drives located on remote nodes. In [61], the authors present LambdaML, a benchmarking platform that enables comparing performances of distributed machine learning model training across IaaS and FaaS offerings. They find out that using FaaS for ML training can be profitable as long as the models present reduced communication patterns.

In [89], the authors show that FaaS can be profitable when running sporadic, interactive queries on gigabytes to a terabyte of cold data. By providing serverless-specific data operators with Lambada, they achieve interactive queries on more than 1 TB of Amazon S3-stored data in approximately 15 s, which is in the same ball park as commercial Query-as-a-Service solutions.

In [105], the authors argue that serverless offerings lack an in-memory, per-application data caching layer that would allow auto-scaling and work transparently. Faa$T can form a strongly consistent distributed caching layer when multiple instances of an application are spun up, with the latest caching node vanishing as the application scales down to zero, effectively enabling billing on the basis of effective resources usage. The experimentations show that Faa$T can improve performance of varying applications by 57% on average while being 99% cheaper than serverful alternatives.

SAND [2] introduces a hierarchy in communication buses. In SAND, an application's functions are always deployed on the same node. A node-local bus serves as a shortcut for inter-function communications, allowing for fast sequential execution. A global bus, distributed across nodes, ensures reliability through fault tolerance. In SAND, the local bus achieves almost 3× speedup for message delivery compared to the global bus.

As serverless functions are ephemeral by nature, and given the isolation mechanisms rolled out by providers to meet privacy and security objectives, minimizing the overhead of inter-function communication seems to be a two-fold problem: on the one hand, serverless platforms need both domain-specific solutions that factor in the characteristics of the data that are fed to and returned by the functions; on the other hand, there is room for general improvements in the field of distributed caches.

4.3   Durable State and Statefulness

"State", or "local state", refers to data usually read and written from and to variables or disk by a process during its execution. FaaS offers no guarantees as to the availability of such storage across multiple executions. That is why serverless functions are referred to as "stateless": data that need to be persisted have to be stored externally, and functions should be idempotent in order to prevent state corruption.

Furthermore, FaaS offerings present arbitrary limitations including a function's execution time, payload size and allocated memory (cf. Table 3). This is a problem when designing "real world" applications that consist of long-lived jobs, and/or that comprise functions that need to communicate or synchronize, e.g. to pass on intermediate results depending on transient state.

Given the ephemeral nature of function instances, one must be aware of fault tolerance and data consistency in their application when they deploy to FaaS.

In [133], the authors address I/O latency in the context of serverless function composition, where an application is divided into multiple functions that may run concurrently on different nodes while accessing remote storage. They propose HydroCache, a system that implements their idea of Multisite Transactional Causal Consistency (causal consistency in the scope of a single transaction distributed across multiple nodes). They observe improvements up to an order of magnitude in performance while achieving consistency. HydroCache outperforms state-of-the-art solutions such as Anna [132] and ElastiCache [5].

In [97], the authors argue that serverless database analytics would allow data analysts to avoid upfront costs by achieving elasticity. However, as these kinds of workloads are by nature unpredictable, cloud providers tend to have difficulty in provisioning adequate resources, leading to solutions that are elastic but sometimes suffer minutes of latency during scaling phases. They present Starling, a query execution engine built on FaaS: three stage of functions (Producers, Combiners and Consumers) can scale independently to handle datasets stored on remote cold storage. Their evaluation shows that Starling is cost-effective on moderate query volumes (under 120 queries per hour on a 10 TB TPC-H dataset), while showing good latency results for ad-hoc analytics on cold data in Amazon S3 and being able to scale on a per-query basis.

In serverless, scaling from zero when activity returns after an idle period is usually event-driven. This poses a problem when no hardware resources are immediately available to resume workloads, inducing high latency. In [99], the authors investigate proactive auto-scaling for their serverless Azure SQL database offering. The contribution focuses on the prediction of pause and resume patterns in order to avoid the latency issue when resuming activity, and to minimize resources reclamation in the first place when idle periods are short. Using samples from thousands of production databases, they found that only 23% of databases are unpredictable, and trained machine learning models on three weeks of historical data to build a prediction system. The approach has been successfully used in production at Azure, achieving 80% of proactive resumes and avoiding up to 50% less pauses.

In [116], the authors build on the Anna KVS [132] to propose a stateful FaaS platform. Cloudburst achieves low-latency mutable state and communication with minimal programming effort. Leveraging Anna's capabilities, they provide essential building blocks to allow statefulness in an FaaS context: direct communication between functions, low-latency access to shared mutable state with distributed session consistency, and programmability to transparently implement Cloudburst consistency protocols. In their evaluation against real-world applications, Cloudburst outperforms both commercial and state-of-the-art solutions by at least an order of magnitude while maintaining auto-scaling capabilities.

4.3.1 Distributed Data Stores

Event-driven invocation implies that functions of a single application are not always executed on the same node, thus these functions cannot make use of shared memory or inter-process communications. Moreover, given the nature of serverless offerings that allow scaling to zero, functions are not always in an execution state and as such are not network addressable. Given these constraints, developers have to rely on increased communications through slow storage such as S3 buckets to handle statefulness within their applications.

There are hard challenges in scaling a database to zero: coming up with a database design that allows *true* serverless is an ongoing engineering and research effort. Microsoft recently proposed auto-scaling capabilities in their Azure SQL database [99]. In 2022, Cloudflare introduced D1 [32], which is based on SQLite.

Indeed, serverless applications are often deployed alongside a key-value store that scales much more naturally than a database, as key-value stores (KVS) are essentially stateless and thus can be distributed across nodes [65]. Given that KVS systems are central to serverless statefulness, implementing consistent, efficient and elastic KVS is a lively research subject. However, industry-grade storage systems were not designed with serverless properties in mind, resulting in impaired elasticity and thus costs that grow faster than linearly with the infrastructure's size, and inconsistent performance depending on the scale.

In [132], the authors set out to design a KVS for any scale: the store should be extremely efficient on a single node and has to be able to scale up elastically to any cloud deployment. Their design requirements include partitioning the key space (starting at the multi-core level to ensure performance) with multi-master replication to achieve concurrency; wait-free execution to minimize latency and coordination-free consistency models to avoid bottlenecks during communications across cores and nodes. Using a state-of-the-art data structure, lattices, Anna can efficiently merge state in an asynchronous (or wait-free) fashion. The evaluation shows that Anna outperforms Cassandra by a $10\times$ factor when used in a distributed setting, across four 32-core nodes in different geographical locations.

In [65], the authors argue that existing store services have objectives orthogonal or contradictory to those of a serverless KVS: they sacrifice performance or cost for durability or high availability of data. In particular, they find that these systems are inherently not suitable for intermediate (or "ephemeral") data in the context of inter-functions communications, as they require a long-running agent to achieve communication among tasks. The authors present Pocket, a distributed data store designed for intermediate data sharing in the context of serverless analytics, with sub-second response times, automatic resource rightsizing and intelligent data placement across multiple storage tiers (DRAM, Flash, disk). This is achieved by dividing responsibilities between three planes that scale independently: a control plane that implements data placement policies, a metadata plane that allows distributing data across nodes, and the data storage plane. When evaluated against Redis for MapReduce operations on a 100 GB dataset, Pocket shows comparable performance while saving close to 60% in cost. It is also significantly faster than Amazon S3, with a $4.1\times$ speedup on ephemeral I/O.

4.3.2 Ephemeral Storage

Cloud storage is devised as a tiered service: data is disaggregated across fast, but costly medium and slow, but cheap medium, according to frequency of use, size, age, etc (Table 5).

Intel Optane are persistent memory (PM) modules that target a tier in-between Flash SSDs and DRAM: their latency and bandwidth are slightly worse than DRAM, but they offer SSD-level capacities of non-volatile memory at an affordable price [18, 19, 59].

In [25], the authors aim at delivering a key-value storage engine that would take advantage of persistent memory (PM, or NVM for non-volatile memory) technology to achieve greater performance than on spinning disks or Flash memory. They focus on write-intensive, small-sized workloads: indeed, previous studies [12, 90] have shown that Memcached pools in the wild are mainly used to store small objects,

Table 5 A simplified overview of media choice in tiered infrastructures

Capacity	TB			GB	
Addressability	Block			Byte	
Consideration	Cost			Data	
Latency	s	ms	μs	μs	ns
Data	Cold	Warm	Hot	Hot	Mission critical
Medium	Tape	HDD	SSD (Flash)	NVRAM	DRAM

e.g. 70% of them are smaller than 300 bytes at Facebook. Moreover, serverless analytics exchange short-lived data and thus are very write-intensive, while object stores have historically been used as a read-dominated caching layer. Building on these observations, and characteristics specific to PM devices, they present FlatStore, a KVS engine with minimal write overhead, low latency and multi-core scalability. As persistent memories present fine-grain addressability and show low latency as compared to HDDs and SSDs, the authors designed FlatStore for minimal batching of write operations so as to avoid contention. When benchmarked on Facebook data with tiny (1–13 bytes) to large (> 300 bytes) items, evaluation shows that FlatStore performs 2.5–6.3 times faster than state-of-the-art solutions.

Statefulness is a major problem for serverless platforms. Service providers are deploying a variety of BaaS software to bridge the gap between traditional service models and FaaS and allow developers to deploy their full applications to their serverless offerings. Serverless functions present intrinsically disaggregated storage and compute, as they are deployed on-the-fly to multiple geographic zones, on hardware resources that are dynamically allocated by the provider. They need a means to operate on data that is fast enough, offers consistency guarantees, and scales in coherence with the pay-as-you-go pricing model. There is room for research in the field of distributed data stores, and using emerging non-volatile memory to accelerate throughput.

4.4 Hardware Heterogeneity

Cloud customers are expected to book different resources depending on their applications' needs, be it a specific CPU architecture, hardware accelerators, ad-hoc storage. A striking example is distributed machine learning, in which many GPUs are used to speed up the training of models—furthermore, cloud providers are starting to generalize access to specialized hardware such as TPUs [28] in VMs.

Manual selection of hardware resources ("instance type"), expected from customers in IaaS offerings such as Amazon EC2, does not make sense in the serverless paradigm. Hardware acceleration should be decided by the provider on a per-application or per-request basis. To date, that possibility is not available in FaaS offerings such as AWS Lambda.

In [61], the authors set out to compare IaaS and FaaS configurations for machine learning training on Amazon Web Services offerings (resp. EC2 and Lambda). They propose an implementation of FaaS-based learning, LambdaML, and benchmark it against state-of-the-art frameworks running on EC2 instances. They measured that serverless training can be cost-effective as long as the model converges quickly enough so that inter-function communications do not dominate the total run time. Otherwise, an IaaS configuration using GPUs will outperform any FaaS configuration, yielding better performance while being more cost-effective.

In [14], the authors explore multitenancy in FPGAs to achieve higher board usage rate. They propose BlastFunction, a scalable system for FPGA time sharing in a serverless context. Their implementation relies on three building blocks: a library that allows transparent access to remote shared devices, a distributed control plane that monitors the FPGAs to achieve time sharing, and a central registry that handles allocating the boards to each compute node. This design allows reaching higher utilization rates on the boards and thus processing a higher number of requests, especially under high load, although at the cost of a 36% increase in latency due to the added concurrency.

In [35], the authors focus on a financial services use case and propose FPGAs to decrease end-to-end response time and increase scalability in a microservices architecture. The application they studied is computationally intensive and has real-time characteristics. They propose CloudiFi, a cloud-native framework that exposes hardware accelerators as microservices through a RESTful HTTP API. CloudiFi allows offloading workloads to hardware accelerators at a function level. An evaluation of the application's performance under CloudiFi shows $485\times$ gains in response time when using network-attached FPGAs against a vanilla configuration.

ARM and RISC CPUs, GPUs and FPGAs are increasingly used in datacenters to address demand for performance, power efficiency and reduced form factor. In [56], the authors argue that since these heterogeneous execution platforms are usually collocated with a general purpose host CPU, being able to leverage their characteristics by migrating workloads could yield significant performance gains. They propose Xar-Trek, a compiler and run-time monitor to enable execution migration across heterogeneous-ISA CPUs and FPGAs according to a scheduling policy. Xar-Trek involves limited programming effort: the application is written once and compiled for different targets thanks to the Xilinx toolchain, without necessary high-level synthesis annotations to guide the compiler. Xar-Trek's runtime system, a user space online scheduler, is able to determine if a migration is effective and to proceed to migrate selected functions that benefit the most from acceleration. Evaluation on machine vision and HPC workloads finds out that as long as the workloads are dominated by compute-intensive functions, Xar-Trek always outperforms vanilla configurations, with performance gains between 26 and 32%.

Even when heterogeneous hardware is collocated on the same node, they are usually interconnected through PCI-Express buses managed by the host's CPU. Communications are achieved with message-passing interfaces that introduce bandwidth and latency costs. In [125], the authors present FractOS, a distributed operating system for heterogeneous, disaggregated datacenters. FractOS allows decentralizing the execution of applications: instead of relying on the CPU to pass on control and data from one execution platform to another, FractOS provides applications with a library that allows direct communications between devices, thanks to an underlying controller that catches system calls and provides direct device-to-device functionality. When benchmarked on a face verification application that leverages GPUs to accelerate computations, their solution shows a speedup of 47% in execution time and an overall network traffic divided by 3.

With the exponential progression and growing interest in the field of machine learning, demand for hardware accelerators in the cloud has never been so prevalent. Commercial serverless offerings are lagging behind traditional IaaS in that regard, as none offer access to GPUs, TPUs nor FPGAs. Furthermore, dynamically allocating such hardware to accelerate select tasks has potential for providers to improve their resource usage and energy consumption.

4.5   Isolation and Security

In order to achieve resource pooling, cloud providers rely on virtualization technologies so as to isolate customers workloads. Furthermore, they offer various models of service ranging from IaaS to FaaS, all of which call for different sandboxing techniques providing a different balance between performance and isolation.

The usual tradeoff happens between the robustness of hypervision-based isolation (VMs) where each sandbox runs a separate OS, and the performance of OS-level virtualization (containers) where sandboxes all share the host's kernel. Ideally, cloud providers should not have to sacrifice one of these two essential characteristics. Efforts have been made to reduce virtualization overhead in order to decrease startup times and reduce the performance gap between these two techniques [77].

4.5.1   MicroVMs

In [1], the authors identify numerous challenges to devise an isolation method specifically suitable for serverless workloads in the context of AWS Lambda—Firecracker must provide VM-level security with container-level sandboxing density on a single host, with close to bare-metal performances for any Linux-compatible application. Firecracker overhead should be small enough so that creating and disposing of sandboxes would be fast enough for AWS Lambda ($\leq 150\,$ms), and the manager should allow over committing hardware resources with sandboxes consuming only the resources it needs. With Firecracker, the authors present a new Virtual Machine Monitor (VMM) based on Linux KVM to run minimal virtual machines (or MicroVMs) that pack an unmodified, minimal Linux kernel and user space. Thanks to sandbox pooling, Firecracker achieves fast boot times and high sandbox density on a single host, for any given Linux application. It has been successfully used in production in AWS Lambda since 2018.

In [10], the authors study the differences in host kernel functionality usage across Linux Containers (LXC), Firecracker MicroVMs and Google's gVisor secure containers. gVisor sandboxes are `seccomp` containers: they are restricted to 4 system calls, namely `exit`, `sigreturn`, and `read` and `write` on already opened file descriptors. Extended functionality relies upon a Go-written user space kernel called Sentry that intercepts and implements system calls, and manages file descriptors. This prevents direct interaction between the sandboxed application and the host OS.

While effectively achieving secure isolation, gVisor's design is complicated and adds overhead: the authors find that gVisor has the largest footprint in CPU and memory usage, with the slowest bandwidth for network operations.

In [129], the authors argue that the virtualization ecosystem lacks a solution tailored for isolation at the granularity of a single function. They present virtines, a lightweight VM isolation mechanism, and Wasp, a type-2, minimal library hypervisor that runs on GNU/Linux and Windows. Virtines are programmer-guided: annotations at function boundaries allow the compiler to automatically package subsets of the application in lightweight VMs with a POSIX-compatible runtime. Wasp works in a client-server fashion: the runtime (client) issues calls to the hypervisor (server) that determines if each individual request is allowed to be serviced according to an administrator-defined policy. In their evaluation with a JavaScript application, the authors find this design introduces limited overhead of 125 μs in boot time compared to baseline, while effectively achieving fine-tunable isolation for selected functions at almost no programmer effort.

4.5.2 Unikernels

The idea behind unikernels is to provide OS functionality as a library that can be embedded in an application sandbox so as to avoid packing and booting a full-fledged operating system to run the application, and to eliminate costly context switches from user space to kernel space. In [67], the authors present Unikraft, a Linux Foundation initiative. Unikraft aims at making the porting process as painless as possible for developers who want to run their applications on top of unikernels. Resulting images for different applications (nginx, SQLite, Redis) come close to the smallest possible size, i.e. Linux user space binary size, with very limited memory overhead during execution ($< 10\,MB$ of RAM) and fast boot times in the milliseconds range. Unikraft-packaged applications achieve 1.7 to $2.7\times$ performance improvements compared to traditional Linux guest VMs.

In [22], the authors present a high-density caching mechanism that leverages unikernels and snapshotting (see Sect. 4.1.1) to speed up deployments. They argue that serverless functions are good candidates for caching: as they usually are written in high-level languages that execute in interpreters, their startup path mainly consists in initializing this interpreter and associated dependencies, which can be shared across different sandboxes. The snapshotting mechanism benefits from the unikernel memory layout, where all functionalities (ranging from filesystem, to network stack, to user application) are combined into a single flat address space. The implement this mechanism in SEUSS to achieve caching over $16\times$ more unikernel-based sandboxes in memory than Linux-based containers. Furthermore, deployment times drop from hundreds of milliseconds to under 10 ms, and platform handling of bursts of requests dramatically improves under high-density caching, leading to reduced numbers of failed requests.

In [118], the authors present Unikernel-as-a-Function (UaaF), a single address space, library OS aimed at deploying serverless functions. UaaF builds on the observation that cross-function invocations are slow in serverless deployments that rely on network-based message passing interfaces (see Sect. 4.2); furthermore, Linux guests suffer from memory usage overhead in sandboxes and their startup latency is not satisfying (see Sect. 4.1). The authors investigate using VMFUNC, an Intel technology for cross-sandboxes function invocations that do not incur latency when exiting from a VM to the hypervisor. It effectively enables remote function invocation, thus giving hardware-supported, secure IPC capabilities to serverless functions. They also propose a new programming model for serverless functions: *session* and *library* functions, with the former being "workflow" (or skeleton) functions and the latter being actual code, uploaded by customers and possibly shared across applications. In their evaluation, the authors implement UaaF with three unikernels (Solo5, MirageOS and IncludeOS) and show that inter-function communication in UaaF is three orders of magnitude lower than native Linux IPC. Their programming model allows for reduced memory overhead and initialization times in several milliseconds thanks to shared functions.

FaaS workloads are orders of magnitude shorter-lived than workloads in traditional offerings. As such, relying on virtualization techniques that were not built for serverless is suboptimal: initialization times may not meet latency requirements when scaling from zero; sandbox sizes may be too high to cache in memory given the increase in multitenancy; isolation might be too weak to collocate different customers' jobs. This assessment sparked interest in research around unikernels and MicroVMs, while commercial providers developed their own approaches such as Firecracker for AWS, or gVisor for Google Cloud.

4.6 Programming Model and Vendor Lock-in

As shown in Fig. 5, FaaS applications tend to rely heavily on BaaS offerings to benefit from costs savings associated to their capability to scale to zero. This tie-in introduces a risk of lock-in with vendor-specific solutions that might not be available across commercial offerings, or available as off-the-shelf, open source software.

Furthermore, some providers will use prohibitive egress bandwidth pricing [55] so as to deter their customers from moving data to a competitor.

Another aspect of that problem is the difficulty to develop, test and debug FaaS applications locally [119]. At the very least, developers will have to simulate the API gateway in order to run test suites; if their application makes use of vendor-specific storage solutions or communication buses, developers will have to deploy similar solutions or mock the specificities of these BaaS building blocks, e.g. their API and performance characteristics.

This amounts to non-negligible engineering efforts and indeed, deploying a full-fledged serverless infrastructure for staging might offset the operations cost benefits of choosing serverless for production. Entry level [33] and seasoned [87] engineers alike report having trouble with tooling, testing and the general ecosystem and added complexity of developing FaaS applications.

In [98], the authors observe that the disaggregation of storage and compute resources in FaaS limits the development of applications that make heavy use of shared mutable state and synchronize a lot between iterations. Indeed, state does not persist between invocations of a same function (see Sect. 4.3), and message passing for inter-function communications induces high overhead (see Sect. 4.2). In particular, they focus on machine learning algorithms (k-means clustering and logistic regression). They present Crucial, a framework aiming at supporting the development of stateful serverless applications. Crucial provides applications with a shared memory layer that guarantees durability through replication, with strong consistency guarantees. Crucial programming model is annotation-based, allowing programmers to port a single-machine, multi-threaded application to a FaaS platform with minimal involvement. Evaluation against a Spark cluster over a 100 GB dataset shows that Crucial running on AWS Lambda introduces very small overhead, enabling it to outperform Spark by 18–40% in performance at similar cost.

In [136], the authors acknowledge that the serverless programming model is challenging for developers. They have the responsibility to correctly partition their code into stateless units of work, to manage coordination mechanisms to achieve a microservices architecture, and to implement consistency models for state retention in case of failures. The complexity might deter customers from deploying general-purpose applications that would greatly benefit from the level of parallelism offered by serverless providers. They present Kappa, a Python framework for serverless applications. Kappa provides a familiar API that achieves checkpointing (by periodically storing the application's state so that the program can resume in case of timeout), concurrency (by supporting spawning tasks, waiting on futures, and cross-function message passing), and fault tolerance (by ensuring idempotent state restoration when resuming from checkpoints). Kappa applications can be deployed to any serverless platform, as the framework requires no change on the server side. In their evaluation, they implement five applications with Kappa and results indicate that the checkpointing mechanism works well when functions time out a lot, with less than 9% response time overhead under heavy (15 s) timeout duration, and a maximum of 3.2% with a more reasonable 60 s timeout period.

In order to limit the increase in latency when scaling from zero, the container or VM images that support serverless applications are usually made as lean and lightweight as possible. This deters developers from including monitoring or debugging tools, making it very hard to inspect a serverless function at runtime. In [119], the authors present VMSH, a mechanism that allows attaching arbitrary guest images to running lightweight VMs in order to instrument it for development or debugging purposes. Evaluation done on KVM—although VMSH is built as a hypervisor-agnostic solution—shows that guest side-loading adds no overhead to the original VM guest, successfully slashing the tradeoff between no-frills, lightweight VMs and functionality.

There is a clear tradeoff in providing sandboxes as small as possible to minimize storage and memory costs in serverless platforms, while shipping adequate tools for developers to build, test, distribute and deploy their functions. Furthermore, the programming model based on stateless functions shed light on a new challenge: provider-side and developer-side tooling for stateful FaaS is needed to enable the serverless deployment of legacy and future applications that make use of long-running services and data persistence.

5 Perspectives and Future Directions

The previous section provided an overview of contributions linked with technical challenges in serverless computing. In this section, we introduce some future directions for research in the field. We present problems investigated in works from the cloud, system and database communities. We argue that contributions building on these insights would have the potential to strengthen serverless platforms for a broader recognition of the serverless paradigm.

5.1 Service Level Agreements

In 2011, Buyya et al. [21] advocated for SLA-oriented resource allocation in cloud computing at the dawn of the microservices era. They identified reliability in utility computing as a major challenge for the next decades: even with reserved resources in traditional service models, the growing complexity of customer applications made meeting Service Level Agreements (SLAs) a hard albeit inescapable problem for cloud providers.

Latency, throughput and continuity of service are difficult to guarantee in cloud computing when using unreserved resources [34]. Due to the transient nature of function sandboxes in serverless computing, auto-scaling platforms face a similar problem of dynamic allocation of resources. However, being able to offer SLAs to customers and meet Quality of Service (QoS) commitments as a provider is necessary for wide-scale adoption of the serverless service model [39].

In [23], the authors argue that serverless auto-scaling platforms are challenged by bursty workloads. In their work, they highlight the importance of workload characterization to rightsize the amount of reserved VMs needed to meet SLAs. When the number of incoming requests drives up the concurrency level in reserved VMs, and makes task latency go past the acceptable threshold negotiated via SLA, they rely on a serverless platform to accommodate for extra tasks and maintain performance. While that framework managed to keep most of response times under the target threshold, the authors still see an incompressible number of violations caused by cold start delays on the serverless platform.

In [27], the authors argue that the task model in serverless computing and the infrastructure view in auto-scaling platforms are inadequate to address customers' needs in terms of service level. Indeed, auto-scalers base their allocation decisions upon generic metrics such as query per second (QPS) that do not reflect application-specific characteristics and do not take into account the heterogeneity of the hardware resources at hand. They propose a framework in which their application metrics (such as request execution time) are fed to the auto-scaler in order for it to allocate resources according to user-specified service level objectives, such as target latency. However, observed response times are non-deterministic due to cold start delays, and user-defined target latencies are subject to violations in an auto-scaling scenario.

In order to meet per-user QoS requirements, the auto-scaling platforms should take into account the characteristics of heterogeneous hardware resources, and SLAs should be negotiated on a per-request basis rather than on a per-function basis. We believe that auto-scaling policies based on workload and platform characterization could be implemented to minimize the impact of cold start latency and allow serverless platforms to meet SLAs with unreserved, heterogeneous resources.

5.2 Energy Efficiency

Power usage in cloud computing is a crucial challenge: in 2010, datacenters totaled between 1.1 and 1.5% of global electricity use [66], and projections for 2030 show that these figures could go up from 3 to 13% of global electricity use [9]. With serverless becoming an increasingly popular service model for the cloud, and many authors considering serverless as the future of cloud computing, there is an opportunity for cloud providers to implement energy policies at scale.

To be efficient in terms of cost and energy consumption, an auto-scaling platform should be able to rightsize the allocated resources in a serverless cloud infrastructure, while being responsive enough to accommodate workload changes without impacting end users with spikes in latency. This highlights a tradeoff between energy and performance: oversubscribing resources can help ensure low latency on function invocation, but will result in higher energy consumption.

Multitenancy helped slow down the growth in server count in datacenters [79]. With promises of massive collocation of short-lived jobs, serverless seems to be a promising direction for cloud infrastructures looking to reduce their energy footprint.

Workload consolidation is a technique that consists in maximizing the number of jobs on the fewest number of nodes [24]. This allows for dynamic power management: nodes that are not solicited can then be powered off, and nodes that observe moderate load can be slowed down, i.e. via CPU throttling [72].

One fundamental problem in the serverless paradigm is the intrinsic data-shipping architecture[2] [26]. Since function sandboxes are deployed on nodes in various geographic regions to achieve load balancing and availability, serverless platforms ship up to terabytes of data from the storage nodes to code which size can range from kilobytes to megabytes within the compute nodes.

---

[2] Moving data from where they are stored to where they need to be processed.

Storage functions allow small units of work to be executed directly on the storage nodes [135], achieving 14–78% speedup against remote storage. Storage functions do not question the physical disaggregation of storage and compute resources that is instrumental in cloud computing, while effectively limiting data movement between nodes and as such reducing energy consumption in a datacenter.

Computational storage is a means to offload workloads from the CPU to the storage controller [15]. When dealing with large amounts of data, such techniques can help decrease data transfers, improve performance and reduce energy consumption. While these technologies are not yet ready for production use, they provide interesting research opportunities for the serverless community.

These techniques could be implemented in serverless platforms to yield further gains in cloud energy consumption. It implies taking into account the diversity of user applications and the heterogeneity of requests and hardware resources.

5.3   AI-Assisted Allocation of Resources

In the serverless paradigm, it is the provider's responsibility to rightsize the allocation of hardware resources so that their customers' workloads are executed in time. Dynamically allocating appropriate hardware resources for event-driven tasks in a heterogeneous infrastructure is a hard problem that may hit a computational complexity barrier at scale, with online scheduler producing sub-optimal solutions [74]. Artificial intelligence (AI) techniques can help overcoming such a challenge.

Some authors expect AI-driven autonomic computing to become the norm in future systems [45]. The idea of autonomic computing is to build self-managed and self-adaptive systems that are resilient to an extremely changing environment at scale [101]. Such systems can be implemented with machine learning (ML) in a cost-effective way, using models that do not require extensive human intervention for supervision.

In [107], the authors show that reinforcement learning (RL) can achieve appropriate scaling on a per-workload basis, resulting in improved performance as compared to baseline configuration. In their contribution, they propose an RL model that effectively determines and adjusts the optimal concurrency level for a given workload.

Serverless platforms require reactive resources allocation and scheduling of tasks under SLAs with per-request QoS requirements [51]. Machine learning techniques can help achieve QoS requirements in traditional cloud computing paradigms [114], and have been used to enforce virtual machine consolidation [111]. Resources management and optimization using AI and ML could further help taking advantage of the heterogeneity of hardware resources in a cloud infrastructure.

6 Conclusion

Serverless is an emerging paradigm in cloud computing, implemented in Function-as-a-Service offerings. Customers devise their applications as compositions of stateless functions and push their code to the serverless provider's function repositories. When an event triggers the execution of a serverless function, the function is reactively instantiated in a virtualized environment.

It is a radical shift in the cloud computing landscape: while traditional offerings such as IaaS or PaaS are based on the reservation of stable resources, FaaS providers propose a demand-driven service model.

This paradigm allows customers to benefit from pay-per-use pricing models to the granularity of a function invocation. On the other hand, serverless providers can maximize the collocation of jobs on nodes and achieve better resource usage.

By freeing customers from the constraint of manual rightsizing of cloud resources, the serverless service model pledges to ease the auto-scaling of applications. Thanks to an event-driven, on-demand resource allocation mechanism, customers can benefit from significant reductions in operations costs as they do not have to pay for idle hardware anymore.

However, current serverless solutions present non-negligible limitations that constrain the model's adoption to specific use cases. This paradigm is based on a programming model in which developers have to design their applications as compositions of pure (stateless, idempotent) functions that cannot rely on side effects. This constitutes a considerable engineering effort.

As with the microservices architecture, serverless software rely on message-passing communications between functions. These functions being non-network addressable in current serverless offerings, these communications have to go through slow storage. This leads to important costs in performances when functions have to synchronize, sometimes outweighing the benefits offered by the virtually infinite level of parallelism in the serverless paradigm.

Furthermore, scaling from zero presents a frequent risk of increased latency during application wake-up, as the provider has to allocate hardware resources and instantiate the application's sandboxes before answering to the event. Serverless providers usually pre-allocate some resources so as to avoid these cold starts, which comes with a cost in resources multiplexing.

Finally, hardware accelerators are not yet available in commercial serverless offerings. With increasing demand in GPUs and FPGAs for massively parallel tasks, such as machine learning training or big data analytics, customers have to turn to conventional cloud offerings such as IaaS if they want to benefit from heterogeneous hardware resources.

For serverless to impose itself as a serious contender among the cloud computing offerings, providers need to be able to guarantee some sort of quality of service through service level agreements. Characterizing the customers' workloads and taking into account the heterogeneity of both the infrastructure and the requests is crucial to that matter, and has the potential to boost both performance and reliability.

Cloud providers have a responsibility to tackle the problem of energy consumption in datacenters. To that end, serverless could prove to be an efficient service model if adequate techniques for workload consolidation are proposed and implemented. Providers could make the most of their new responsibility to allocate resources by devising power-off or slow-down strategies in serverless infrastructures.

AI-assisted resource management appears to be a promising research direction for serverless computing. Indeed, as interactive workloads exhibit hard-to-predict patterns, cloud providers could take advantage of ML models to guide allocation and scheduling decisions in auto-scaling platforms.

References

1. Agache A, Brooker M, Iordache A, Liguori A, Neugebauer R, Piwonka P, Popa DM (2020) Firecracker: lightweight virtualization for serverless applications. In: 17th USENIX symposium on networked systems design and implementation (NSDI 20). USENIX Association, Santa Clara, CA, pp 419–434. https://www.usenix.org/conference/nsdi20/presentation/agache
2. Akkus IE, Chen R, Rimac I, Stein M, Satzke K, Beck A, Aditya P, Hilt V (2018) SAND: towards high-performance serverless computing. In: Proceedings of the 2018 USENIX conference on Usenix annual technical conference, USENIX ATC '18. USENIX Association, USA, pp 923–935. https://doi.org/10.5555/3277355.3277444
3. Alibaba (2022) Alibaba function compute. https://www.alibabacloud.com/product/function-compute
4. Almeida Morais FJ, Vilar Brasileiro F, Vigolvino Lopes R, Araujo Santos R, Satterfield W, Rosa L (2013) Autoflex: service agnostic auto-scaling framework for IaaS deployment models. In: 2013 13th IEEE/ACM international symposium on cluster, cloud, and grid computing. IEEE, Delft, pp 42–49. https://doi.org/10.1109/CCGrid.2013.74
5. Amazon Web Services (2022) Amazon ElastiCache. https://aws.amazon.com/elasticache/
6. Amazon Web Services (2022) AWS Lambda. https://aws.amazon.com/lambda/
7. Amazon Web Services (2022) AWS step functions. https://aws.amazon.com/step-functions/
8. Amazon Web Services (2022) Lambda function scaling. https://docs.aws.amazon.com/lambda/latest/dg/invocation-scaling.html
9. Andrae A, Edler T (2015) On global electricity usage of communication technology: trends to 2030. Challenges 6(1):117–157
10. Anjali, Caraza-Harter T, Swift MM (2020) Blending containers and virtual machines: a study of firecracker and gVisor. In: Proceedings of the 16th ACM SIGPLAN/SIGOPS international conference on virtual execution environments. https://doi.org/10.1145/3381052.3381315
11. Apache (2022) Openwhisk. https://openwhisk.apache.org/
12. Atikoglu B, Xu Y, Frachtenberg E, Jiang S, Paleczny M (2012) Workload analysis of a large-scale key-value store. SIGMETRICS Perform Eval Rev 40(1):53–64
13. Baarzi AF, Kesidis G, Joe-Wong C, Shahrad M (2021) On merits and viability of multi-cloud serverless. In: Proceedings of the ACM symposium on cloud computing. ACM, Seattle WA USA, pp 600–608. https://doi.org/10.1145/3472883.3487002
14. Bacis M, Brondolin R, Santambrogio MD (2020) BlastFunction: an FPGA-as-a-service system for accelerated serverless computing. In: 2020 design, automation and test in Europe conference and exhibition (DATE). IEEE, Grenoble, France, pp 852–857. https://doi.org/10.23919/DATE48585.2020.9116333
15. Barbalace A, Do J (2021) Computational storage: where are we today? In: CIDR, Conference on innovative data systems research 2020, CIDR 2020; Conference date: 11-01-2021 through 15-01-2021, p 6. http://cidrdb.org/cidr2021/index.html

16. Baude B (2019) Basic security principles for containers and container runtimes. https://www.redhat.com/sysadmin/basic-security-principles-containers

17. Bentaleb O, Belloum ASZ, Sebaa A, El-Maouhab A (2022) Containerization technologies: taxonomies, applications and challenges. J Supercomput 78(1):1144–1181

18. Boukhobza J, Olivier P (2017) Flash memory integration. ISTE Press-Elsevier. https://www.elsevier.com/books/flash-memory-integration/boukhobza/978-1-78548-124-6

19. Boukhobza J, Rubini S, Chen R, Shao Z (2017) Emerging nvm: a survey on architectural integration and research challenges. ACM Trans Des Autom Electron Syst 23(2). https://doi.org/10.1145/3131848

20. Burckhardt S, Chandramouli B, Gillum C, Justo D, Kallas K, McMahon C, Meiklejohn CS, Zhu X (2022) Netherite: efficient execution of serverless workflows. Proc VLDB Endow 15(8):1591–1604. https://doi.org/10.14778/3529337.3529344

21. Buyya R, Garg SK, Calheiros RN (2011) SLA-oriented resource provisioning for cloud computing: challenges, architecture, and solutions. In: 2011 international conference on cloud and service computing. IEEE, Hong Kong, China, pp 1–10. https://doi.org/10.1109/CSC.2011.6138522

22. Cadden J, Unger T, Awad Y, Dong H, Krieger O, Appavoo J (2020) SEUSS: skip redundant paths to make serverless fast. In: Proceedings of the fifteenth European conference on computer systems. ACM, Heraklion Greece, pp 1–15. https://doi.org/10.1145/3342195.3392698

23. Chahal D, Palepu S, Mishra M, Singhal R (2020) SLA-aware workload scheduling using hybrid cloud services. In: Proceedings of the 1st workshop on high performance serverless computing. ACM, Virtual Event Sweden, pp 1–4. https://doi.org/10.1145/3452413.3464789

24. Chaurasia N, Kumar M, Chaudhry R, Verma OP (2021) Comprehensive survey on energy-aware server consolidation techniques in cloud computing. J Supercomput 77(10):11682–11737

25. Chen Y, Lu Y, Yang F, Wang Q, Wang Y, Shu J (2020) FlatStore: an efficient log-structured key-value storage engine for persistent memory. In: Proceedings of the twenty-fifth international conference on architectural support for programming languages and operating systems

26. Chikhaoui A, Lemarchand L, Boukhalfa K, Boukhobza J (2021) Multi-objective optimization of data placement in a storage-as-a-service federated cloud. ACM Trans Storage 17(3):1–32

27. Cho J, Tootaghaj DZ, Cao L, Sharma P (2022) SLA-driven ML inference framework for clouds with heterogeneous accelerators, p 13

28. Cloud G (2022) Cloud tpu vms are generally available. https://cloud.google.com/blog/products/compute/cloud-tpu-vms-are-generally-available

29. Cloud Native Computing Foundation (2021) New SlashData report: 5.6 million developers use Kubernetes, an increase of 67% over one year. https://www.cncf.io/blog/2021/12/20/new-slashdata-report-5-6-million-developers-use-kubernetes-an-increase-of-67-over-one-year/

30. Cloud Native Computing Foundation (2022) Kubernetes. https://kubernetes.io/

31. Cloud Native Computing Foundation (2022) Kubevirt. http://kubevirt.io/

32. Cloudflare (2022) Announcing D1: our first SQL database. https://blog.cloudflare.com/introducing-d1/

33. D. J. (2022) Baby's first AWS deployment. https://blog.verygoodsoftwarenotvirus.ru/posts/babys-first-aws/

34. Dartois JE, Ribeiro HB, Boukhobza J, Barais O (2019) Opportunistic MapReduce on ephemeral and heterogeneous cloud resources. In: 2019 IEEE 12th international conference on cloud computing (CLOUD). IEEE, Milan, Italy, pp 396–403. https://doi.org/10.1109/CLOUD.2019.00070

35. Diamantopoulos D, Polig R, Ringlein B, Purandare M, Weiss B, Hagleitner C, Lantz M, Abel F (2021) Acceleration-as-a-μ service: a cloud-native Monte-Carlo option pricing engine on CPUs, GPUs and disaggregated FPGAs. In: 2021 IEEE 14th international conference on cloud computing (CLOUD). IEEE, Chicago, IL, USA, pp 726–729. https://doi.org/10.1109/CLOUD53861.2021.00096

36. Docker Inc. (2022) Docker. https://www.docker.com/

37. Dragoni N, Lanese I, Larsen ST, Mazzara M, Mustafin R, Safina L (2018) Microservices: how to make your application scale. In: Petrenko AK, Voronkov A (eds) Perspectives of system informatics, vol 10742. Springer International Publishing, Cham, pp 95–104. https://doi.org/10.1007/978-3-319-74313-4_8

38. Du D, Yu T, Xia Y, Zang B, Yan G, Qin C, Wu Q, Chen H (2020) Catalyzer: sub-millisecond startup for serverless computing with initialization-less booting. In: Proceedings of the twenty-fifth international conference on architectural support for programming languages and operating systems. ACM, Lausanne, Switzerland, pp 467–481. https://doi.org/10.1145/3373376.3378512

39. Elsakhawy M, Bauer M (2020) FaaS2F: a framework for defining execution-SLA in serverless computing. In: 2020 IEEE cloud summit. IEEE, Harrisburg, PA, USA, pp 58–65. https://doi.org/10.1109/IEEECloudSummit48914.2020.00015

40. Fehling C, Leymann F, Retter R, Schupeck W, Arbitter P (2014) Cloud computing patterns. Springer Vienna, Vienna. https://doi.org/10.1007/978-3-7091-1568-8

41. Foundation C.N.C. (2022). Cloud Native Computing Foundation. https://www.cncf.io/

42. Foundation C.N.C. (2022). Knative. https://knative.dev/

43. Fowler M, Lewis J (2014) Microservices. http://martinfowler.com/articles/microservices.html

44. Fuerst A, Sharma P (2021) FaasCache: keeping serverless computing alive with greedy-dual caching. In: Proceedings of the 26th ACM international conference on architectural support for programming languages and operating systems. ACM, Virtual USA, pp 386–400. https://doi.org/10.1145/3445814.3446757

45. Gill SS, Xu M, Ottaviani C, Patros P, Bahsoon R, Shaghaghi A, Golec M, Stankovski V, Wu H, Abraham A, Singh M, Mehta H, Ghosh SK, Baker T, Parlikad AK, Lutfiyya H, Kanhere SS, Sakellariou R, Dustdar S, Rana O, Brandic I, Uhlig S (2022) AI for next generation computing: emerging trends and future directions. Internet of Things 19:100514

46. Golec M, Ozturac R, Pooranian Z, Gill SS, Buyya R (2022) iFaaSBus: a security- and privacy-based lightweight framework for serverless computing using IoT and machine learning. IEEE Trans Ind Inform 18(5):3522–3529. https://doi.org/10.1109/TII.2021.3095466

47. Google (2022) Google Cloud functions. https://cloud.google.com/functions/

48. Google (2022) Google workflows. https://cloud.google.com/workflows/

49. Google (2022) gVisor. https://gvisor.dev/

50. Greenberger M (1962) Management and the computer of the future, pp 220–248. Published jointly by M.I.T. Press and Wiley, New York. https://archive.org/details/managementcomput00gree/page/220/

51. Gujarati A, Elnikety S, He Y, McKinley KS, Brandenburg BB (2017) Swayam: distributed autoscaling to meet SLAs of machine learning inference services with resource efficiency. In: Proceedings of the 18th ACM/IFIP/USENIX middleware conference. ACM, Las Vegas Nevada, pp 109–120. https://doi.org/10.1145/3135974.3135993

52. Handaoui M, Dartois JE, Boukhobza J, Barais O, d'Orazio L (2020) ReLeaSER: a reinforcement learning strategy for optimizing utilization of ephemeral cloud resources. In: 2020 IEEE international conference on cloud computing technology and science (CloudCom). IEEE, Bangkok, Thailand, pp 65–73. https://doi.org/10.1109/CloudCom49646.2020.00009

53. Hassan HB, Barakat SA, Sarhan QI (2021) Survey on serverless computing. J Cloud Comput 10(1):39. https://doi.org/10.1186/s13677-021-00253-7

54. Hellerstein JM, Faleiro JM, Gonzalez J, Schleier-Smith J, Sreekanti V, Tumanov A, Wu C (2019) Serverless computing: one step forward, two steps back. In: 9th biennial conference on innovative data systems research, CIDR 2019, Asilomar, CA, USA, Jan 13–16, 2019, Online proceedings. www.cidrdb.org. http://cidrdb.org/cidr2019/papers/p119-hellerstein-cidr19.pdf

55. Holori (2022) Holori GCP pricing calculator. https://holori.com/gcp-pricing-calculator/

56. Horta E, Chuang HR, VSathish NR, Philippidis C, Barbalace A, Olivier P, Ravindran B (2021) Xar-Trek: run-time execution migration among FPGAs and heterogeneous-ISA CPUs. In: Proceedings of the 22nd international middleware conference. ACM, Québec City, Canada, pp 104–118. https://doi.org/10.1145/3464298.3493388

57. IBM (2022) IBM cloud functions. https://cloud.ibm.com/functions/
58. Ionescu V (2022) Scaling containers on AWS in 2022. https://www.vladionescu.me/posts/scaling-containers-on-aws-in-2022/
59. Izraelevitz J, Yang J, Zhang L, Kim J, Liu X, Memaripour A, Soh YJ, Wang Z, Xu Y, Dulloor SR, Zhao J, Swanson S (2019) Basic performance measurements of the Intel Optane DC persistent memory module. http://arxiv.org/abs/1903.05714. https://dblp.uni-trier.de/rec/journals/corr/abs-1903-05714.html
60. Jia Z, Witchel E (2021) Nightcore: efficient and scalable serverless computing for latency-sensitive, interactive microservices. In: Proceedings of the 26th ACM international conference on architectural support for programming languages and operating systems, ASPLOS '21. Association for Computing Machinery, New York, NY, USA, pp 152–166. https://doi.org/10.1145/3445814.3446701
61. Jiang J, Gan S, Liu Y, Wang F, Alonso G, Klimovic A, Singla A, Wu W, Zhang C (2021) Towards demystifying serverless machine learning training. In: Proceedings of the 2021 international conference on management of data. https://doi.org/10.1145/3448016.3459240
62. Jonas E, Schleier-Smith J, Sreekanti V, Tsai C, Khandelwal A, Pu Q, Shankar V, Carreira J, Krauth K, Yadwadkar NJ, Gonzalez JE, Popa RA, Stoica I, Patterson DA (2019) Cloud programming simplified: a Berkeley view on serverless computing. http://arxiv.org/abs/1902.03383
63. Khandelwal A, Kejariwal A, Ramasamy K (2020) Le Taureau: deconstructing the serverless landscape and a look forward. In: Proceedings of the 2020 ACM SIGMOD international conference on management of data. ACM, Portland, OR, USA, pp 2641–2650. https://doi.org/10.1145/3318464.3383130
64. Kivity A, Kamay Y, Laor D (2007) Kvm: the linux virtual machine monitor. In: Proceedings of the 2007 Ottawa linux symposium, OLS'-07, p 8. https://www.kernel.org/doc/ols/2007/ols2007v1-pages-225-230.pdf
65. Klimovic A, Wen Wang Y, Stuedi P, Trivedi AK, Pfefferle J, Kozyrakis CE (2018) Pocket: elastic ephemeral storage for serverless analytics. Usenix Magazine, vol 44. https://doi.org/10.5555/3291168.3291200. https://www.usenix.org/conference/osdi18/presentation/klimovic
66. Koomey JG (2011) Growth in data center electricity use 2005 to 2010. Analytics Press for the New York Times. https://www.koomey.com/post/8323374335
67. Kuenzer S, Bădoiu VA, Lefeuvre H, Santhanam S, Jung A, Gain G, Soldani C, Lupu C, Teodorescu Ş, Răducanu C, Banu C, Mathy L, Deaconescu R, Raiciu C, Huici F (2021) Unikraft: fast, specialized unikernels the easy way. In: Proceedings of the sixteenth European conference on computer systems. ACM, Online Event, United Kingdom, pp 376–394. https://doi.org/10.1145/3447786.3456248
68. Leite L, Rocha C, Kon F, Milojicic D, Meirelles P (2019) A survey of DevOps concepts and challenges. ACM Comput Surv 52(6). https://doi.org/10.1145/3359981
69. Linux Foundation Projects (2022) Open container initiative. https://opencontainers.org/
70. Linux Foundation Projects (2022) Xen project. https://xenproject.org/
71. Linux Kernel: KVM (2022). https://www.linux-kvm.org/page/Main_Page
72. Liu N, Li Z, Xu J, Xu Z, Lin S, Qiu Q, Tang J, Wang Y (2017) A hierarchical framework of cloud resource allocation and power management using deep reinforcement learning. In: 2017 IEEE 37th international conference on distributed computing systems (ICDCS). IEEE, Atlanta, GA, USA, pp 372–382. https://doi.org/10.1109/ICDCS.2017.123
73. Lloyd W, Vu M, Zhang B, David O, Leavesley G (2018) Improving application migration to serverless computing platforms: latency mitigation with keep-alive workloads. In: 2018 IEEE/ACM international conference on utility and cloud computing companion (UCC Companion). IEEE, Zurich, pp 195–200. https://doi.org/10.1109/UCC-Companion.2018.00056
74. Lopes RV, Menasce D (2016) A taxonomy of job scheduling on distributed computing systems. IEEE Trans Parallel Distrib Syst 27(12):3412–3428
75. Mackey K (2022) Fly machines: an API for fast-booting VMs. https://fly.io/blog/fly-machines/

76. Magoulas R, Swoyer S (2020) Cloud adoption in 2020. https://www.oreilly.com/radar/cloud-adoption-in-2020/
77. Manco F, Lupu C, Schmidt F, Mendes J, Kuenzer S, Sati S, Yasukata K, Raiciu C, Huici F (2017) My VM is lighter (and safer) than your container. In: Proceedings of the 26th symposium on operating systems principles, SOSP '17. Association for Computing Machinery, New York, NY, USA, pp 218–233. https://doi.org/10.1145/3132747.3132763
78. Marshall P, Keahey K, Freeman T (2010) Elastic site: using clouds to elastically extend site resources. In: 2010 10th IEEE/ACM international conference on cluster, cloud and grid computing. IEEE, Melbourne, Australia, pp 43–52. https://doi.org/10.1109/CCGRID.2010.80
79. Masanet E, Shehabi A, Lei N, Smith S, Koomey J (2020) Recalibrating global data center energy-use estimates. Science 367(6481):984–986. https://doi.org/10.1126/science.aba3758
80. Matei O, Skrzypek P, Heb R, Moga A (2020) Transition from serverfull to serverless architecture in cloud-based software applications. In: Silhavy R, Silhavy P, Prokopova Z (eds) Software engineering perspectives in intelligent systems, vol 1294. Springer International Publishing, Cham, pp 304–314. https://doi.org/10.1007/978-3-030-63322-6_24
81. McGrath G, Brenner PR (2017) Serverless computing: design, implementation, and performance. In: 2017 IEEE 37th international conference on distributed computing systems workshops (ICDCSW). IEEE, Atlanta, GA, USA, pp 405–410. https://doi.org/10.1109/ICDCSW.2017.36
82. Mell P, Grance T (2011) The NIST definition of cloud computing. National Institute of Standards and Technology Special Publication 800-145. https://doi.org/10.6028/NIST.SP.800-145
83. Microsoft (2022) Azure functions. https://azure.microsoft.com/products/functions/
84. Microsoft (2022) GitHub. https://github.com/
85. Microsoft (2022) Introduction to hyper-V on windows. https://learn.microsoft.com/en-us/virtualization/hyper-v-on-windows/about/
86. Microsoft (2022) What are durable functions? https://learn.microsoft.com/en-us/azure/azure-functions/durable/durable-functions-overview
87. Mitchell B (2022) After 5 years, I'm out of the serverless compute cult. https://dev.to/brentmitchell/after-5-years-im-out-of-the-serverless-compute-cult-3f6d
88. Mohan A, Sane H, Doshi K, Edupuganti S, Nayak N, Sukhomlinov V (2019) Agile cold starts for scalable serverless. In: 11th USENIX workshop on hot topics in cloud computing (HotCloud 19). USENIX Association, Renton, WA, p 6. https://www.usenix.org/conference/hotcloud19/presentation/mohan
89. Müller I, Marroquín R, Alonso G (2020) Lambada: interactive data analytics on cold data using serverless cloud infrastructure. In: Proceedings of the 2020 ACM SIGMOD international conference on management of data. ACM, Portland, OR, USA, pp 115–130. https://doi.org/10.1145/3318464.3389758
90. Nishtala R, Fugal H, Grimm S, Kwiatkowski M, Lee H, Li HC, McElroy R, Paleczny M, Peek D, Saab P, Stafford D, Tung T, Venkataramani V (2013) Scaling memcache at facebook. In: 10th USENIX symposium on networked systems design and implementation (NSDI 13). USENIX Association, Lombard, IL, pp 385–398. https://www.usenix.org/conference/nsdi13/technical-sessions/presentation/nishtala
91. Oakes E, Yang L, Zhou D, Houck K, Harter T, Arpaci-Dusseau A, Arpaci-Dusseau R (2018) SOCK: rapid task provisioning with serverless-optimized containers. In: 2018 USENIX annual technical conference (USENIX ATC 18). USENIX Association, Boston, MA, pp 57–70. https://www.usenix.org/conference/atc18/presentation/oakes
92. Oracle (2022) Fn. https://fnproject.io/
93. Oracle (2022) Oracle cloud functions. https://www.oracle.com/cloud/cloud-native/functions/
94. Oracle (2022) VirtualBox. https://www.virtualbox.org/
95. Owens K (2018) CNCF WG-serverless whitepaper v1.0. Technical report. Cloud Native Computing Foundation (2018)

96. Passwater A (2018) Serverless community survey: huge growth in serverless usage. https://www.serverless.com/blog/2018-serverless-community-survey-huge-growth-usage/

97. Perron M, Fernandez RC, DeWitt DJ, Madden S (2020) Starling: a scalable query engine on cloud functions. In: Proceedings of the 2020 ACM SIGMOD international conference on management of data. https://doi.org/10.1145/3318464.3380609

98. Pons DB, Artigas MS, París G, Sutra P, López PG (2019) On the FaaS track: building stateful distributed applications with serverless architectures. In: Proceedings of the 20th international middleware conference. https://doi.org/10.1145/3361525.3361535

99. Poppe O, Guo Q, Lang W, Arora P, Oslake M, Xu S, Kalhan A (2022) Moneyball: proactive auto-scaling in Microsoft Azure SQL database serverless. In: VLDB. ACM, pp 1279–1287. https://doi.org/10.14778/3514061.3514073. https://www.microsoft.com/en-us/research/publication/moneyball-proactive-auto-scaling-in-microsoft-azure-sql-database-serverless/

100. Project (2022) Fission. https://fission.io/

101. Puviani M, Frei R (2013) Self-management for cloud computing. In: Science and information conference, p 7. https://ieeexplore.ieee.org/document/6661855?arnumber=6661855

102. QEMU team (2022) QEMU. https://www.qemu.org/

103. Reeve J (2018) Kubernetes: a cloud (and data center) operating system? https://blogs.oracle.com/cloud-infrastructure/post/kubernetes-a-cloud-and-data-center-operating-system

104. Roberts M (2018) Serverless architectures. https://martinfowler.com/articles/serverless.html

105. Romero F, Chaudhry GI, Goiri Í, Gopa P, Batum P, Yadwadkar NJ, Fonseca R, Kozyrakis CE, Bianchini R (2021) Faa$T: a transparent auto-scaling cache for serverless applications. In: Proceedings of the ACM symposium on cloud computing

106. Schleier-Smith J, Sreekanti V, Khandelwal A, Carreira J, Yadwadkar NJ, Popa RA, Gonzalez JE, Stoica I, Patterson DA (2021) What serverless computing is and should become: the next phase of cloud computing. Commun ACM 64(5):76–84

107. Schuler L, Jamil S, Kuhl N (2021) AI-based resource allocation: reinforcement learning for adaptive auto-scaling in serverless environments. In: 2021 IEEE/ACM 21st international symposium on cluster, cloud and internet computing (CCGrid). IEEE, Melbourne, Australia, pp 804–811 (2021). https://doi.org/10.1109/CCGrid51090.2021.00098

108. Shafiei H, Khonsari A, Mousavi P (2022) Serverless computing: a survey of opportunities, challenges, and applications. ACM Comput Surv (Just Accepted). https://doi.org/10.1145/3510611

109. Shahin M, Ali Babar M, Zhu L (2017) Continuous integration, delivery and deployment: a systematic review on approaches, tools. Challenges and practices. IEEE Access 5:3909–3943

110. Shahrad M, Fonseca R, Goiri Í, Chaudhry G, Batum P, Cooke J, Laureano E, Tresness C, Russinovich M, Bianchini R (2020) Serverless in the wild: characterizing and optimizing the serverless workload at a large cloud provider, p 14. https://www.usenix.org/conference/atc20/presentation/shahrad

111. Shaw R, Howley E, Barrett E (2022) Applying reinforcement learning towards automating energy efficient virtual machine consolidation in cloud data centers. Inf Syst 21. https://doi.org/10.1016/j.is.2021.101722

112. Shillaker S, Pietzuch P (2020) FAASM: lightweight isolation for efficient stateful serverless computing. USENIX Association, USA. https://doi.org/10.5555/3489146.3489174

113. Singhvi A, Balasubramanian A, Houck K, Shaikh MD, Venkataraman S, Akella A (2021) Atoll: a scalable low-latency serverless platform. In: Proceedings of the ACM symposium on cloud computing. ACM, Seattle, WA, USA, pp 138–152. https://doi.org/10.1145/3472883.3486981

114. Soni D, Kumar N (2022) Machine learning techniques in emerging cloud computing integrated paradigms: a survey and taxonomy. J Netw Comput Appl 205:103419

115. SPEC Research Group (2022) https://research.spec.org/

116. Sreekanti V, Wu C, Lin XC, Schleier-Smith J, Faleiro JM, Gonzalez JE, Hellerstein JM, Tumanov A (2020) Cloudburst: stateful functions-as-a-service. Proc VLDB Endow 13:2438–2452. https://doi.org/10.14778/3407790.3407836

117. Taibi D, El Ioini N, Pahl C, Niederkofler J (2020) Patterns for serverless functions (function-as-a-service): a multivocal literature review. In: Proceedings of the 10th international conference on cloud computing and services science. SCITEPRESS—Science and Technology Publications, Prague, Czech Republic, pp 181–192. https://doi.org/10.5220/0009578501810192
118. Tan B, Liu H, Rao J, Liao X, Jin H, Zhang Y (2020) Towards lightweight serverless computing via unikernel as a function. In: 2020 IEEE/ACM 28th international symposium on quality of service (IWQoS). IEEE, Hang Zhou, China, pp 1–10. https://doi.org/10.1109/IWQoS49365.2020.9213020
119. Thalheim J, Okelmann P, Unnibhavi H, Gouicem R, Bhatotia P (2022) VMSH: hypervisor-agnostic guest overlays for VMs. In: Proceedings of the seventeenth European conference on computer systems. ACM, Rennes, France, pp 678–696. https://doi.org/10.1145/3492321.3519589
120. The Linux Foundation (2022) https://www.linuxfoundation.org/
121. Ustiugov D, Petrov P, Kogias M, Bugnion E, Grot B (2021) Benchmarking, analysis, and optimization of serverless function snapshots. In: Proceedings of the 26th ACM international conference on architectural support for programming languages and operating systems. ACM, Virtual USA, pp 559–572. https://doi.org/10.1145/3445814.3446714
122. Vahidinia P, Farahani B, Aliee FS (2020) Cold start in serverless computing: current trends and mitigation strategies. In: 2020 international conference on omni-layer intelligent systems (COINS). IEEE, Barcelona, Spain, pp 1–7. https://doi.org/10.1109/COINS49042.2020.9191377
123. van Eyk E, Iosup A, Abad CL, Grohmann J, Eismann S (2018) A SPEC RG cloud group's vision on the performance challenges of FaaS cloud architectures. In: Companion of the 2018 ACM/SPEC international conference on performance engineering. ACM, Berlin, Germany, pp 21–24. https://doi.org/10.1145/3185768.3186308
124. Vaquero LM, Rodero-Merino L, Morán D (2011) Locking the sky: a survey on IaaS cloud security. Computing 91(1):93–118
125. Vilanova L, Maudlej L, Bergman S, Miemietz T, Hille M, Asmussen N, Roitzsch M, Härtig H, Silberstein M (2022) Slashing the disaggregation tax in heterogeneous data centers with FractOS. In: Proceedings of the seventeenth European conference on computer systems. ACM, Rennes, France, pp 352–367. https://doi.org/10.1145/3492321.3519569
126. Villamizar M, Garces O, Castro H, Verano M, Salamanca L, Casallas R, Gil S (2015) Evaluating the monolithic and the microservice architecture pattern to deploy web applications in the cloud. In: 2015 10th computing Colombian conference (10CCC). IEEE, Bogota, Colombia, pp 583–590. http://doi.org/10.1109/ColumbianCC.2015.7333476
127. VMware (2022) ESXi. https://www.vmware.com/products/esxi-and-esx.html
128. VMware (2022) Openfaas. https://www.openfaas.com/
129. Wanninger NC, Bowden JJ, Shetty K, Garg A, Hale KC (2022) Isolating functions at the hardware limit with virtines. In: Proceedings of the seventeenth European conference on computer systems. ACM, Rennes, France, pp 644–662. https://doi.org/10.1145/3492321.3519553
130. Weissman CD, Bobrowski S (2009) The design of the Force.Com multitenant internet application development platform. In: Proceedings of the 2009 ACM SIGMOD international conference on management of data. ACM, Providence Rhode Island, USA, pp 889–896. https://doi.org/10.1145/1559845.1559942
131. Wiggins A (2017) The twelve-factor app. https://12factor.net/
132. Wu C, Faleiro JM, Lin Y, Hellerstein JM (2018) Anna: a KVS for any scale. In: 2018 IEEE 34th international conference on data engineering (ICDE), pp 401–412. https://doi.org/10.1109/TKDE.2019.2898401
133. Wu C, Sreekanti V, Hellerstein JM (2020) Transactional causal consistency for serverless computing. In: Proceedings of the 2020 ACM SIGMOD international conference on management of data. ACM, Portland, OR, USA, pp 83–97. https://doi.org/10.1145/3318464.3389710
134. Yalles S, Handaoui M, Dartois JE, Barais O, d'Orazio L, Boukhobza J (2022) Riscless: a reinforcement learning strategy to guarantee SLA on cloud ephemeral and stable resources. In: 2022 30th Euromicro international conference on parallel, distributed and network-based processing (PDP), pp 83–87. https://doi.org/10.1109/PDP55904.2022.00021

135. Zhang T, Xie D, Li F, Stutsman R (2019) Narrowing the gap between serverless and its state with storage functions. In: Proceedings of the ACM symposium on cloud computing. ACM, Santa Cruz, CA, USA, pp 1–12. https://doi.org/10.1145/3357223.3362723
136. Zhang W, Fang V, Panda A, Shenker S (2020) Kappa: a programming framework for serverless computing. In: Proceedings of the 11th ACM symposium on cloud computing, SoCC '20. Association for Computing Machinery, New York, NY, USA, pp 328–343. https://doi.org/10.1145/3419111.3421277
137. Zomer J (2022) Escaping privileged containers for fun. https://pwning.systems/posts/escaping-containers-for-fun/

Printed in the United States
by Baker & Taylor Publisher Services